Lecture Notes in Mathematics 2024

Editors:
J.-M. Morel, Cachan
B. Teissier, Paris

For further volumes:
http://www.springer.com/series/304

Mariarosaria Padula

Asymptotic Stability
of Steady Compressible
Fluids

Springer

Mariarosaria Padula
University of Ferrara
Faculty of Engineering
Department of Mathematics
via Machiavelli 35
44124 Ferrara
Italy
pad@unife.it

ISBN 978-3-642-21136-2   e-ISBN 978-3-642-21137-9
DOI 10.1007/978-3-642-21137-9
Springer Heidelberg Dordrecht London New York

Lecture Notes in Mathematics ISSN print edition: 0075-8434
ISSN electronic edition: 1617-9692

Library of Congress Control Number: 2011933136

Mathematics Subject Classification (2010): 35-XX, 76-XX

Cover design: deblik, Berlin

Printed on acid-free paper

Springer is part of Springer Science+Business Media (www.springer.com)

*Dedicated to my children
and grandchildren.*

Preface

Qui non se canta al modo de le rane,
qui non se canta al modo del poeta
che finge imaginando cose vane;
ma qui resplende e luce onne natura,
che a chi intende fa la mente leta;
qui non se gira per la selva oscura.

Dall'"Acerba", Cecco D'Ascoli

This book aims to introduce a systematic approach for solving mathematical problems arising in the study of hyperbolic-parabolic systems governing the motions of thermodynamic fluids. It appeals to a wide range of theoretical and applied mathematicians whose research interests may include compressible flows, capillarity theory, and control of motions. It aims to acquaint mathematicians and fluid dynamics researchers with recent results achieved in the investigation and of nonlinear asymptotic stability under no smallness assumption on initial data, and of loss of control from initial data for some steady flows of compressible fluids as well.

The main ideas are illustrated using three different model problems. The following classes of motions are studied:

(i) Barotropic viscous gases in rigid domains with compact, either impermeable or porous boundaries, and exterior domains.
(ii) Isothermal viscous gases with free boundaries.
(iii) Heat conducting viscous polytropic gases.

Equations governing non-steady flows of the fluids described in (i), (ii), (iii) are classified as follows:

(1) One vector parabolic equation for the velocity, and one scalar hyperbolic equation for the density.
(2) One vector parabolic equation for the velocity, and two scalar hyperbolic equations for the density and the free boundary.

(3) Two parabolic equations with one vectorial for the velocity, one scalar for the temperature, and one scalar hyperbolic equation for the density.

The main goal of this text is to reduce the study of stability, linear and nonlinear of basic motion S_b, to the sign of a suitable functional $\mathbb{E}$ *called the **modified energy functional**.* To this end, we employ the classical Lyapunov direct method[1] where the Lyapunov functional is identified with the *modified energy of perturbations* $\mathbb{E}(t)$. The modified energy $\mathbb{E}(t)$ consists of the difference between the total energies of unsteady $E(t)$ and basic E_b motions, plus an extra energy term given by the functional $\mathcal{I}(t)$, named *free work functional.* Namely $\mathbb{E}(t)$ is given by

$$\mathbb{E}(t) = E(t) - E_b + \gamma\mathcal{I},$$

where γ is an arbitrary parameter.

To apply the Lyapunov method using the modified energy as functional we are led to *formulate a differential equation governing the time evolution of the free work functional $\mathcal{I}(t)$, called the free work equation*, which is the key ingredient for the proof of uniqueness, asymptotic nonlinear stability, and loss of control from initial data, cf. [104]. Specifically, in coupled conservative-dissipative systems, the free work equation provides an artificial dissipative term for the variables satisfying conservative equations. As such, the free work equation, in combination with the energy equation for perturbations, results in a differential equation for the modified energy $\mathbb{E}(t)$, which then allows for a new a priori estimate that provides control of perturbations in particular norms that we call "natural norms"; cf. [4].

The two main tools that we will introduce in this book are:

(1) *A variant of Lyapunov's second method* (a generalization of Dirichlet method) to prove nonlinear stability;
(2) The *free work equation*, useful in proving asymptotic nonlinear stability, and loss of initial data control.

In our stability problems the control occurs on the employed "natural" norms for perturbations, that reduce only in particular cases to the L^2 norms of perturbations. In most cases, for regular flows, they will be equivalent to the

---

[1]Sometime the Lyapunov method is erroneously confused with the energy method. Actually in the energy method the Lyapunov functional is identified with the L^2 norm $K(t)$ of the difference $\mathbf{u} = \mathbf{v} - \mathbf{v}_b$ between the velocities $\mathbf{v}$, $\mathbf{v}_b$ of unsteady $S(t)$ and basic S_b motions. Of course $K(t) = \int_\Omega \mathbf{u}^2 dx$ doesn't coincide with the difference between the kinetic energies $E(t) = \int_\Omega \mathbf{v}^2 dx$, $E_b(t) = \int_\Omega \mathbf{v}_b^2 dx$ of the two motions,

$$K(t) \neq E(t) - E_b.$$

L^2 norm of perturbations. Furthermore, the basic flow S_b will represent either an equilibrium position or a steady motion.

In this text we will confine ourselves to the listed three cases (i), (ii) and (iii), as our aim is to explain a new algorithm, named the *free work identity*. It is introduced for the study of nonlinear stability of basic flows. We deem that our three examples of fluid motions, each containing an elastic behavior, cover a sufficiently wide mathematical set of partial differential equations (PDE), without being so many as to give rise to confusion.

Below is a summary of how the text will approach the subject matter.

(1) Chapter 1 is a prologue to thermo-fluid dynamics. In this chapter we introduce the equations for compressible fluids, with related boundary and initial-boundary value problems for the cases (i), (ii) and (iii) defined in the preface.

(2) In Chap. 2 we recall direct methods in the study of nonlinear stability, with the aid of three simple applications of the Dirichlet method. We end the chapter listing the main theorems proven in the paper.

(3) In Chap. 3 we consider barotropic fluids filling domains Ω with fixed, rigid, and compact closed boundaries $\partial\Omega$. The fluid may fill either the region Ω interior to $\partial\Omega$ where Ω is a bounded domain, or the region exterior to $\partial\Omega$, where Ω is an unbounded domain. For the exterior region we distinguish between cases when the fluid has finite mass, and when it has an infinite mass. We prove uniqueness theorems for several steady flows in a given regularity class of steady motions corresponding to the same data, thus we prove asymptotic decay of regular perturbed unsteady motions to these steady flows.

(4) In Chap. 4 we consider the rest state of an isothermal fluid in a section of horizontal layer with free boundaries. Periodicity in the horizontal direction is assumed. First we prove a uniqueness theorem of the rest state in the class of steady motions corresponding to the same data, thus we prove asymptotic decay to the rest. We also study the instability problem that occurs when the fluid is below the rigid plane of the layer. We extend this study by constructing rest states which are linearly stable, but not physically observable for large initial data. This result is achieved by introducing the concept of loss of initial data control.

(5) In Chap. 5 we consider the rest state of a polytropic viscous gas in a rigid, bounded domain, with perfectly heat conducting walls and periodicity in the horizontal direction. We then prove a uniqueness theorem for the rest state in the class of steady motions corresponding to the same data, and asymptotic decay to the rest.

Note that in Chaps. 2, 3 and 5:

(1) The initial perturbations may be large;
(2) The class of perturbations is quite large despite the fact that we will not be covering the problem of optimizing the regularity class of solutions.

This book deals with a single approach, in order not to obscure the main ideas.

To restrict the length of the book, the discussion of materials that require more extensive coverage, such as problems relating to fluid motions in pipes [57, 77, 111] and of heat conducting fluids with free boundaries [45], has been omitted.

The results described within the book are obtained using simple form, and the reader is required to know only the basic elements of classical and functional analysis.

Also note that each chapter is self-contained, and can be read independently.

Acknowledgements

The results in this book represent the summa of a long research period that began 10 years ago. The work presented was performed, in part, during the visit of the author to TIFR, where she delivered an academic course on Fluid Dynamics.

The author wishes to express her deep gratitude to professors G. Bizhanova, E. Frolova, T. Nishida, and V.A. Solonnikov, for their friendship, and for the endless discussions concerning the organization of the matter.

Furthermore she acknowledges the 60% and GNFM of Italian CNR-INDAM. She would like to thank to professors A. Borrelli, V. Coscia, G.P. Galdi, J. Heywood, P. Maremonti, F. Mollica, C. Patria and to dr. E.M. Galdi for scientific support in the compilation of the text.

Contents

1   **Topics in Fluid Mechanics**   1
    1.1   Introduction   1
    1.2   Mathematical Notations   2
        1.2.1   Geometrical Notations   2
        1.2.2   Analytical Notations   5
    1.3   Kinematics   9
        1.3.1   Vectorial Notations   10
        1.3.2   Eulerian and Lagrangean Descriptions   11
        1.3.3   Reynolds Transport Theorem   11
        1.3.4   Frames   13
        1.3.5   From Macroscopic to Local Laws   14
        1.3.6   Mass, Momentum, Rotational Momentum,
               Kinetic Energy   14
    1.4   Mechanics   15
        1.4.1   Forces   16
        1.4.2   Conservation of Mass   17
        1.4.3   Balance Laws of Momentum, Rotational Momentum ...   18
        1.4.4   Stress Tensor   19
        1.4.5   Balance Equations in Local Form   20
        1.4.6   Incompressible Euler Equations   21
        1.4.7   Constitutive Equations   22
        1.4.8   Linearly Viscous Fluids   23
    1.5   Thermodynamics   25
        1.5.1   Thermodynamical Variables   25
        1.5.2   The First Law of Thermodynamics   27
        1.5.3   The Second Law of Thermodynamics   29
        1.5.4   Clausius–Duhem Inequality   30
        1.5.5   Phenomenological Laws   32
        1.5.6   Compressible Euler Equations   33

       1.5.7   Linearly Viscous Fluids 34

       1.5.8   Heat-Conducting Fluids.................................. 35

  1.6 Side Conditions 36

       1.6.1   Classes of Boundary Conditions 37

       1.6.2   Boundary Conditions on Velocity 38

       1.6.3   Boundary Conditions on Stress.......................... 40

       1.6.4   Boundary Conditions on Temperature................... 43

       1.6.5   Side Conditions on Density 44

  1.7 Three Model Problems 46

       1.7.1   Incompressible Fluids 47

       1.7.2   Barotropic Fluids 48

       1.7.3   Polytropic Fluids 50

       1.7.4   Bibliographical Notes.................................... 51

2 **Topics in Stability** 53

  2.1 Introduction.. 53

  2.2 Nonlinear Stability 54

       2.2.1   Abstract Settings 55

       2.2.2   Initial Data Control 58

       2.2.3   Lyapunov Method 59

       2.2.4   Energy Method 60

  2.3 Lagrange–Dirichlet Method...................................... 62

       2.3.1   Hyperbolic First Order Systems......................... 62

       2.3.2   Second Order ODE 64

       2.3.3   Stability of Barotropic Inviscid Fluids $\mathbf{v}_b \neq 0$ 66

       2.3.4   Isothermal Viscous Fluids $\mathbf{v}_b = 0$ 71

  2.4 Main Theorems 74

       2.4.1   Case (a) Barotropic Fluid, Rigid Boundary 74

       2.4.2   Case (b) Isothermal Fluid, Deformable Boundary 80

       2.4.3   Case (c) Polytropic Fluid, Rigid Boundary.............. 85

  2.5 Bibliographical Notes 86

3 **Barotropic Fluids with Rigid Boundary** 87

  3.1 Introduction.. 87

  3.2 Ω Bounded, Potential Forces 89

       3.2.1   Uniqueness of the Rest State 90

       3.2.2   Nonlinear Stability 91

       3.2.3   Nonlinear Exponential Stability 96

  3.3 Ω Bounded, Non-Potential Forces.............................. 100

       3.3.1   Uniqueness $f \neq \nabla U$ 100

       3.3.2   Nonlinear Stability 106

       3.3.3   Nonlinear Exponential Stability 109

  3.4 Ω Bounded, Non Zero Boundary Data.......................... 113

       3.4.1   Uniqueness 113

       3.4.2   Nonlinear Exponential Stability 115

3.5 Ω Exterior, Fixed Compact Region 116
    3.5.1 Uniqueness 117
    3.5.2 Nonlinear Exponential Stability 119
3.6 Instability 122
3.7 Auxiliary Lemmas 126
    3.7.1 Function $\mathbf{V}$ for Uniqueness 126
    3.7.2 Function $\mathbf{V}$ for Stability 128
    3.7.3 Bibliographical Notes....................................... 131

4 Isothermal Fluids with Free Boundaries 133
4.1 Introduction... 133
4.2 Position of the Problem 136
    4.2.1 Geometrical Tools 136
    4.2.2 Equations of Motion....................................... 138
    4.2.3 Rest State and Equilibrium Configurations 139
    4.2.4 First and Second Variations of Energy.................. 141
4.3 Definitions Related to Stability 147
    4.3.1 Initial Data Control 149
    4.3.2 Present Results 152
4.4 Uniqueness 154
4.5 Nonlinear Stability 158
    4.5.1 Energy Equation... 158
    4.5.2 Nonlinear Stability 162
    4.5.3 Nonlinear Exponential Stability 164
4.6 Loss of Stability 171
    4.6.1 Energy Inequality... 171
    4.6.2 Equivalence of Norms 173
    4.6.3 Instability 175
    4.6.4 Loss of Initial Data Control 179
4.7 Sign of E_0 in Terms of Size of Perturbation.................... 179
    4.7.1 Proof of Theorem 4.7.1.................................... 180
    4.7.2 Proof of Theorem 4.7.2.................................... 183
4.8 Auxiliary Lemmas 187
    4.8.1 Function $\mathbf{V}$ for Uniqueness 187
    4.8.2 Functions $\mathbf{V}$ for Stability and Instability 190
    4.8.3 Bibliographical Notes....................................... 194

5 Polytropic Fluids with Rigid Boundary 197
5.1 Introduction... 197
    5.1.1 Equations of Motion....................................... 198
5.2 Uniqueness of the Rest State 201
5.3 Stability Problem 207
    5.3.1 Energy Equation... 208
    5.3.2 Energy Equation of Perturbation 209
    5.3.3 Nonlinear Exponential Stability 213

5.4 Auxiliary Lemmas 217
     5.4.1 Some Inequalities 217
     5.4.2 Function **V** for Uniqueness 218
     5.4.3 Auxiliary Function for Stability 219
     5.4.4 Bibliographical Notes..................................... 221

Bibliography 223

Index 231

Chapter 1
Topics in Fluid Mechanics

Concreato fu ordine construtto
alle sustanzie, e quelle furon cima
nel mondo, in che puro atto fu produtto.
31, XXIX Paradiso A. Dante

Experience only decides on the truth.

1.1 Introduction

The aim of this chapter is, after some mathematical preliminaries, to introduce the physical equations governing steady and unsteady motions of barotropic and polytropic fluids. The field of macroscopic thermodynamics provides us with a general framework for the description of irreversible continuum processes. Certain areas of macroscopic physics have connections with fluid dynamics, and the first part of the chapter is devoted to a systematic development of this theory, referencing [143]. More specifically, we will be treating state parameters as field variables, thus formulating the basic equations of continuous thermodynamics in the form of local equations. This will allow us to formulate correct well-posed problems.

In this chapter we will distinguish three model problems:

(a) *Barotropic viscous fluid*, fluid filling a domain Ω with rigid boundaries
(b) *Isothermal viscous fluid*, fluid filling a domain Ω with deformable boundaries
(c) *Polytropic viscous fluid*, fluid filling a domain Ω with rigid, perfectly heat-conducting boundaries

It is clear that in cases (a), (b) and (c), the unknown functions correspond to different physical variables. In addition, as observed in the preface, each

M. Padula, *Asymptotic Stability of Steady Compressible Fluids*,
Lecture Notes in Mathematics 2024, DOI 10.1007/978-3-642-21137-9_1,
© Springer-Verlag Berlin Heidelberg 2011

unknown function satisfies a different PDE. The character of these PDEs changes by changing the physical model, thus steady or unsteady problems are governed by different mathematical problems (see cases (i), (ii), (iii) defined in the preface). In all cases that we are dealing with, problems of unsteady fluid flow are all described by a coupled parabolic-hyperbolic system.

Chapter 1 will proceed as follows:

Section 1.2 Geometric and analytical tools are introduced.

Section 1.3 Kinematical tools are introduced.

Section 1.4 Systems of equations governing motions of general continua are written in local form.

Section 1.5 Equations of thermodynamics are introduced, and constitutive equations for a fluid are given.

Section 1.6 General physical boundary conditions are introduced for different domains and material boundaries.

Section 1.7 The exact mathematical position is given for three model problems.

1.2   Mathematical Notations

In this section we will introduce some notations concerning the geometry of the region where the motion occurs, and the functional spaces, with relative norms, where stability is studied. Finally, some elementary topics in kinematics are covered.

1.2.1   Geometrical Notations

The domain

Let a fluid fill the region Ω_t at time t, with boundary $\partial\Omega_t$.

If the domain is fixed we omit the subscript t.

In subsequent sections we will use the symbol Ω to analyze motions occurring only in one of the following regions Ω:

1. Ω is a bounded, fixed, simply connected domain.
2. Ω is exterior to a bounded, fixed, simply connected domain.
3. Ω is a portion of fluid layer contained between a rigid flat surface Π, and a deformable stress free surface that do not intersect each other.
4. Ω is a portion of fluid layer contained between two rigid, parallel, flat surfaces Π, Π_1.

The reference frame

Consider rectilinear coordinates in a reference frame $\mathcal{R} =: \{O, \mathbf{i}, \mathbf{j}, \mathbf{k}\}$, with the basis $\{\mathbf{i}, \mathbf{j}, \mathbf{k}\}$ ortho-normal.

The origin O is:
in Ω in the first case itemized above;
in B in the second case;
on one rigid surface Π in the latter two cases.

In the latter two cases, the x and y axes, parallel to $\mathbf{i}$, $\mathbf{j}$ respectively, are horizontal lines on Π, crossing orthogonally to each other at O, while the z axis, parallel to $\mathbf{k}$, is orthogonal to the rigid plane Π.

The rectangle $\Sigma = (0, a) \times (0, b)$ on Π denotes a periodicity cell.

We add the index $'$ to denote quantities calculated on Σ, specifically to denote vector functions which are linear combinations of $\mathbf{i}$, $\mathbf{j}$.

The symbol ∇' represents derivatives along the coordinate lines x, y. Thus given a differentiable function ψ defined on Σ, we set

$$\nabla'\psi \equiv \left(\partial_x\psi, \partial_y\psi\right), \qquad \psi_{,x} := \partial_x\psi \quad \psi_{,y} := \partial_y\psi.$$

The Cartesian representation of a domain and of a surface

By Ω_t we denote the subset of R^3 that has the **Cartesian representation**

$$\Omega_t =: \{(x, y, z) \in R^3 : \ (x, y) \in \Sigma, \ z \in (0, \zeta) \ \ \zeta = \zeta(x, y, t)\}, \ \ t \in (0, \infty).$$

Ω_t is partially bounded by the free surface Γ_t expressed by the **Cartesian representation**

$$\Gamma_t = \{(x, y, z) \in R^3 : \ (x, y) \in \Sigma, \ \ z = \zeta(x, y, t)\}, \ \ \ t \in (0, \infty),$$

where ζ is an *unknown* scalar differentiable function.

The Cartesian representation of Γ_t implies the absence of **reversal flows**.[1]

To have a **simply connected domain** we assume ζ to be strictly positive. Precisely, given a positive constant ζ_b, (discussed further in upcoming sections), we assume that the deformable surface may be represented at any time by the equation

$$\zeta = \zeta_b + \eta(x, y, t), \quad |\eta| > \frac{\zeta_b}{2}.$$

---

[1] In the Cartesian representation of Γ_t, by reversal flows we mean flows for which there exists a $x' \in \Sigma$ and two points of Γ_t, $(x', \zeta_1(x', t))$, $(x', \zeta_2(x', t))$ having velocities $\mathbf{V}(x', \zeta_1(x', t))$, $\mathbf{V}(x', \zeta_2(x', t))$ satisfying

$$\mathbf{V}'(x', \zeta_1(x', t)) = -\mathbf{V}'(x', \zeta_2(x', t)).$$

Definition 1.2.1 *The surface Γ_t has the **Lipschitz (1832-1903) property** if there exist two positive numbers δ, L such that*

$$|\zeta(x'_1,t) - \zeta(x'_2,t)| \le L|x'_1 - x'_2|, \qquad \forall |x'_1 - x'_2| < \delta.$$

*The domain Ω has the **strong local Lipschitz property** provided that there exists a finite open cover U_i of the boundary $\partial\Omega$ such that the surface $\Gamma_i = U_i \cap \partial\Omega$ has the **Lipschitz property**.*

Definition 1.2.2 *The surface Γ_t is of class $C^1(\Sigma \times (0,T))$ if the function $\zeta(x',t)$ is of class C^1. There are two unit normals that can be introduced at each point of Γ_t. Each of such vectors introduces an **orientation** over Γ_t. We orientate Γ_t by fixing the exterior unit normal.*

The **unit normal n** has components $(-\nabla'\zeta, 1)/\sqrt{1 + |\nabla'\zeta|^2}$, where $\sqrt{1 + |\nabla'\zeta|^2}$ is the metric element. We recall that on Γ_t the two vectors

$$\mathbf{t}_1 := \frac{1}{\sqrt{1+\zeta_{,x}^2}}\partial_x\left(x\mathbf{i} + y\mathbf{j} + \zeta\mathbf{k}\right) \equiv \frac{1}{\sqrt{1+\zeta_{,x}^2}}\left(\zeta_{,x}\mathbf{k} + \mathbf{i}\right), \qquad (1.2.1)$$

$$\mathbf{t}_2 := \frac{1}{\sqrt{1+\zeta_{,y}^2}}\partial_y\left(x\mathbf{i} + y\mathbf{j} + \zeta\mathbf{k}\right) \equiv \frac{1}{\sqrt{1+\zeta_{,y}^2}}\left(\zeta_{,y}\mathbf{k} + \mathbf{j}\right),$$

represent two linearly independent, unitary vectors tangent to Γ_t, expressed by $\mathbf{t}_1 = \frac{1}{\sqrt{1+\zeta_{,x}^2}}\left(1, 0, \zeta_{,x}\right)$, $\mathbf{t}_2 = \frac{1}{\sqrt{1+\zeta_{,y}^2}}\left(0, 1, \zeta_{,y}\right)$. The normal vector $\mathbf{n}$ directed outward Ω_t, is given by

$$\mathbf{n} = \frac{1}{\sqrt{1 + |\nabla'\zeta|^2}}\left[-\zeta_{,x}\mathbf{i} - \zeta_{,y}\mathbf{j} + \mathbf{k}\right]. \qquad (1.2.2)$$

The curvature

Let γ be a curve with curvilinear abscissa $s \in (0,1)$, of class $C^1(0,1)$,

$$\mathbf{x} = \mathbf{x}(s).$$

Definition 1.2.3 *The curvature $1/r(s)$ of γ at s is the inverse of the radius of the circumference tangent to γ at s. If the curvature is considered as positive when the curve is bending in the direction of its principal normal $\mathbf{N}$, the analytical notation requires*

$$\frac{1}{r(s)} := \left|\frac{d^2\mathbf{x}}{ds^2}\right|.$$

Definition 1.2.4 *Let the surface Γ_t be of class $C^1(\Sigma \times (0,T))$. Fixed (x',t) there are infinite curves $\gamma \in C^1(0,1)$ on Γ_t crossing at $(x', \zeta(x',t))$. Let us*

*call $1/r$ their curvatures. We define principal curvatures the maximum $1/r_1$
and minimum $1/r_2$ of these curvatures. The **doubled mean curvature** $\mathcal{H}(\zeta)$
of Γ_t at $(x', \zeta(x', t))$ is given by the sum of the principal curvatures*

$$\frac{2}{r} = \frac{1}{r_1} + \frac{1}{r_2}.$$

*The double curvature $2/r = \mathcal{H}(\zeta)$ in analytical notation is expressed
through the nonlinear positive **Laplace (1749-1827)-Beltrami (1836-
1900) operator** requires*

$$\mathcal{H}(\zeta) = \nabla' \cdot \left(\frac{\nabla' \zeta}{\sqrt{1 + |\nabla' \zeta|^2}} \right). \tag{1.2.3}$$

1.2.2 Analytical Notations

The mathematical notations for the norms of general functions may change
from chapter to chapter. Below are some main notations.

Notations

First, let us introduce some classical notation relating to functional spaces.
For any function defined in a domain of R^n, for given integers l, n, we let
$(i_1, \ldots, i_n)$ be the multi-index such that $i_1 + \ldots + i_n = l$. Let $\partial_{x_k}^{i_k}$ denotes the
partial derivative of order i_k with respect to the variable x_k, $k = 1, \ldots, n$,
and we introduce the notation

$$D^l f = \partial_{x_1}^{i_1} \ldots \partial_{x_n}^{i_n} f.$$

We use the Einstein convention, whereby repeated indices in one term indicate
summation, thus for the vector functions $\mathbf{u} \equiv (u_j)$, $\mathbf{v} \equiv (v_i)$, for $i = 1, 2, 3$,
it will hold that

$$u_j \partial_j v_i = u_1 \partial_1 v_i + u_2 \partial_2 v_i + \partial_3 v_i = \mathbf{u} \cdot \nabla \, v_i.$$

Lebesgue (1875-1941) spaces

We denote by $L^p(\Omega)$, $1 \le p \le \infty$, the Lebesgue space of all measurable
functions f, defined on Ω, for which

$$\int_\Omega |f(x)|^p dx < \infty.$$

The functional $\| \cdot \|_{L^p(\Omega)}$ defined by

$$\|f\|_{L^p(\Omega)} := \left(\int_\Omega |f(x)|^p dx < \infty \right)^{1/p},$$

is a norm on $L^p(\Omega)$, provided $1 \leq p < \infty$.[2] Finally, $L^\infty(\Omega)$ is the Lebesgue space of all measurable functions f, defined on Ω, for which

$$\sup_\Omega |f(x)| < \infty.$$

The functional $\| \cdot \|_{L^\infty(\Omega)}$ defined by

$$\|f\|_{L^\infty(\Omega)} := \sup_\Omega |f(x)|,$$

is a norm on $L^\infty(\Omega)$. If the domain is bounded it holds

$$\|f\|_{L^\infty(\Omega)} := \sup_\Omega |f(x)| = \lim_{p \to \infty} \|f\|_{L^p(\Omega)}.$$

The numbers p, p' satisfying

$$\frac{1}{p} + \frac{1}{p'} = 1,$$

are called conjugate numbers. For a pair p, p' of conjugate numbers, and for all functions $f \in L^p(\Omega)$, $g \in L^{p'}(\Omega)$ it holds the **Hölder (1859-1937) inequality**

$$\left| \int_\Omega f g \, dx \right| \leq \|f\|_{L^p(\Omega)} \|g\|_{L^{p'}(\Omega)}. \tag{1.2.4}$$

For $p = p' = 2$ Hölder inequality implies the **Schwartz (1843-1921) inequality**

$$\left| \int_\Omega f g \, dx \right| \leq \|f\|_{L^2(\Omega)} \|g\|_{L^2(\Omega)}. \tag{1.2.5}$$

For $p = 2$, the space $L^2(\Omega)$ becomes an **Hilbert (1862-1943) space** where it is defined the scalar product

$$(f, g) := \int_\Omega f g \, dx, \quad \forall f, g \in L^2(\Omega). \tag{1.2.6}$$

Notice that by Hölder inequality for all $g \in L^{p'}(\Omega)$, it is meaningful the application

$$(., g): \quad f \in L^p(\Omega) \quad \longrightarrow \quad (f, g) \in R.$$

It defines a linear functional on $L^p(\Omega)$. Thus for a pair p, p' of conjugate numbers, $L^{p'}(\Omega)$ is contained in the dual space $(L^p(\Omega))'$ [3] of $L^p(\Omega)$.

---

[2]The functional $\| \cdot \|_p$ is not a norm if $0 < p < 1$.

[3]The set of all continuous, linear functionals on a linear space X is called **Dual Space** of X and is denoted by X'.

As a consequence for $f \in L^p(\Omega)$, $g \in L^{p'}(\Omega)$ it is meaningful the bilinear functional (f, g) defined in (1.2.6).

Indeed for $1 \leq p < \infty$ it is possible to prove that $L^{p'}(\Omega)$ **is the dual space** of $L^p(\Omega)$, $L^{p'}(\Omega) = (L^p(\Omega))'$.

For a pair p, p' of conjugate numbers, and for all functions $f \in L^p(\Omega)$, $g \in L^{p'}(\Omega)$ it holds the **Young (1863-1942) inequality**

$$|(f, g)| \leq \frac{\epsilon}{p}\|f\|_{L^p(\Omega)}^p + \frac{1}{\epsilon\, p'}\|g\|_{L^{p'}(\Omega)}^{p'} \qquad \forall \epsilon > 0. \tag{1.2.7}$$

When $p = p' = 2$ the inequality is known as **Cauchy (1789-1857) inequality**

We recall also the **Minkowski (1864-1909) inequality**

$$\|f + g\|_{L^p(\Omega)} \leq \|f\|_{L^p(\Omega)} + \|g\|_{L^p(\Omega)}, \quad \forall f, g \in L^p(\Omega). \tag{1.2.8}$$

Finally let $1 \leq s \leq p \leq r < \infty$, $f \in L^s(\Omega) \cap L^r(\Omega)$, with

$$\frac{1}{p} = \theta\frac{1}{s} + (1 - \theta)\frac{1}{r}, \quad \theta \in [0, 1],$$

thus the **interpolation inequality** holds

$$\|f\|_{L^p(\Omega)} \leq \|f\|_{L^s(\Omega)}^\theta \|f\|_{L^r(\Omega)}^{1-\theta}. \tag{1.2.9}$$

Sobolev spaces

For $m \geq 0$ and $1 \leq q \leq \infty$, as usual **Sobolev (1908-1989) space** we set

$$W^{m,q}(\Omega) = \{f \in L_{loc}^1(\Omega) : \quad D^l f \in L^q(\Omega), \quad |l| \leq m\}.$$

Since elements f of $W^{m,q}(\Omega)$ are equivalence class of functions equal up to set of zero measure, surfaces S, it must specified sense the value of f on S can be considered. The value of f on S in called **trace** of f on S. Indeed the trace of a function f in $W^{m,q}(\Omega)$ on $W^{j,p}(S)$, for suitably indices, may be considered as limit value of a sequence of regular functions f_n converging to f in $W^{m,q}(\Omega)$.

$$W_0^{m,q}(\Omega) = \{f \in L_{loc}^1(\Omega) : \quad D^l f \in L^q(\Omega), \quad 1 \leq |l| \leq m; \quad D^{l-1} f|_{\partial\Omega} = 0\},$$

where $D^{l-1} f|_{\partial\Omega}$ denotes the trace of $D^{l-1} f$ on the boundary. For $q = 2$ we use the notation

$$H^m(\Omega) = W^{m,2}(\Omega) \qquad H_0^m(\Omega) = W_0^{m,2}(\Omega),$$

to note Hilbert spaces.

We define the functionals

$$\|f\|_{W^{m,p}(\Omega)} := \Big(\sum_{0 \le |l| \le m} \|D^l f(x)\|_p^p dx < \infty \Big)^{1/p}, \quad 1 \le p < \infty,$$

$$\|f\|_{W^{m,\infty}(\Omega)} := \max_{0 \le |l| \le m} \|D^l f(x)\|_\infty,$$

as norms on $W^{m,p}(\Omega), W^{m,\infty}(\Omega)$.

Embedding inequalities

Below we recall some important inequalities, without repeating the proofs.

Let $\Omega \subseteq R^n$, then for all $f \in L^n(\Omega)$ the **Niremberg (1925) inequality** yields

$$\|f\|_{L^n(\Omega)} \le \frac{1}{2\sqrt{n}} \|\nabla f\|_{L^1(\Omega)}. \tag{1.2.10}$$

Let

$$r \in \Big[p, \frac{np}{n-p}\Big], \qquad 1 \le p < n,$$

$$r \in [p, \infty), \qquad n \le p.$$

Then if $f \in W_{1,p}(\Omega)$ it holds the **embedding inequality**

$$\|f\|_{L^r(\Omega)} \le \Big(\frac{c}{2\sqrt{n}}\Big)^\lambda \|f\|_{L^p(\Omega)}^{1-\lambda} \|\nabla f\|_{L^p(\Omega)}^\lambda, \tag{1.2.11}$$

with

$$c = max\Big\{p, \frac{r(n-1)}{n}\Big\}, \qquad \lambda := n\Big(\frac{1}{p} - \frac{1}{r}\Big).$$

When $1 \le p < n$, $r = np(n-p)$, and $f \in W^{1,p}(\Omega)$, it holds the **Sobolev inequality**

$$\|f\|_{L^r(\Omega)} \le \frac{p(n-1)}{2\sqrt{n}(n-p)} \|\nabla f\|_{L^p(\Omega)}. \tag{1.2.12}$$

When $f \in W^{1,2}(\Omega)$, $\Omega \subseteq R^n$, $n = 2, 3$, it holds the **Ladyzhenskaja (1922-2004) inequality**

$$\|f\|_{L^4(\Omega)} \le \Big(\frac{2(n-1)}{n\sqrt{n}}\Big)^{n/4} \|f\|_{L^2(\Omega)}^{(4-n)/4} \|\nabla f\|_{L^2(\Omega)}^{n/4}. \tag{1.2.13}$$

For Sobolev spaces $W^{m,p}(\Omega)$ the general **embedding inequality** holds

$$\|D^j f\|_{L^s(\Omega)} \le c\|f\|_{L^p(\Omega)}^{1-\lambda} \|D^m f\|_{L^p(\Omega)}^\lambda, \tag{1.2.14}$$

where $c = c(j, p, r, \lambda, n, m)$, $j \ne 0$, and

$$\frac{1}{s} = \frac{j}{n} + \lambda\left(\frac{1}{r} - \frac{m}{n}\right) + (1-\lambda)\frac{1}{p},$$

with $j/m \leq \lambda \leq 1$.

If $1 < r < \infty$, $m - j - n/r > 0$ then it must hold $j/m \leq \lambda < 1$.

The dual space of $H_0^1(\Omega)$ will be denoted by $H^{-1}(\Omega)$.

Let us denote by $J(\Omega)$, $J_1(\Omega)$ the subsets of vector functions defined in $L^2(\Omega)$, $W^{1,2}(\Omega)$ respectively, having zero divergence.

Vectorial Notations

Let $\mathbf{u}$ and $\mathbf{v}$ be two vector functions in dual Lebesgue spaces; we use the notation $(\mathbf{u}, \mathbf{v})$ to denote the integral over Ω of the scalar product between these two functions,

$$(\mathbf{u}, \mathbf{v}) := \int_\Omega \mathbf{u} \cdot \mathbf{v} \, dx.$$

In Chaps. 3 and 5, the domains are fixed. Since this negates the possibility of confusion, we use the symbols

$$\|\cdot\|_{L^p}, \qquad p > 1, \qquad \|\cdot\|_{W^{m,p}}, \qquad m \geq 0. \quad p > 1,$$

to denote the norms in the Lebesgue, and Sobolev spaces $L^p(\Omega)$, $W^{m,p}(\Omega)$ respectively.

In Chap. 4, the domain is unknown, and its boundary admits Cartesian representation in the reference frame $\mathcal{R} = \{O, \mathbf{i}, \mathbf{j}, \mathbf{k}\}$. In $\mathcal{R}$ the domain Ω_t and its moving boundary Γ_t are expressed by:

$$\Omega_t = \left\{(x', z) \in R^2 \times R : \quad x' \in \Sigma, \quad z < \zeta(x', t)\right\},$$

$$\Gamma_t = \left\{(x', z) \in R^2 \times R : \quad x' \in \Sigma, \quad z = \zeta(x', t)\right\},$$

where $\zeta(x', t)$ is an unknown function. In this case, we use the symbols

$$\|\cdot\|_{L^2(\Omega_t)}, \qquad \|\cdot\|_{L^2(\Sigma)}, \qquad \|\cdot\|_{L^2(\Gamma_t)},$$

to denote the L^2 norms in Ω_t, Σ, Γ_t respectively.

Let X be a Banach (1892-1945) space, with $L^p(0, T; X)$ denoting the set of functions f defined in $(0, T) \times X$, with finite **Bochner (1821-1894) integrals**:

$$\|f\|_{X,p} := \left(\int_0^T \|f\|_X^p(t) \, dt\right)^{1/p} < \infty, \qquad \|f\|_{X,\infty} := \sup_{t \in (0,\infty)} \|f\|_X(t).$$

We may use the notation $f \in L^p(0, T; X)$.

1.3 Kinematics

In this section we introduce some kinematical notation for the study of a continuum.

1.3.1 Vectorial Notations

The following conventions are used

1. $t \in (0, T)$ signifies a time instant
2. $\mathbf{x} \in \Omega$ signifies a position vector
3. $\mathbf{X}$ is a material particle, $\mathbf{x}$ is the position of the space occupied by the particle $\mathbf{X}$ at time t
4. $\mathbf{v}(\mathbf{x}, t)$ is the velocity of a particle at a specified position in the space $\mathbf{x} \in \Omega$ and time $t \in (0, T)$
5. The operator

$$\frac{d}{dt}\mathbf{w} = \partial_t \mathbf{w} + \mathbf{v} \cdot \nabla \mathbf{w}$$

denotes the **material time derivative**[4] of a function $\mathbf{w}(\mathbf{x}, t)$, or the rate of change at a point $(\mathbf{x}, t)$, when movement occurs along the trajectory of a particle of fluid moving with velocity $\mathbf{v}$

6. The components (a_i) of acceleration $\mathbf{a}$ of a fluid particle at position $\mathbf{x}$ at time t are given by

$$a_i = \frac{d}{dt} v_i = \partial_t v_i + v_k \partial_k v_i$$

7. A macroscopic quantity Q function of the n-dimensional domain C, $n = 2, 3$, is called **absolutely continuous function** of C if

$$\lim_{C \to \mathbf{x}} \frac{|Q(C)|}{|C|} < \infty, \qquad \mathbf{x} \in C. \qquad (1.3.1)$$

In kinematics we work with material volumes and surfaces, thus the macroscopic mass M is an absolutely continuous function of C, and it is reasonable to introduce the material density function ρ as

$$M(C) := \int_C \rho(\mathbf{x}) dx.$$

For a generic quantity Q the limit is function of $\mathbf{x}$ and we may set

$$\lim_{C \to \mathbf{x}} \frac{Q(C)}{|C|} = q_1(\mathbf{x}) = \rho(\mathbf{x}) q(\mathbf{x}),$$

where q_1 is the density function of Q per unit of volume, q is the specific density of Q, namely density per unit of mass

8. A system of applied vector functions $(\mathbf{x}, \mathbf{f}(\mathbf{x})$, $\mathbf{x} \in C$, is characterized by its **resultant** $\mathbf{R}(C)$, and a **torque of pole** $\mathbf{x}_O$ $\mathbf{M}_O(C)$ given by

---

[4]Material time derivative means the derivative along the motion of the fluid particle $\mathbf{X}$, such as $\mathbf{X}$ where fixed.

$$\mathbf{R}(C) := \int_C \mathbf{f}(\mathbf{x})dx, \qquad \mathbf{M}_O(C) := \int_C (\mathbf{x} - \mathbf{x}_O) \times \mathbf{f}(\mathbf{x})dx.$$

Notice that in our notation C denotes either a volume or a surface
9. The flux of q entering in C across its boundary ∂C is defined by

$$-\int_{\partial\Omega} q \cdot \mathbf{n}dS.$$

The flux of q across a oriented surface S is defined by

$$\int_S q \cdot \mathbf{n}dS.$$

Thèse conventions will be used throughout the book unless otherwise stated.

1.3.2 Eulerian and Lagrangean Descriptions

It is customary to distinguish between Lagrangean $(\mathbf{X}, t)$, Lagrange (1736-1813), and Eulerian $(\mathbf{x}, t)$, Euler (1707-1783), representations of flow.

Eulerian coordinates refer to $(\mathbf{x}, t) \in \Omega \times (0, T)$ as a given point $\mathbf{x}$ in the spatial domain Ω at a given instant t in the time interval $(0, T)$; $(\mathbf{x}, t)$ refers to fixed points in space and not to fixed particles of fluid.

Lagrangian coordinates refer to $(\mathbf{X}, t)$ as a given particle $\mathbf{X}$ of fluid filling Ω at a given instant $t \in (0, T)$; $(\mathbf{X}, t)$ refers to fixed material particle and not to a fixed point of the domain Ω.

The mathematical model describing the physical properties of fluids is customarily written in Eulerian form.

1.3.3 Reynolds Transport Theorem

We recall that macroscopic balance laws can be formulated in at least three different ways since the volume C, over which density, momentum, total energy and entropy are calculated, may be either a material region, constituted always by the same particles, or a fixed geometrical region, or a moving physical region. In any case, to formulate local balance laws, we must move the time derivative of a macroscopic quantity, integrated over the region C, inside the integral sign. This operation is known as the change between the time derivative and spatial integration if the domain is fixed, and as the *transport theorem*, or the *Reynolds transport theorem* if the domain is material.

We prove the theorem of change between time derivative and spatial integration, over fixed and moving domains, for a regular function.

To this end we introduce the scalar function ρ, which in a moving domain Ω_t, satisfies the transport equation

$$\partial_t \rho + \nabla \cdot (\rho \mathbf{v}) = 0, \qquad (1.3.2)$$

where $\mathbf{v}$ is a given regular vector function.

If $\mathbf{v}$ is the velocity vector then ρ will represent the material density, see Sect. 1.4.2.

Where Domain C Is Fixed

The region C is a fixed, domain of R^n, and the results that follow are well known.

Lemma 1.3.1 *Given a function f defined in C, if C is sufficiently regular $C \in C^1$, and $f \in C^1(0,T; L^1(C))$, then the change between the time derivative and spatial integration holds*

$$\frac{d}{dt} \int_C f(\mathbf{x},t)dx = \int_C \partial_t f(\mathbf{x},t)dx. \qquad (1.3.3)$$

Where Domain Ω_t Is Moving

Lemma 1.3.2 *Let Ω_t be a regular domain with boundary $\Gamma_t \in C^1(\Sigma \times (0,T))$ which moves with velocity $\mathbf{V} \in C^0(0,T; L^1(\Sigma))$. Given a function f defined in Ω_t and $f \in C^1(0,T; L^1(C))$, Then the change between the time derivative and spatial integration of the function $f(\mathbf{x},t)$, density per unit of volume, yields*

$$\frac{d}{dt} \int_{\Omega_t} f(\mathbf{x},t)dx = \int_{\Omega_t} \partial_t f(\mathbf{x},t)dx + \int_{\Gamma_t} f(\mathbf{x},t)\mathbf{V} \cdot \mathbf{n}\, dS, \qquad (1.3.4)$$

where $\mathbf{n}$ is normal to Γ_t oriented toward the exterior of Ω_t.

Where Domain Ω_t Is Material

The following Lemma, called the Reynolds transport theorem, elaborates upon the possibility of carrying out the time derivative in spatial integrations for functions that are density functions per unit of volume and unit of mass.

Lemma 1.3.3 *Let $\mathbf{v}$ be a regular vector function defined in the regular domain Ω_t, and ρ a scalar function solution to (1.3.2). If $\mathbf{V}$ is the velocity of points of the boundary, then*

$$\frac{d}{dt} \int_{\Omega_t} \rho f dx = \int_{\Omega_t} \rho \frac{df}{dt} dx + \int_{\Gamma_t} \rho f(\mathbf{V} - \mathbf{v}) \cdot \mathbf{n} dS. \qquad (1.3.5)$$

Proof. Any regular function f satisfies the following identity

$$\int_{\Omega_t} \rho \frac{df}{dt} dx = \int_{\Omega_t} \rho(\partial_t f + \mathbf{v} \cdot \nabla f) dx = \int_{\Omega_t} \Big(\partial_t(\rho f) + \nabla \cdot (\rho f \mathbf{v}) \Big) dx$$

$$= \frac{d}{dt} \int_{\Omega_t} \rho f dx - \int_{\Gamma_t} \rho f \mathbf{V} \cdot \mathbf{n} dS + \int_{\Gamma_t} \rho f \mathbf{v} \cdot \mathbf{n} dS, \qquad (1.3.6)$$

where $\mathbf{V}$ is the velocity of points of the boundary, and where ρ is the solution of (1.3.2). Then we find that (1.3.6) is equivalent to (1.3.5).   □

Remark 1.3.1 *When Ω_t is a material volume it is $\mathbf{V} \cdot \mathbf{n} = \mathbf{v} \cdot \mathbf{n}$. Hence from (1.3.5) we deduce*

$$\frac{d}{dt} \int_{\Omega_t} \rho f dx = \int_{\Omega_t} \rho \frac{df}{dt} dx. \qquad (1.3.7)$$

In other words, the time derivative is taken inside the integral sign as though the domain and ρ were constants.

1.3.4   Frames

A phenomenon is independent of its frame; as such, its description is reported differently by different observers. Let $\mathcal{R} = \{O, \mathbf{e}_i, t\}$, $\mathcal{R}^* = \{O^*, \mathbf{e}_i^*, t^*\}$ be two frames representing two observers and two watches. A change of frame has the representation

$$\mathbf{x}^* = \mathbf{x}_O^*(t) + \mathbf{Q}(t)(\mathbf{x} - \mathbf{x}_O), \qquad\qquad t^* = a + t,$$

where $\mathbf{Q}$ is the **matrix of transformation of the basis** $\{\mathbf{e}_i\}$ **into** $\{\mathbf{e}_i^*\}$. A motion $\mathbf{x}(\mathbf{X}, t)$ in $\mathcal{R}$, represented by $\mathbf{x}^*(\mathbf{X}, t)$ in $\mathcal{R}^*$, is subject to the **transformation rule**

$$\mathbf{x}^*(\mathbf{X}, t) = \mathbf{x}_O^*(t) + \mathbf{Q}(t)(\mathbf{x}(\mathbf{X}, T) - \mathbf{x}_O), \qquad\qquad t^* = t. \qquad (1.3.8)$$

A change of frame induces the following transformations in tensor spaces

$$\alpha = \alpha^*, \qquad \mathbf{w}^* = \mathbf{Q}(t)\mathbf{w}, \qquad\qquad \mathbf{T}^* = \mathbf{Q}(t)\mathbf{T}\mathbf{Q}^T(t), \qquad (1.3.9)$$

where α is a scalar, $\mathbf{w}$ a vector, and $\mathbf{T}$ a tensor. As known, velocity and acceleration do not follow the rule of change of frame (1.3.9). In particular, for velocity we have

$$\mathbf{v}^*(\mathbf{X}, t) = \mathbf{v}_O^*(t) + \mathbf{Q}(t)\mathbf{v}(\mathbf{x}, t) + \dot{\mathbf{Q}}(t)(\mathbf{x}(\mathbf{X}, t) - \mathbf{x}_O), \qquad (1.3.10)$$

where the dot sign ˙ over the tensor $\mathbf{Q}$ denotes time derivative.

A **physical event** must be considered **independent from its observer**, though we may describe it only by introducing the frame through which it is observed. Given the motion of a continuum in the time interval $(0, T)$, and given two frames $\mathcal{R}$, $\mathcal{R}^*$, there are two ways to proceed:

(1) Find the motion laws for each frame, directly in $\mathcal{R}$, and $\mathcal{R}^*$
(2) Find the motion laws in the frame $\mathcal{R}$, and translate these laws in $\mathcal{R}^*$

Definition 1.3.1 *A quantity is called **frame indifferent** if it satisfies* (1.3.9).

One important example of this is given by the scalar density function ρ.

For the transformation of tensorial functionals, please refer to Sect. 1.4.6.

The velocity is not frame indifferent, actually points moving with $\mathcal{R}$ have zero velocity in $\mathcal{R}$, and non zero velocity in $\mathcal{R}^*$ given by (1.3.10).

1.3.5   From Macroscopic to Local Laws

Physical laws relate various macroscopic quantities that are functions of the volume or the surface of a medium when time is varying. In classical continuum thermodynamics, macroscopic balance laws are postulated that describe the evolution of the global physical characteristic of a system.

Regularity Assumption I

Each macroscopic quantity can be expressed through its density function, of either volume, mass, or surface.

A typical balance law for the macroscopic quantity $Q(t, A)$ in the domain A relates the evolution in time of $Q(t, A)$ to its flux through the boundary ∂A, plus the macroscopic long and short range actions between A and its environment A^c.

Regularity Assumption II

The basic unknowns involved in macroscopic balance laws must have smoothness properties in order to ensure the passage to local form.

These regularity assumptions may lead one to consider the models unrealistic. However, we stress the fact that regularity properties can be proven, at least for small data; cf. bibliographical notes. Moreover, the estimate derived using the free work method has frequently allowed for the proofs of existence theorems; cf. [115, 116].

1.3.6   Mass, Momentum, Rotational Momentum, Kinetic Energy

To state the balance laws, we begin by introducing the definitions of mass, momentum, and rotational momentum. Kinetic energy is also defined.

Definition 1.3.2 *The **mass of a portion of continuum** in the material region C is given by*

$$\int_C \rho(\mathbf{x}, t)\, dx,$$

where ρ is the material density of the particle $\mathbf{X}$ that occupies the position $\mathbf{x}$ at time t,

$$\mathbf{x} = \mathbf{x}(\mathbf{X}, t).$$

Definition 1.3.3 *The **momentum of a portion of continuum** in the material region C is given by*

$$\int_C \rho \mathbf{v}(\mathbf{x}, t)\, dx,$$

where ρ is the density, and $\mathbf{v}$ the velocity of the particle. The quantity $\rho \mathbf{v}(x, t)$ denotes the momentum of the particle $\mathbf{X}$ occupying the position $\mathbf{x}$ at time t.

Definition 1.3.4 *The **rotational, or angular, momentum of a portion of continuum** in the material region C with respect to the point $\mathbf{x}_O$ is given by*

$$\int_C (\mathbf{x} - \mathbf{x}_O) \times \rho \mathbf{v}(\mathbf{x}, t)\, dx.$$

The quantity $(\mathbf{x} - \mathbf{x}_O) \times \rho \mathbf{v}(x, t)$ denotes the angular momentum of the particle $\mathbf{X}$ in the position $\mathbf{x}$ at time t.

Definition 1.3.5 *The **kinetic energy of a portion of continuum** in the material region C is given by*

$$\frac{1}{2} \int_C \rho \mathbf{v}^2(\mathbf{x}, t)\, dx.$$

1.4   Mechanics

A systematic macroscopic approach to non-equilibrium processes must be built upon a certain number of phenomenological laws, true in every three-dimensional region having a non zero Lebesgue measure. These laws are of two kinds: (1) balance equations including the conservation of the mass, the balance of momentum, rotational momentum, and the first and second law of thermodynamics; (2) constitutive equations. In the next subsections global and local formulations of these laws will be derived.

Notation

The continuum fills the domain Ω. By C we denote any regular subset of Ω, $C \subseteq C$.

In Sects. 1.4.2, 1.4.5, 1.4.6 we deduce the indefinite partial differential equations governing the motion of a continuum. To these PDEs initial and boundary conditions have still to be added, see Sect. 1.6.

1.4.1 Forces

External actions are basic elements of mechanics. They are mathematical quantities introduced a priori and subject to mathematical axioms.

Traditionally, we represent the actions on a body as being distinguished by two types of forces: body forces, also called long-range actions, and surface forces, or tractions, called short range actions. Both categories of forces are characterized by a resultant and a torque. By abuse of notations we continue to call force also the resultant.

Body forces are vector fields defined in any region $C \subseteq \Omega$, and include *long-range forces* such as gravity, fictitious forces and electromagnetic forces. Long-range forces may have density per unit mass, as gravity, fictitious forces and per unit volume, like electromagnetic forces. The body resultant acting on the portion of continuum in the region C is expressed by

$$\mathbf{F}(C, t) = \int_C \rho \mathbf{f}(\mathbf{x}, t)\, dx + \int_C \mathbf{f}_1(\mathbf{x}, t)\, dx,$$

where $\mathbf{f}$ is the *density of force per unit of mass*, also said **specific force**, and $\mathbf{f}_1$ is the *density of force per unit of volume*.

The **body torques** of *long-range torques* acting on the portion of continuum $\mathcal{C}$ in the region C, with pole $\mathbf{x}_O$ are given by

$$\mathbf{M}_O(C, t) = \int_C (\mathbf{x} - \mathbf{x}_O) \times \rho \mathbf{f}(\mathbf{x}, t)\, dx + \int_C (\mathbf{x} - \mathbf{x}_O) \times \mathbf{f}_1(\mathbf{x}, t)\, dx.$$

In subsequent sections we will refer only to forces and torques per unit of mass.

The surface forces, or **tractions** are defined on any portion of oriented regular surface S contained in Ω. The tractions include *short-range forces*, such as capillary forces and stress, which have direct molecular origin and decrease rapidly with an increase in distance between interacting elements. The surface *forces acting on the portion of oriented surface $\mathcal{S}$* of continuum in the region S are expressed by

$$\mathbf{F}(S, t) = \int_S \widetilde{\mathbf{t}}(\mathbf{x}, t)\, dS,$$

where $\widetilde{\mathbf{t}}$ is the density of *short-range force*, called *traction* per unit of surface.

The **surface torques** acting on the portion of oriented surface S of continuum in the region S, with pole $\mathbf{x}_O$ are given by

$$\mathbf{M}_O(S,t) = \int_S (\mathbf{x} - \mathbf{x}_O) \times \tilde{\mathbf{t}}(\mathbf{x},t)\, dS.$$

Classical mechanics assumes the following:

The Cauchy Postulate[5] *The tractions on all like-oriented surfaces of contact with a common tangent plane π at $\mathbf{x}$ are the same at $\mathbf{x}$. That is, $\tilde{\mathbf{t}}$ at $\mathbf{x}$ is assumed to depend upon S only through the normal $\mathbf{n}$ of S at $\mathbf{x}$:*

$$\tilde{\mathbf{t}}(\mathbf{x},t) = \mathbf{t}(\mathbf{x},t,\mathbf{n}). \tag{1.4.1}$$

S is oriented so that its normal $\mathbf{n}$ points out of $\mathcal{C}$.

The corresponding torques with respect to a point $\mathbf{x}_O$ are given by

$$\mathbf{M}_O(C,t) \int_C (\mathbf{x} - \mathbf{x}_O) \times \rho \mathbf{f}(\mathbf{x},t)\, dx = \int_C (\mathbf{x} - \mathbf{x}_O) \times \mathbf{b}(\mathbf{x},t)\, dx,$$

$$\mathbf{M}_O(S,t) := \int_S (\mathbf{x} - \mathbf{x}_O) \times \mathbf{t}(\mathbf{x},t,\mathbf{n}) dS.$$

1.4.2 Conservation of Mass

In this treatment, $C \subseteq \Omega$ represents any fixed three-dimensional region contained in the domain of motion Ω.

The law of conservation of mass holds true for all continua. It states that the increase in time of the mass of the continuum $\mathcal{F}$ in the volume C, for all C,

$$\frac{d}{dt} \int_C \rho(\mathbf{x},t) dx,$$

equals the total mass of $\mathcal{F}$ entering the volume C

$$-\int_{\partial C} \rho \mathbf{v} \cdot \mathbf{n}\, dS,$$

where dS is the infinitesimal element of area of ∂C, and $\mathbf{n}$ is the normal to ∂C directed outside C.

Thus the **axiom of conservation of mass** claims that for sufficiently regular density and velocity fields it holds

$$\frac{d}{dt} \int_C \rho(\mathbf{x},t) dx + \int_{\partial C} \rho \mathbf{v} \cdot \mathbf{n}\, dS = 0, \qquad \forall C \subseteq \Omega. \tag{1.4.2}$$

---

[5]The first statement of stress principle traces back to 1823, cf. [17].

Using the Gauss theorem, the transport theorem, and owing the arbitrariness of C, for sufficiently regular density and velocity fields, (1.4.2) at any time t and point x yields the local form of the **continuity equation**

$$\partial_t \rho + \nabla \cdot (\rho \mathbf{v}) = 0, \qquad (\mathbf{x}, t) \in \Omega \times (0, T). \qquad (1.4.3)$$

1.4.3 Balance Laws of Momentum, Rotational Momentum

The **momentum balance law** states that the increase in time of the momentum of the portion of the continuum $\mathcal{F}$ in the volume C, for all C,

$$\frac{d}{dt} \int_C \rho \mathbf{v}(\mathbf{x}, t) \, dx, \qquad (1.4.4)$$

equals the total momentum of $\mathcal{F}$ flowing in the volume C

$$- \int_{\partial C} (\rho \mathbf{v}\, \mathbf{v})(\mathbf{x}, t) \cdot \mathbf{n} \, dS, \qquad (1.4.5)$$

plus the long range force due to sources acting on volume C

$$\int_C \rho \mathbf{f}(\mathbf{x}, t) dx, \qquad (1.4.6)$$

plus the short range force, due to internal friction, acting on the oriented surface ∂C

$$\int_{\partial C} \mathbf{t}(\mathbf{x}, t, \mathbf{n}) dS. \qquad (1.4.7)$$

The **axiom of balance of momentum** claims that for sufficiently regular density, velocity, and stress fields it holds

$$\frac{d}{dt} \int_C \rho \mathbf{v}(\mathbf{x}, t) dx + \int_{\partial C} (\rho \mathbf{v}\, \mathbf{v})(\mathbf{x}, t) \, dS = \int_C \mathbf{f}(\mathbf{x}, t) dx + \int_{\partial C} \mathbf{t}(\mathbf{x}, t\, \mathbf{n}) dS. \qquad (1.4.8)$$

The **rotational momentum balance law** states that the increase in time of the angular momentum of pole $\mathbf{x}_O$ of the portion of the continuum $\mathcal{F}$ in the volume C, for all C,

$$\frac{d}{dt} \int_C (\mathbf{x} - \mathbf{x}_O) \times \rho \mathbf{v}(\mathbf{x}, t) \, dx, \qquad (1.4.9)$$

equals the angular momentum of $\mathcal{F}$ flowing in the volume C

$$- \int_{\partial C} (\mathbf{x} - \mathbf{x}_O) \times \rho \mathbf{v} \, \mathbf{v} \cdot \mathbf{n} \, dS, \tag{1.4.10}$$

plus the body torques of long-range forces acting on the volume C,

$$\int_C (\mathbf{x} - \mathbf{x}_O) \times \rho \mathbf{f}(\mathbf{x}, t) dx, \tag{1.4.11}$$

plus the surface torques, due to internal friction, acting on the oriented surface ∂C

$$\int_{\partial C} (\mathbf{x} - \mathbf{x}_O) \times \mathbf{t}(\mathbf{x}, t, \mathbf{n}) d\mathcal{S}. \tag{1.4.12}$$

The **axiom of balance of rotational momentum**, when $\mathbf{x}_O$ is a fixed point, claims that for all region C, for sufficiently regular density, velocity, and stress fields it holds

$$\frac{d}{dt} \int_C (\mathbf{x} - \mathbf{x}_O) \times \rho \mathbf{v}(\mathbf{x}, t) dx + \int_{\partial C} (\mathbf{x} - \mathbf{x}_O) \times (\rho \mathbf{v} \mathbf{v})(\mathbf{x}, t) \cdot \mathbf{n} \, dS$$
$$= \int_C (\mathbf{x} - \mathbf{x}_O) \times \rho \mathbf{f}(\mathbf{x}, t) dx + \int_{\partial C} (\mathbf{x} - \mathbf{x}_O) \times \mathbf{t}(\mathbf{x}, t, \mathbf{n}) d\mathcal{S}. \tag{1.4.13}$$

Currently, we are not yet in a position to derive local equations, because integral terms of volume and surface appear in the same balance equation. In order to deduce local equations we must therefore create a postulate concerning the regularity of the surface force.

1.4.4 Stress Tensor

To deduce local equations of motion we first state the **Cauchy–Noll theorem** [17, 143]. Such a theorem restricts the class of surfaces forces $\mathbf{t}(\mathbf{x}, t, \mathbf{n})$ to those vectors which are linear functions of $\mathbf{n}$, precisely it states

Theorem 1.4.1 *If* $\mathbf{t}(\cdot, \mathbf{n})$ *is a continuous vectorial function satisfying* (1.4.8) *for all volumes* C, *then there exists a second order tensor function* $\mathbf{T}(\cdot)$ *such that*

$$\mathbf{T}(\mathbf{x}, t)\mathbf{n} = \mathbf{t}(\mathbf{x}, t, \mathbf{n}). \tag{1.4.14}$$

The function $\mathbf{T}(\mathbf{x}, t)$ is known as the **Chauchy stress tensor**.

By the Cauchy–Noll theorem it follows the **action-reaction theorem** that claims

$$\mathbf{t}(\mathbf{x}, t, \mathbf{n}) = -\mathbf{t}(\mathbf{x}, t, -\mathbf{n}). \tag{1.4.15}$$

A fluid is said **isotropic** at $\mathbf{x}$ if its traction in $\mathbf{x}$ is expressed as a pressure independent of the orientation of the surface crossing $\mathbf{x}$ in the direction of the normal $\mathbf{n}$.

Fluids at rest are normally in a state of compression, and it is therefore convenient to write the stress tensor of a fluid at rest as i.e.,

$$\mathbf{t}(\mathbf{x}, \mathbf{n}) = -p(\mathbf{x})\mathbf{n}. \qquad (1.4.16)$$

In (1.4.16) the function $p = p(\mathbf{x})$ may be termed **static-fluid pressure**, because only normal stresses are exerted, and the intensity of normal stress is independent of the normal to surface element across which it acts.

For fluids in motion observation shows that the result (1.4.16) ceases to be valid. The tangential stresses are non zero in general, and the normal component of the stress acting across a surface element depends upon the direction of the normal to the surface element.

Sometime for fluids in slow motion, or in motions where internal friction does not occur, the expression (1.4.16) may still be valid; such fluids are called *ideal fluids*.

For general fluid motions internal friction occurs, thus it is convenient to regard the stress tensor $\mathbf{T}$ as the sum of an isotropic part $-p\mathbf{I}$, analogous to that of the stress tensor in a fluid at rest, and a remaining non isotropic part $\mathbb{V}$ called the **deviatoric stress tensor** contributing the tangential stresses, namely we set

$$\mathbf{T}(\mathbf{x}, t)\mathbf{n} = -p(\mathbf{x}, t)\mathbf{n} + \mathbb{V}(\mathbf{x}, t)\mathbf{n}. \qquad (1.4.17)$$

1.4.5 Balance Equations in Local Form

Appealing to the balance of momentum (1.4.8), true for every volume C, employing the Cauchy–Noll theorem (1.4.16), allows the expression of the momentum equation in local form:

$$\partial_t(\rho\mathbf{v}) + \nabla \cdot (\rho\mathbf{v} \otimes \mathbf{v}) = \nabla \cdot \mathbf{T}(\mathbf{x}, t) + \rho\mathbf{f}, \qquad x, t \in \Omega \times (0, T), \quad (1.4.18)$$

where suitable regularity on density ρ, velocity $\mathbf{v}$, stress tensor $\mathbf{T}$ is assumed. We compute the cross product of (1.4.18) times $(\mathbf{x} - \mathbf{x}_O)$, integrate over C and add the resulting equation to (1.4.13). Owing to the variability of volume C, we deduce the equation for the balance of angular momentum in local form, namely the symmetry of the stress tensor

$$\mathbf{T}(\mathbf{x}, t) = \mathbf{T}^T(\mathbf{x}, t), \qquad x, t \in \Omega \times (0, T). \qquad (1.4.19)$$

If we substitute (1.4.17), (1.4.19) in (1.4.18) we obtain the following system

$$\partial_t \rho + \nabla \cdot (\rho\mathbf{v}) = 0, \qquad (1.4.20)$$

$$\partial_t(\rho\mathbf{v}) + \nabla \cdot (\rho\mathbf{v} \otimes \mathbf{v}) = -\nabla p + \nabla \cdot \mathbb{V}(\mathbf{x}, t) + \rho\mathbf{f},$$

$$\mathbb{V}(\mathbf{x}, t) = \mathbb{V}^T(\mathbf{x}, t), \qquad x, t \in \Omega \times (0, T).$$

System (1.4.20) contains ten equations. Unknown of system (1.4.20) are density ρ, velocity $\mathbf{v}$, pressure p and deviatoric tensor $\mathbb{V}$ which are totally eleven. Therefore, to solve the problem of motion we need one more equation.

1.4.6 Incompressible Euler Equations

The Euler equations are derived when **processes of internal friction do not occur**, precisely the Euler equations suppose

$$\mathbb{V}(x,t) = 0.$$

Hence for Euler fluids the stress acts as a pure pressure in the direction normal to the surface:

$$\mathbf{Tn} = \mathbf{t}(\mathbf{x}, \mathbf{tn}) = -p\mathbf{n}. \qquad (1.4.21)$$

The continuity and momentum equations (1.4.20) yield the Euler equations

$$\partial_t \rho + \nabla \cdot (\rho \mathbf{v}) = 0, \qquad (1.4.22)$$

$$\partial_t (\rho \mathbf{v}) + \nabla \cdot (\rho \mathbf{v} \otimes \mathbf{v}) = -\nabla p + \rho \mathbf{f}, \qquad x, t \in \Omega \times (0, T).$$

The Euler equations hold when thermal conductivity and viscosity are not relevant to motion. We can see at this point that the **number of unknowns** $(\rho, \mathbf{v}, p)$, five, is **still greater than the number of equations** (1.4.22), four.

Toward a new equation

When *processes of thermal conduction* are not relevant to the motion of the fluid, we can assume the kinematic constraint of isocoric motions in the Euler equations. Since the material volume is constant the fluid cannot be compressed, thus we call the fluid **incompressible**.

Incompressible Fluids

The kinematic constraint entails that the fluid is incompressible, this is mathematically expressed by the equation

$$\nabla \cdot \mathbf{v} = 0, \qquad x, t \in \Omega \times (0, T). \qquad (1.4.23)$$

In this case the equations governing the motion of an ideal fluid become five and equate the number of unknown. The mathematical model governing the motion of a inviscid incompressible fluid is given by (1.4.22), and (1.4.23)

$$\partial_t \rho + \mathbf{v} \cdot \nabla \rho = 0,$$

$$\partial_t (\rho \mathbf{v}) + \nabla \cdot (\rho \mathbf{v} \otimes \mathbf{v}) = -\nabla p + \rho \mathbf{f}$$

$$\nabla \cdot \mathbf{v} = 0, \qquad x, t \in \Omega \times (0, T). \qquad (1.4.24)$$

For incompressible fluids $p = p(\mathbf{x}, t)$ is a dynamical variable. The pressure represents the reaction of liquid to compression in order to not change its volume.

If at initial time we prescribe a uniform density $\rho(\mathbf{x}, 0) = r_0$, then the density will not vary for all time. Thus homogeneous initial data furnish **homogeneous fluids**. In this case we obtain the **Euler equations for incompressible fluids**

Unsteady homogeneous fluids

$$\partial_t \mathbf{v} + \nabla \cdot (\mathbf{v} \otimes \mathbf{v}) = -\nabla \pi + \mathbf{f}, \qquad \pi = \frac{p}{r_0},$$

$$\nabla \cdot \mathbf{v} = 0, \qquad x, t \in \Omega \times (0, T). \qquad (1.4.25)$$

Steady homogeneous fluids

$$\nabla \cdot (\mathbf{v} \otimes \mathbf{v}) = -\nabla \pi + \mathbf{f}, \qquad \pi = \frac{p}{r_0},$$

$$\nabla \cdot \mathbf{v} = 0, \qquad x \in \Omega. \qquad (1.4.26)$$

Notice that in this model of fluid the number of unknown $\mathbf{v}$, π equals the number of equations.

1.4.7   Constitutive Equations

Let $\mathcal{R} = \{O, e_i, t\}$, $\mathcal{R}^* = \{O^*, e_i^*, t^*\}$ be two frames representing two observers and two periods of observation.

Definition 1.4.1 *The pair* $\left(\mathbf{x}(\mathbf{X}, t), \mathbf{T}(\mathbf{x}(\mathbf{X}, t), t) \right)$ *defines a* ***mechanical process*** *if it satisfies the balance laws of momentum and rotational momentum.*

We have seen in Sect. 1.3.3 that the stress tensor $\mathbf{T}$ and the vector body force $\mathbf{f}$ must satisfy equations (1.3.9). However, in the physical case the force and the traction are functions of motion, hence we should write

$$\mathbf{f}^*(\mathbf{x}^*, t) = \mathbf{Q}(t)\mathbf{f}(\mathbf{x}, t) \qquad \mathbf{T}^*(\mathbf{x}^*, t) = \mathbf{Q}(t)\mathbf{T}(\mathbf{x}, t)\mathbf{Q}^T(t), \qquad (1.4.27)$$

where $\mathbf{x}^*$ is determined by (1.3.8). In (1.4.27), the functions for $\mathbf{T}(\mathbf{x}, t)$ defined in $\mathcal{R}$, $\mathbf{T}^*(\mathbf{x}^*, t)$ defined in $\mathcal{R}^*$ are different.

Consider an event occurring in the time interval (0,T). Until now, we have presented properties common to all continua and to all motions in a given reference frame. **Constitutive equations** allow for a diversity of answer (motion) for different materials corresponding to the same external data.

The general theory of constitutive equations restricts constitutive equations via **constitutive axioms**.

In the class of constitutive equations satisfying the constitutive axioms there are added further axioms, called **phenomenological laws**, used to describe different models of continua.

We define history of a function of time $\Psi(t)$ whose value is a scalar, a vector, or a tensor. For convenience t is the present time, and we shall represent the past time $\tau \geq t$ by the positive quantity $s := t - \tau$. The history of Ψ up to time t is denoted by the application Ψ^t

$$\Psi^t : s \in [0, \infty) \longrightarrow \Psi^t(s) := \Psi(t - s).$$

Below are listed the axioms of mechanical nature that restrict the dependence law of Cauchy stress on motion:

The Principle of Determinism The stress at $\mathbf{x}$ of the particle $\mathbf{X}$, at time t, $\mathbf{x}(\mathbf{X}, t)$, is determined by the history $\mathbf{X}^t$ of the particle $\mathbf{X}$.

$$\mathbf{T}(\mathbf{x}(\mathbf{X}, t), t) = \mathcal{F}(\mathbf{X}^t; \mathbf{X}, t). \tag{1.4.28}$$

By the principle of determinism, the stress at $\mathbf{x}$ of the particle $\mathbf{X}$ at time t, $\mathbf{x}(\mathbf{X}, t)$, should be determined by the motion $\mathbf{y}(\mathbf{Y}, t)$ of all particles $\mathbf{Y}$ in a continuum, which contradicts the short range assumption. Hence we set

The Principle of Local Action *The motion of material points at a finite distance from the particle $\mathbf{X}$ at any given time t is negligible in calculating the stress at $\mathbf{X}$.*

The Principle of Material Indifference states that

$$\mathbf{f}(\mathbf{x}^{*t}; \mathbf{X}, t) = \mathbf{Q}(t)\mathbf{f}(\mathbf{x}^t; \mathbf{X}, t) \qquad \mathcal{T}(\mathbf{x}^{*t}; \mathbf{X}, t) = \mathbf{Q}(t)\mathcal{T}(\mathbf{x}^t; \mathbf{X}, t)\mathbf{Q}(t).$$
$$\tag{1.4.29}$$

Consequently, we see that the stress is invariant under superposed rigid motions.

Note that in (1.4.29), the functionals $\mathbf{f}(\mathbf{x}^t; \mathbf{X}, t)$, $\mathcal{T}(\mathbf{x}^t; \mathbf{X}, t)$, $\mathbf{f}(\mathbf{x}^{*t}; \mathbf{X}, t)$, $\mathcal{T}(\mathbf{x}^{*t}; \mathbf{X}, t)$ defined respectively in $\mathcal{R}$, $\mathcal{R}^*$ are the same!

1.4.8 Linearly Viscous Fluids

If *internal friction is relevant to the motion,* and *processes of thermal conduction and internal friction do not occur,* the state of a fluid is still described by the five variables given by the velocity (3), the pressure (1), and the density (1), plus the six components of the symmetric deviatoric stress $\mathbb{V}$. Concerning the governing equations, the continuity equation (1.4.3) and the balance equation (1.4.20) continue to hold. However, due to internal friction, the stress tensor contains the deviatoric tensor $\mathbb{V}$ responsible for the irreversible transfer of momentum from points where velocity is large to those where it is small. $\mathbb{V}$ *is an unknown function of motion.*

By the principle of material indifference, for simple materials it follows that $\mathbb{V}(\mathbf{x}, t) = \widehat{\mathbb{V}}(\mathbf{D}, t)$, cf. [143]. Assuming linear state equations, we obtain **Newtonian fluids**, or **linearly viscous fluids**

$$\mathbf{T}(\mathbf{v}, p) = -p\mathbf{I} + \lambda \nabla \cdot \mathbf{v}\mathbf{I} + 2\mu\mathbf{D}(\mathbf{v}), \qquad (1.4.30)$$

where $\mathbf{D}(\mathbf{v})$ is the symmetric part of the velocity gradient tensor, also called velocity deformation tensor, and the shear μ and bulk λ coefficients are unknown constants; cf. Batchelor [6].

When *processes of thermal conduction are not relevant* to the motion of viscous fluids, we assume the kinematic constraint of isocoric motions

$$\nabla \cdot \mathbf{v} = 0,$$

for the last equation.

Assumption

Let us now assume that at initial time the fluid is homogeneous, resulting in a density that is uniform in space and time.

In this case, we chose the unknown coefficient $\mu = r_0 \nu$ as a constant.[6]

If in (1.4.20) we take expression (1.4.30) and limit ourselves to the class of homogeneous incompressible fluids, we obtain the complete system of **Naiver–Stokes equations**

Unsteady flows

$$\partial_t \mathbf{v} + \nabla(\mathbf{v} \otimes \mathbf{v}) = -\nabla \pi + \nu \Delta \mathbf{v} + \mathbf{f},$$
$$\nabla \cdot \mathbf{v} = 0, \qquad x, t \in \Omega \times (0, T), \qquad (1.4.31)$$
$$\mathbf{v}(x, 0) = \mathbf{v}_0(x), \qquad x \in \Omega,$$

where $\pi = p/r_0$ is a dynamical unknown.

Steady flows

$$\nabla(\mathbf{v} \otimes \mathbf{v}) = -\nabla \pi + \nu \Delta \mathbf{v} + \mathbf{f},$$
$$\nabla \cdot \mathbf{v} = 0, \qquad x \in \Omega.$$

Notice that the number of unknown $\mathbf{v}$, π equals the number of equations.

---

[6]This choice may be verified by experience as a phenomenological law.

1.5 Thermodynamics

So far, the mathematical description of the motion of a non-heat conducting fluid has been given by the distribution of the fluid velocity $\mathbf{v}(\mathbf{x}, t)$, of the deviatoric tensor $\mathbb{V}$, and of the two scalar functions the pressure $p(\mathbf{x}, t)$, and density $\rho(\mathbf{x}, t)$.

Thermodynamics studies the effect of heat on motions of a continuum body. The heat may influence the motion of general fluids only indirectly through a functional law of the pressure and the deviatoric tensor in terms of the density.

For fluids in which *processes of thermal conduction occurs*, the state is described by the previous variables, plus one variable describing the thermal state of the fluid. In this book we use the *temperature* as the new independent thermodynamical variable. As equations of mechanics we continue to use the continuity and momentum equations. To deal with concrete models, next we shall adopt the Newtonian stresses (1.4.30) that will bring to the Euler or Navier–Stokes as momentum equations.

1.5.1 Thermodynamical Variables

Thermodynamics studies effects of heat on physical systems analyzing heat transfers and joint heat-work actions on materials. Specifically, a thermal system exchanges heat and work with the environment through mechanical and thermal interactions. To this study it is need the introduction of a primary independent variable to describe the thermal state of a body given by the **empirical temperature**. We begin with the introduction of an empirical scale of temperature which signs the heatness level of the body.

First request is that heatness level is one to one with numbers of an oriented line, each number furnishes an empirical temperature, which depends on the choice of thermometer. A justification on the possibility of construction of a thermometer is provided by the experimental statement by Fowler, *Two systems in thermal equilibrium with a third one are also in thermal equilibrium each other.*

We now define an ideal instrument.

Definition 1.5.1 *The thermostat $\mathcal{T}$ is a body with uniform temperature. The body $\mathcal{B}$ is said thermometer if, in contact with the thermostat $\mathcal{T}$, assumes the same temperature of $\mathcal{T}$, while $\mathcal{T}$ doesn't change its temperature.*

We postulate

Zero Principle of Temperature Do exist perfect termometers, equals for all observers. It exists a infimum for the empirical temperature and it is possible to choose a scale of temperature where the infimum is zero. The temperature measured with such scale is said **absolute temperature** and it is denoted by θ.

To study thermodynamics of irreversible processes, we work in space-time reference frame $\{\mathcal{R}, t\}$.

Definition 1.5.2 *The **thermal state** of a continuous body at a given time t in the position* $\mathbf{x}$*, is characterized by the particle* $\mathbf{X} = \mathbf{X}(\mathbf{x}, t)$*, the time t, and the empirical temperature* $\theta(\mathbf{x}, t)$*, and is said the **thermo-kinetic process**, given by*

$$\Theta(\mathbf{x}, t) := (\mathbf{x}(\mathbf{X}, t), t, \theta(\mathbf{x}(\mathbf{X}, t), t)).$$

For fluids the thermo-kinetic process simplifies as

Definition 1.5.3 *The motion of a fluid is given by*

$$\Theta(\mathbf{x}, t) := (\mathbf{x}, t, \rho(\mathbf{x}, t), \mathbf{v}(\mathbf{x}, t), \theta(\mathbf{x}, t)).$$

Notice that the measure unities for time, space, mass and temperature are dimensionally independent.

We are now in the position to introduce thermodynamical potentials.

Definition 1.5.4 *The **internal energy** $\mathcal{E}(C)$ of the portion of continuum filling the region C is given by*

$$\mathcal{E}(C) = \int_C \rho \epsilon(\mathbf{x}, t) dx, \tag{1.5.1}$$

*where ϵ is the **density of internal energy per unit of mass**, or **specific internal energy**.*

For all thermodynamical variables, such as density, momentum, internal energy, etc., we have started with the balance laws of macroscopic quantities to bring us to a consistent theory; we have been able to do this because we have a measurement instrument for each macroscopic quantity, and as such we may control the truth of our axioms. The only macroscopic quantity for which there is no known instrument of measurement is the **entropy**, therefore along with Callen, we postulate the existence of a functional, called the entropy. We set

Definition 1.5.5 *The **entropy** of a portion of continuum filling the region C $S(C)$ is given by*

$$S(C) = \int_C \rho \, s(\mathbf{x}, t) \, dx, \tag{1.5.2}$$

*where s is the **density of entropy per unit of mass**, or **specific entropy**.*

Thermodynamics is based on the balance laws of total energy and entropy, and assumes that the increase in time of the total energy and entropy for a continuum $\mathcal{F}$ contained in the region C are due to total energy and entropy fluxes through the boundary ∂C, plus the total energy and entropy sources acting in C. The entropy sources are due to irreversible phenomena inside C. *The main goal here is to quantify explicitly entropy sources created by irreversible processes* [21].

1.5.2 The First Law of Thermodynamics

Concerning state equations we begin with the definition of total energy.

Definition 1.5.6 *The **total energy** e per unit mass is the sum of kinetic energy $v^2/2$ plus the internal energy ϵ per unit mass, i.e.*

$$e = v^2/2 + \epsilon.$$

The **first law of thermodynamics**, states that, for all fixed volume C, the increase in time of total energy $\mathcal{E}(C)$

$$\frac{d}{dt} \int_C (\rho\, e)(\mathbf{x}, t)\, dx, \tag{1.5.3}$$

equals the flux of the total energy of fluid flowing in the volume C

$$\int_{\partial C} \rho\, e\, \mathbf{v} \cdot \mathbf{n}\, dS, \tag{1.5.4}$$

plus the long and short range sources of total energy acting on, and through, the volume C

$$\int_C r_e(\mathbf{x}, t) dx + \int_{\partial C} \mathbf{h} \cdot \mathbf{n}\, dS, \tag{1.5.5}$$

where $\mathbf{h}$ is the surface production of total energy, and r_e is the total energy source. In fact, the first law of thermodynamics postulates the **equation governing the evolution of total energy in macroscopic form**

$$\frac{d}{dt} \int_C (\rho\, e)(\mathbf{x}, t)\, dx + \int_{\partial C} \rho\, e\, \mathbf{v} \cdot \mathbf{n}\, dS = \int_C r_e(\mathbf{x}, t) dx + \int_{\partial C} \mathbf{h} \cdot \mathbf{n}\, dS. \tag{1.5.6}$$

Multiplying (1.4.18) by $\mathbf{v}$, integrating over C, and applying the transport theorem, we get

$$\frac{1}{2}\frac{d}{dt}\int_C \rho\,\mathbf{v}^2\,dx + \frac{1}{2}\int_{\partial C}\rho\,\mathbf{v}^2\,\mathbf{v}\cdot\mathbf{n}\,dS = \int_C \rho\mathbf{f}\cdot\mathbf{v}\,dx - \int_C \mathbf{T}\cdot\nabla\mathbf{v}\,dx + \int_{\partial C}\mathbf{v}\cdot\mathbf{T}\mathbf{n}\,dS.$$

$$(1.5.7)$$

Subtracting (1.5.7) from (1.5.6), and recalling the definition of internal energy, we deduce

$$\frac{d}{dt}\int_C(\rho\,\epsilon)(\mathbf{x},t)\,dx + \int_{\partial C}\rho\,\epsilon\,\mathbf{v}\cdot\mathbf{n}\,dS = \int_C r(\mathbf{x},t)dx + \int_C\mathbf{T}\cdot\nabla\mathbf{v}\,dx + \int_{\partial C}\mathbf{q}\cdot\mathbf{n}\,dS,$$

$$(1.5.8)$$

where

$$r = \Big(r_e - \rho\mathbf{f}\cdot\mathbf{v}\Big),$$

is the internal energy source, or the energy source due to heat, and

$$-\mathbf{q} = \Big(\mathbf{h} - \mathbf{v}\cdot\mathbf{T}\Big),$$

is the heat flux of the internal energy, or the external surface production of internal energy.

We can now derive the equation governing the evolution of the specific internal energy ϵ.

As with previous macroscopic laws resulting from the variable nature of C, and the continuity equation we may easily derive the local form of the energy equation. The first law of thermodynamics introduces the following **equation for the internal energy ϵ**

$$\rho\frac{d\epsilon}{dt} = \nabla\cdot\mathbf{q} - p\nabla\cdot\mathbf{v} + \Xi + \rho\,r, \qquad p = p(\rho,\theta), \qquad \epsilon = \epsilon(\rho,\theta), \quad (1.5.9)$$

where $\frac{d}{dt} = \partial_t + \mathbf{v}\cdot\nabla$ is the material derivative, r is the energy source per unit of mass, $\mathbf{q} = \mathbf{q}(\theta)$ the heat flux vector, Ξ the **kinematical dissipation**

$$\Xi = 2\mu D(\mathbf{v}):D(\mathbf{v}) + \lambda(\nabla\cdot\mathbf{v})^2. \qquad (1.5.10)$$

Assuming linear state equations for $\mathbf{q}$ in terms of $\nabla\theta$, we obtain the following constitutive equation for the heat flux, known as the **Fourier law (1768-1830)**,

$$\mathbf{q} = -\chi\,\nabla\theta, \qquad (1.5.11)$$

where the constant χ denotes the heat diffusivity if it is positive. In the next section we shall see that $\chi > 0$ follows from the Clausius–Duhem inequality. The precise dependence of χ on the motion of continuum, or its value if constant, is given by phenomenological laws.

Some time we use

$$\frac{d\varphi}{dt} = \partial_t\varphi + \mathbf{v} \cdot \nabla\varphi = \dot\varphi.$$

1.5.3 The Second Law of Thermodynamics

Recalling the definition of entropy (1.5.2), following up on the work outlined in Sect. 1.3.1, for any macroscopic portion filling a generic region C, we introduce as a state equation the entropy $S(C)$ of a portion of continuum filling the region C. We experimentally observe that

The variation in time of entropy $\dot S(C)$ in the continuum filling C is due to two variable influences (1) the variation of external entropy $\dot S_e$, supplied to C by its surroundings, and (2) the variation of internal entropy $\dot S_i(C)$, produced inside the system

$$\dot S(C) = \dot S_i(C) + \dot S_e(C).$$

The variation in time of internal entropy $\dot S_i$ must vanish at equilibrium and along reversible transformations

$$\dot S_i(C) = 0, \qquad \forall C,$$

while $\dot S_i(C)$ must be positive along irreversible processes

$$\dot S_i(C) \geq 0, \qquad \forall C.$$

The supply for the variation in time of external entropy $\dot S_e(C)$ has no definite sign, however *adiabatic processes* in an *isolated system* must be $\dot S_e(C) = 0$, which yields

$$\dot S(C) \geq 0.$$

In the thermodynamics of **irreversible processes**, one objective is to relate the entropy variations $\dot S(C)$ to various irreversible phenomena occurring inside and outside the continuum body. The **second law of thermodynamics** states that

$$\frac{dS}{dt} \geq \int_{\partial C} \mathbf{J}\, d\mathcal{S} + \int_C r_1\, d\,x, \qquad (1.5.12)$$

$\mathbf{J}$ is the entropy flux per unit area and unit time, and r_1 is the entropy body source per unit volume and unit time.

Again owing to the variability of volume C, we may derive the entropy equation in local form. Unfortunately, this equation introduces more unknowns than equations. Here we use a simplified version of the second law of thermodynamics; specifically, we assume that the entropy flux and entropy source are given by

$$\mathbf{J} = \frac{\boldsymbol{q}}{\theta},$$

$$r_1 = \frac{\rho\, r}{\theta}. \tag{1.5.13}$$

Assuming sufficient regularity in the constitutive functionals, the variability of C, we get the **second law of thermodynamics** in local form

$$\rho\frac{ds}{dt} \geq -\nabla \cdot \left(\frac{\boldsymbol{q}}{\theta}\right) + \rho\frac{r}{\theta}. \tag{1.5.14}$$

1.5.4   Clausius–Duhem Inequality

Let us write (1.5.9) in the equivalent form

$$-\nabla \cdot \boldsymbol{q} + \rho r = \rho\frac{d\epsilon}{dt} + p\nabla \cdot \mathbf{v} - \Xi, \tag{1.5.15}$$

where Ξ is defined in (1.5.10). Next we rewrite (1.5.14) in the following form

$$\rho\theta\frac{ds}{dt} \geq -\nabla \cdot \boldsymbol{q} + \frac{\boldsymbol{q}\cdot\nabla\theta}{\theta} + \rho r. \tag{1.5.16}$$

Substituting the energy equation (1.5.15) into (1.5.14) multiplied by θ, we deduce the well known **Clausius–Duhem inequality**

$$\rho\theta\frac{ds}{dt} - \rho\frac{d\epsilon}{dt} - p\nabla \cdot \mathbf{v} + \Xi - \frac{\boldsymbol{q}\cdot\nabla\theta}{\theta} \geq 0. \tag{1.5.17}$$

This equation is true for all regular motions and allows us to deduce further constraints on the constitutive laws. Actually the Clausius–Duhem inequality, is called **dissipation principle**, when it is employed the arbitrariness of thermodynamical processes. The condition of validity of (1.5.17) for all thermodynamical processes infers severe constraints on constitutive relations. Below we furnish two examples.

Compressible fluids

We assume the following constitutive hypotheses

$$\epsilon = \epsilon(\rho,\theta), \qquad s = s(\rho,\theta), \qquad p = p(\rho,\theta), \tag{1.5.18}$$

$$\mathbf{T} = -p\mathbf{I} + \lambda\nabla \cdot \mathbf{v} + 2\mu\mathbf{D}, \qquad \boldsymbol{q} = -\chi\nabla\theta.$$

Substituting constitutive hypotheses (1.5.18) in (1.5.17) we obtain

$$\rho\left(\theta\frac{\partial s}{\partial\rho}-\frac{\partial\epsilon}{\partial\rho}+\frac{p}{\rho^2}\right)\frac{d\rho}{dt}+\rho\left(\theta\frac{\partial s}{\partial\theta}-\frac{\partial\epsilon}{\partial\theta}\right)\frac{d\theta}{dt}$$

$$+\lambda(\nabla\cdot\mathbf{v})^2+2\mu\mathbf{D}:\mathbf{D}+\chi\left(\frac{(\nabla\theta)^2}{\theta}\right)\geq 0. \qquad (1.5.19)$$

Notice that

$$\frac{d\rho}{dt},\qquad \frac{d\theta}{dt},\qquad \nabla\theta,\qquad \mathbf{D},$$

are independent variables, moreover the coefficients of these variables are independent of these values. Therefore by letting suitably varying the above variables in (1.5.21) we deduce

$$\frac{p}{\rho^2}=\frac{\partial\epsilon}{\partial\rho}-\theta\frac{\partial s}{\partial\rho},\qquad \frac{\partial\epsilon}{\partial\theta}=\theta\frac{\partial s}{\partial\theta},$$

$$\mu\geq 0,\qquad 3\lambda+2\mu\geq 0,\qquad \chi\geq 0.$$

Incompressible, homogeneous fluids

For incompressible homogeneous fluids, we assume initially uniform density, and this infers $\rho(\mathbf{x},t) = const. = 1$. Thus the following constitutive hypotheses hold

$$\epsilon=\epsilon(\theta),\qquad s=s(\theta),\qquad p=p(\mathbf{x},t), \qquad (1.5.20)$$

$$\mathbf{T}=-p\mathbf{I}+2\mu\mathbf{D},\qquad q=-\chi\nabla\theta.$$

Substituting constitutive hypotheses (1.5.20) in (1.5.17) we obtain

$$\rho\left(\theta\frac{ds}{d\theta}-\frac{d\epsilon}{d\theta}\right)\frac{d\theta}{dt}+2\mu\mathbf{D}:\mathbf{D}+\chi\left(\frac{(\nabla\theta)^2}{\theta}\right)\geq 0. \qquad (1.5.21)$$

Set $\mathbf{D}_0$ the symmetric tensor with zero trace. Notice that

$$\frac{d\theta}{dt},\qquad \nabla\theta,\qquad \mathbf{D}_0,$$

are independent variables, moreover the coefficients of these variables are independent of these values. Therefore by letting suitably varying the above variables in (1.5.21) we deduce

$$\frac{d\epsilon}{d\theta}=\theta\frac{ds}{d\theta},\qquad \mu\geq 0\qquad \chi\geq 0.$$

In this case no informations on the pressure are available, and the pressure becomes a dynamical unknown.

1.5.5   *Phenomenological Laws*

Let us recall some basic state equations.

Important phenomenological laws include the *constitutive equations* for pressure $p = p(\rho, \theta)$, internal specific energy $\epsilon(\rho, \theta)$, internal specific entropy $s(\rho, \theta)$ as given positive functions of ρ and θ. The thermodynamic potentials $p(\rho, \theta)$, $\epsilon(\rho, \theta)$, $s(\rho, \theta)$ determine the thermodynamical state of system. The reduced dissipation principle (1.5.20) has demonstrated that for any motion, the following local equation must be satisfied:

$$p(\rho, \theta) = \rho^2 \Big(\epsilon_\rho(\rho, \theta) - \theta s_\rho(\rho, \theta) \Big), \qquad (1.5.22)$$

where the subscript ρ denotes the partial derivative with respect to ρ.

We are using as independent variables temperature θ and density ρ. The fluid $\mathcal{F}$ is said to be **polytropic** if along its motions its internal energy is a linear function of temperature, i.e.

$$\left. \frac{\partial \epsilon}{\partial \theta} \right|_{\rho = const.} = c_V, \qquad \qquad \epsilon = c_v \theta,$$

where the positive constant c_V denotes the *specific heat at constant volume*.

Along all stationary thermal processes, almost all gases experience the Boyle–Mariot law, which states that

$$p(\rho, \theta) = R_* \rho \theta, \qquad (1.5.23)$$

where $R_* > 0$ is the universal gas constant. A gas satisfying this law is called a *perfect gas*; cf. [65] Sect. 80. Equation (1.5.23) is customarily adopted for wide classes of gas. There is a safety range for $r_1 < \rho < r_2$ and $\tau_1 < \theta < \tau_2$ where the gas satisfies (1.5.23). However, for each gas there is a threshold θ_c for absolute temperature, below which it starts to become liquid. When a phase change occurs, other state equations become more appropriate, e.g. *the Van der Waals law* (1837-1923). More general phenomenological state equations for the pressure may be proposed, all satisfying the natural requests; cf. [23, 26–28, 39]. The instability of phase changes will be studied in Chap. 3.

We will end this subsection by reviewing some definitions that will be used in the next chapter.

Definition 1.5.7 *The **enthalpy** per unit of mass and per unit of volume is the thermodynamical potential $\Phi(\rho)$ given by*

$$\Phi(\rho) := \int^\rho \frac{p'(s)}{s} ds; \qquad \rho \Phi(\rho) := \rho \int^\rho \frac{p'(s)}{s} ds. \qquad (1.5.24)$$

*The **Helmholtz (1821-1894) free energy** per unit of mass and per unit of volume is the thermodynamical potential $\Psi(\rho)$ by*

$$\Psi(\rho) := \int^{\rho} \frac{p(s)}{s^2} ds; \qquad \rho\Psi(\rho) := \rho \int^{\rho} \frac{p(s)}{s^2} ds. \qquad (1.5.25)$$

From these given definitions it follows that

$$\rho \frac{d^2}{d\rho^2}(\rho\Psi(\rho)) = p'(\rho),$$

$$\rho\left(\Psi(\rho) - \Phi(\rho_b)\right) = \frac{1}{2} \frac{p'(\overline{\rho})}{\overline{\rho}} \sigma^2, \qquad (1.5.26)$$

where $\overline{\rho}$ is a point between ρ, and ρ_b. Equation $(1.5.26)_2$ seems to be new, it will play a crucial role in the study of stability of barotropic fluids.

1.5.6 Compressible Euler Equations

For isotropic fluids in which *processes of thermal conduction and internal friction do not occur*, stress acts as a pure pressure in the direction normal to the surface

$$\mathbf{Tn} = \mathbf{t}(\mathbf{x}, t, \mathbf{n}) = -p(\mathbf{x}, t)\mathbf{n}. \qquad (1.5.27)$$

The continuity (1.4.3) and Euler (1.4.22) equations still hold, however we now have density, velocity and pressure as unknowns, and the number of unknowns once again exceeds the number of equations.

Towards a new equation

If thermodynamics modifies the motion of a fluid only through pressure, we can assume either the thermodynamic constraint of isothermal motions, or the thermodynamic constraint of adiabatic motions.

Isothermal flows

Assume the motion occurs at constant temperature Θ_*.

Let R_* be the universal gas constant, then the Boyle–Mariot law states that

$$p = R_* \rho \Theta_*. \qquad (1.5.28)$$

Fluids satisfying (1.5.28) are called **isothermal**. Thus, the equations governing the motion of a ideal fluid are (1.4.3), (1.4.22) and (1.5.28).

Adiabatic flows

Take the independent thermodynamic variables ρ and s, and assume $s = s_* = const.$. Then the temperature Θ verifies $\Theta = \Theta(\rho, s_*)$; with regards to pressure for the Gibbs relations, we can deduce

$$p = k\rho^m, \tag{1.5.29}$$

where m is the polytropy index. Fluids satisfying (1.5.29) are called **isentropic**.

Barotropic flows

If one of the classical thermodynamic relations isn't used, for pressure we can assume

$$p = p(\rho). \tag{1.5.30}$$

Fluids satisfying (1.5.30) are called *barotropic*.

Thus, equations governing the motion of a ideal fluid are (1.4.3), (1.4.22) and (1.5.30), and are called **Compressible Euler equations**.

Unsteady flows

$$\partial_t \rho + \nabla \cdot (\rho \mathbf{v}) = 0,$$
$$\partial_t (\rho \mathbf{v}) + \nabla(\rho \mathbf{v} \otimes \mathbf{v}) = -\nabla p + \rho \mathbf{f},$$
$$p = p(\rho), \qquad (x, t) \in \Omega \times (0, T), \qquad (1.5.31)$$
$$\rho(x, 0 = \rho_0(x), \qquad \mathbf{v}(\mathbf{x}, \mathbf{0}) = \mathbf{v_0}(\mathbf{x}), \qquad \mathbf{x} \in \mathbf{\Omega}.$$

Steady flows

$$\nabla \cdot (\rho \mathbf{v}) = 0,$$
$$\nabla(\rho \mathbf{v} \otimes \mathbf{v}) = -\nabla p + \rho \mathbf{f}, \qquad x \in \Omega,$$
$$p = p(\rho), \qquad x \in \Omega.$$

These equations must be completed with boundary conditions.

1.5.7   *Linearly Viscous Fluids*

If *internal friction is relevant to motion* and *processes of thermal conduction do not directly occur*, then following the line of Sect. 4.8, we can deduce that the state of a fluid is described by five unknown variables: velocity (3), pressure (1), density (1) plus unknown coefficients λ and μ. In fact, again by constitutive axioms, for simple materials, it follows that $\mathbf{S}(\mathbf{x}, t) = S(\mathbf{D}, t)$; cf. [6,143]. For linear state equations we obtain the Newtonian equation (1.4.30) for the stress tensor. In this case, the shear μ and bulk λ coefficients are still assumed constants, however they must obey (in three-dimensional domains) the Clausisus–Duhem inequality. Hence we deduce the conditions

$$\mu \geq 0, \qquad 2\mu + 3\lambda \geq 0, \qquad (1.5.32)$$

where the number 3 refers to the dimension of space in which the motion occurs. This result follows from thermodynamics, however in next section we shall see that they follow more exactly from the Clausius–Duhem inequality. Viscosity coefficients are given by phenomenological laws.

Remark 1.5.1 *The most important difference between the mechanical properties of liquids and gases lies in their bulk elasticity, given by the viscosity coefficient λ, which expresses the compressibility of the fluid.*

Thus we obtain the compressible Navier–Stokes equations.
Compressible Navier–Stokes equations
Unsteady flows

$$\partial_t \rho + \nabla \cdot (\rho \mathbf{v}) = 0,$$
$$\partial_t (\rho \mathbf{v}) + \nabla(\rho \mathbf{v} \otimes \mathbf{v}) = -\nabla p + (\lambda + \mu)\nabla(\nabla \cdot \mathbf{v}) + \mu \Delta \mathbf{v} + \rho \mathbf{f},$$
$$p = p(\rho), \qquad\qquad (x, t) \in \Omega \times (0, T), \qquad (1.5.33)$$
$$\rho(x, 0 = \rho_0(x), \qquad \mathbf{v}(\mathbf{x}, \mathbf{0}) = \mathbf{v_0}(\mathbf{x}), \qquad\qquad \mathbf{x} \in \mathbf{\Omega}.$$

Steady flows

$$\nabla \cdot (\rho \mathbf{v}) = 0,$$
$$\nabla(\rho \mathbf{v} \otimes \mathbf{v}) = -\nabla p + (\lambda + \mu)\nabla(\nabla \cdot \mathbf{v}) + \mu \Delta \mathbf{v} + \rho \mathbf{f},$$
$$p = p(\rho), \qquad\qquad x \in \Omega. \qquad (1.5.34)$$

These equations must be completed with boundary conditions.

1.5.8   Heat-Conducting Fluids

If *processes of thermal conduction and internal friction do occur*, then we must consider the full set of equations governing thermodynamical flows.

Unsteady flows

$$\frac{\partial \rho}{\partial t} + \nabla \cdot (\rho \mathbf{v}) = 0, \qquad\qquad x, t \in \Omega \times (0, T),$$

$$\rho \frac{d\mathbf{v}}{dt} = \nabla \cdot \mathbf{T} + \rho \mathbf{f}, \qquad\qquad x, t \in \Omega \times (0, T),$$

$$\rho \frac{d\epsilon}{dt} = \mathbf{T} : \mathbf{D} - \nabla \cdot \mathbf{q} + \rho r, \qquad x, t \in \Omega \times (0, T),$$

$$\rho(x, 0) = \rho_0(x), \qquad \mathbf{v}(x, 0) = \mathbf{v}_0(x), \; \theta(x, 0) = \theta_0(x), \qquad x \in \Omega.$$

(1.5.35)

with constitutive equations

$$p = R_* \theta \rho, \qquad \epsilon = \epsilon(\rho, \theta), \qquad\qquad (1.5.36)$$

$$\mathbf{T} = -p\mathbf{I} + \lambda \nabla \cdot \mathbf{v} + 2\mu \mathbf{D}, \qquad \mathbf{q} = -\chi \nabla \theta,$$

$$\mu \geq 0, \qquad\qquad 3\lambda + 2\mu \geq 0, \qquad\qquad \chi \geq 0, \qquad\qquad (1.5.37)$$

(1.5.35) represents a system of five scalar equations in the five unknown functions ρ, v_1, v_2, v_3, θ of $(\mathbf{x}, t)$.

Steady flows

$$\nabla \cdot (\rho \mathbf{v}) = 0,$$

$$\rho \mathbf{v} \cdot \nabla \mathbf{v} = \rho \mathbf{b} + \nabla \cdot \mathbf{T} \qquad\qquad (1.5.38)$$

$$\rho \mathbf{v} \cdot \nabla \epsilon = \mathbf{T} : \mathbf{D} + r - \nabla \cdot \mathbf{q}.$$

Equation (1.5.38) represents a system of five scalar equations in the five unknown functions ρ, v_1, v_2, v_3, θ of $\mathbf{x}$.

In order to close systems (1.5.35), (1.5.38) we must control boundary terms.

1.6   Side Conditions

In order to determine solutions to either one of the initial boundary value problems (1.4.25), or (1.4.31), or (1.5.31) or (1.5.33), (1.5.35) or one of the boundary value problems (1.4.26), or (1.4.8), or (1.5.6) or (1.5.34), (1.5.38) we must prescribe the side and boundary conditions that describe the environment where the motion occurs.

Conditions occurring at boundary between a compressible fluid and some other medium warrant special consideration because they *give rise to several important phenomena.* The boundary may separate different phases, solid, liquid or gaseous, or it may separate two media of the same phase but different constitution or at different flow regimes.

Notice that the correct boundary conditions, BC, must satisfy three criteria: (1) on one side, the BCs must be physically controllable; (2) on the other side, the BCs must be not so many that they make existence impossible; (3) the BC must be enough so as to ensure uniqueness. In this section we will cover several kinds of correct boundary conditions.

In the sequel we study only viscous fluids. For solutions to PDE systems of second order in space, it is customary to prescribe either Dirichlet or Neumann conditions at the boundaries, as such:

(1) *Steady flows*, systems (1.4.8), and (1.5.34), or (1.5.38) become of elliptic type in $\mathbf{v}$, or in $\mathbf{v}$, θ, respectively. It is not, however, clear which type of condition one should prescribe on density, as the steady continuity equation results just a first order PDE not characterized in canonical way; cf. [57, 89, 128].

(2) *Unsteady flow* systems (1.4.31), and (1.5.33), or (1.5.35) become of parabolic type in $\mathbf{v}$ and in $\mathbf{v}$, θ respectively, and of hyperbolic type in ρ.

1.6.1 Classes of Boundary Conditions

In order to set boundary conditions on $\mathbf{v}$, distinctions are made between:
(**A**) the type of boundary, and (**B**) the type of domain.

(**A**) Depending on the type of boundary, $\partial\Omega$ can be union of surfaces that are either:

(**A-i**) *Rigid moving*[7]: boundary is rigid, impermeable, moving
(**A-ii**) *Rigid moving*: boundary is rigid, porous, moving
(**A-iii**) *Stress free boundary*: *known deformable boundary* boundary is in contact with a known external medium
(**A-iv**) *Free boundary with, or without, capillary effect*: an *unknown deformable boundary* in contact with an external medium

B) Depending on the type of domain, Ω may have one of the following characteristics:

(**B-i**) *Bounded*
(**B-ii**) *Exterior to a compact region*
(**B-iii**) *With a non-compact boundary*

Remark 1.6.1 Motion in a layer *with a free boundary may occur when the fluid moves over an inclined plane (see Couette, Poisuelle flows). Stability*

---

[7]If $\partial\Omega$ is a rigid, impermeable moving domain in the reference $\mathcal{R}$, by suitably changing the external forces, it may be transformed to a rigid fixed domain in another reference $\mathcal{R}'$.

properties of these flows depend on the thickness of the layer, and on the inclination of the plane; cf. [75, 76, 103].

1.6.2   Boundary Conditions on Velocity

Concerning the value of the velocity of fluid particles at a boundary, we can apply either the Dirichlet or Neumann conditions depending upon the type of walls in question, either perfect heat conductors or totally adiabatic, respectively.

(A) Type of boundary

(A-i) $\mathcal{S} \subseteq \partial\Omega$: rigid and impermeable.
   We assume the *adherence condition* of the viscous fluid to $\mathcal{S}$, that is

$$\mathbf{v}(x,t) = \mathbf{w}(x,t), \qquad (x,t) \in \mathcal{S} \times (0,\infty),$$

where $\mathbf{w}$ is a given vector function representing the velocity of the points of $\mathcal{S}$, and $\mathbf{v}$ is the velocity of fluid particles at boundary $\partial\Omega$. In particular, in the reference $\mathcal{R}'$ where $\mathcal{S}$ is fixed we assume $\mathbf{v}$ to vanish.

(A-ii) $\mathcal{S}' \subseteq \partial\Omega$: rigid and porous.
   We continue to apply

$$\mathbf{v}(x,t) = \mathbf{w}'(x,t), \qquad (x,t) \in \mathcal{S}' \times (0,\infty),$$

where $\mathbf{w}'$ is still a given vector function, however $\mathbf{w}'$ is no longer the velocity of the points of $\mathcal{S}'$, while $\mathbf{v}$ is the velocity of fluid particles at boundary $\partial\Omega$. In order to adhere to the law of conservation of mass, the following condition must be satisfied[8]

$$\int_{\partial\Omega} \rho\,\mathbf{w}' \cdot \mathbf{n}\,dx = 0. \tag{1.6.1}$$

Note that, despite incompressible fluids, where density is known, (1.6.1) doesn't represent a compatibility condition for compressible fluids, because ρ is unknown.

(A-iii) $\partial\Omega_t$: deformable, *outside Ω_t there is a known body $\mathcal{C}(t)$.*
   On the geometric surface $S(t) = \partial\Omega_t \cap \partial\mathcal{C}(t)$ we assume impermeability

$$\mathbf{v}(x,t) \cdot \mathbf{n}(x,t) = \mathbf{w}(x,t) \cdot \mathbf{n}(x,t), \qquad (x,t) \in S(t) \times (0,\infty), \quad (1.6.2)$$

where $\mathbf{n}$ is the exterior normal to $S(t)$, $\mathbf{w}$ is the velocity of the material points at $\partial\mathcal{C}(t)$, and $\mathbf{v}$ is the velocity of fluid particles at boundary $\partial\Omega_t$. We remark that in this case, $S(t)$ is a *known material surface*.

---

[8]The total mass may vary in a porous region, where $(1.6.1)$ doesn't hold.

(A-iv) $\partial\Omega_t$: deformable, *outside Ω_t there is a unknown body $\mathcal{C}(t)$.*
On the geometric surface $\Gamma_t = \partial\Omega_t \cap \partial\mathcal{C}(t)$ we assume impermeability

$$\mathbf{v}(x,t)\cdot\mathbf{n}(x,t) = \mathbf{w}(x,t)\cdot\mathbf{n}(x,t), \qquad\qquad (x,t)\in\Gamma_t\times(0,\infty), \qquad (1.6.3)$$

$$\mathbf{w}_{\mathcal{C}}(x,t)\cdot\mathbf{n}(x,t) = \mathbf{w}(x,t)\cdot\mathbf{n}(x,t), \qquad\qquad (x,t)\in\Gamma_t\times(0,\infty),$$

where $\mathbf{n}$ is the exterior normal to Γ_t, $\mathbf{w}$ is the velocity of the points of Γ_t, $\mathbf{v}(x,t)$ is the velocity of fluid particles at boundary $\partial\Omega_t$, and $\mathbf{w}_{\mathcal{C}}$ is the velocity of material points at $\partial\mathcal{C}(t)$. However now Γ_t is no longer a known material surface, but rather itself an unknown.

(A-iv)-(I) When only the surface Γ_t is unknown, and the exterior is a vacuum, we use only the kinematical condition $(1.6.3)_1$, considering the motion of Γ_t described by the particle x, (i.e. $\frac{dx}{dt} = \mathbf{w}$)

$$\frac{dx}{dt}\cdot\mathbf{n}(x,t) = \mathbf{v}(x,t)\cdot\mathbf{n}(x,t), \qquad\qquad (x,t)\in\Gamma_t\times(0,\infty).$$

This condition allows for the sleeping of the fluid on the surface, and constitutes a more difficult problem.

(A-iv)-(II) On the part Γ_t' of the boundary of Ω_t where there is an unknown body $\mathcal{C}(t)$, we prescribe continuity on the tangential component of the velocity fields on Γ_t'; cf. [19]. That is:

$$\frac{dx}{dt} = \mathbf{v}(x,t) = \mathbf{w}(x,t), \qquad\qquad (x,t)\in\Gamma_t'\times(0,\infty).$$

In general, the boundary may be the union of the parts referred to above, say $\partial\Omega_t = \Sigma\cup\Sigma'\cup S(t)\cup\Gamma_t\cup\Gamma_t'$, the interior of each part has an empty intersection with the other parts, and for each part the corresponding condition is assumed.

To the kinetic condition (1.6.2) we must add a *dynamical condition* which prescribes a condition on the jump of stresses $\mathbf{Tn}$ on both sides of Γ_t, see (1.6.5).

(B) Type of domain The following distinctions apply to domains:

(B-i) Ω *interior* to a *bounded* container $\mathcal{C}$
(B-ii) Ω is *exterior* to a bounded region $\mathcal{C}$
(B-iii) Ω with a non compact boundary

(B-i) Let $\mathcal{C}(t)$ be a rigid container (a shell), in motion in reference frame $\mathcal{R} = \{O, \mathbf{i}, \mathbf{j}, \mathbf{k}\}$. As known, if $\mathbf{v}_G$ is the velocity of the center of mass of $\mathcal{C}(t)$, and $\omega(t)$ denotes its angular velocity at time t, then the velocity $\mathbf{w}$ of its generic particle $x\in\mathcal{C}(t)$ is given by

$$\mathbf{w}(x,t) = \mathbf{v}_G(t) + \omega(t)\times x, \qquad\qquad (x,t)\in\mathcal{C}(t)\times(0,\infty).$$

On $\partial \mathcal{C}(t) = \partial \Omega_t$ we prescribe the adherence condition

$$\mathbf{v}(x,t) = \mathbf{w}(x,t) = \mathbf{v}_G(t) + \omega(t) \times x, \qquad (x,t) \in \partial \Omega_t \times (0,\infty).$$

In **(B-ii)** we apply in the reference frame $\mathcal{R}$ the conditions at infinity

$$\lim_{|x| \to \infty} \mathbf{v}(x,t) = 0.$$

Notice that in the frame $\mathcal{R}'$ moving together with $\mathcal{C}$ with angular speed $\omega(t) \neq 0$, the *velocities of fluid particles grow to infinity at a linear rate!*

Note also that if $\omega(t) = 0$ and $\mathbf{v}_G \neq 0$ in the frame with origin attached to $G \in \mathcal{C}$, we find that

$$\widehat{\mathbf{v}}(x,t) = 0, \quad (x,t) \in \partial \Omega \times (0,\infty), \qquad \lim_{|x| \to \infty} \widehat{\mathbf{v}}(x,t) = -\mathbf{v}_G(t), \quad t \in (0,\infty),$$
$$(1.6.4)$$

where $\widehat{\mathbf{v}}(x,t)$ is the velocity of the fluid particles in the new frame. In this frame, the momentum equation will differ from equations (1.4.3), (1.4.20) and (1.5.9) since it presents new linear terms. The resulting linearized equations for incompressible fluids have been proposed by Oseen (1879-1944); cf. [88], and are known as *Oseen equations*.

Analogous equations can be introduced for compressible fluids and will be called **compressible Oseen equations**.

The compressible Oseen equations have been studied for compressible fluids, in the steady state we direct the reader to the results of Novotny and the author; cf. [85, 86].

(B-iii) In this text we will consider only thin horizontal layers of fluid motion with capillary effects. Note that we will take into account the surface tension, as it has a dissipative effect. Moreover, assuming periodicity on the flat variable, the problem will be reduced to a bounded domain.

1.6.3 Boundary Conditions on Stress

To (1.6.2) we must add a *dynamic condition*, whereby we assume a jump condition on the stress $\mathbf{Tn}$ on both sides of Γ_t, as follows

$$\mathbf{Tn} = \mathbf{T}_C \mathbf{n} + \mathbf{c}, \qquad (\mathbf{y},t) \in \Gamma_t \times (0,\infty), \qquad (1.6.5)$$

where $\mathbf{T}_C$ is the known stress tensor of C, and $\mathbf{c}$ is the vectorial operator responsible for capillary effects, relevant only within the microscopic scale.

The effect of an 'elastic' structure could be taken into account in order to introduce a discontinuity in normal stress proportional to the operator K_1 describing the dynamic behavior of the structure. In the linear case one may assume that *any deformation creates stresses only in the **normal** direction,*

$$\mathbf{T}_e = \Big(K_1(\mathbf{y}) - p_e\Big)\mathbf{n},$$

where p_e is the constant external pressure. On microscopic scale the stress due to capillary effects depends on the curvature of the surface through a factor κ called **surface tension**. The surface tension may depend on density, temperature, and has no fixed sign. For volatile gases κ is negative. In the sequel we take as a positive constant, and we deduce

$$\mathbf{c}(\mathbf{y}, t) = \kappa \, \mathcal{H}(\mathbf{y}) \mathbf{n}(\mathbf{y}, t), \qquad \mathbf{y} \in \Gamma_t, \ t \in (0, T).$$

The jump condition between the stress tensors of fluid $\mathbf{T}_f(\mathbf{u}, p)$ and of elastic walls $\mathbf{T}_e$ is expressed by

$$(\mathbf{T}_f - \mathbf{T}_e)\mathbf{n} = \kappa \mathcal{H}(\mathbf{y})\mathbf{n}, \tag{1.6.6}$$

where the constant positive κ denotes surface tension. Let us set

$$S(\mathbf{u}) = (\nabla \mathbf{u} + \nabla \mathbf{u}^T) + \frac{\lambda}{\mu} \nabla \cdot \mathbf{u} \, \mathbf{I},$$

where $\mathbf{I}$ is the unitary tensor; we have introduced the constants $\mu > 0$, λ, shear and bulk viscosities verifying $3\lambda + 2\mu \geq 0$. To further our work, it is useful to set

$$T_f(\mathbf{u}, p) = -p\mathbf{I} + \mu \mathbf{S}(\mathbf{u}),$$

where p is pressure.

Assume locally Γ_t has cartesian representation $\zeta = \zeta(\mathbf{x}', t)$, $\mathbf{x}' \in \Sigma$, Σ is a rectangle. Thus we have

$$\mathbf{y} = \mathbf{y}(\mathbf{x}', \zeta(\mathbf{x}', t), \qquad \mathbf{x}' \in \Sigma, \quad t \in (0, T),$$

and we substitute $\mathbf{y}$ with the scalar variable ζ.

We can write the dynamic balance of stresses (1.6.6) in the form

$$-p + \mu \mathbf{n} \cdot S(\mathbf{v}) \cdot \mathbf{n} = K(\zeta) - p_e, \qquad \text{on } \Gamma_t,$$

$$\mathbf{t}_1 \cdot S(\mathbf{v}) \cdot \mathbf{n} = 0, \tag{1.6.7}$$

$$\mathbf{t}_2 \cdot S(\mathbf{v}) \cdot \mathbf{n} = 0, \qquad \text{on } \Gamma_t,$$

where $\mathbf{t}_\alpha$, $\alpha = 1, 2$ are the unit vectors tangent to the free surface Γ_ζ, and p_e is a constant pressure. Moreover, we introduce the *leading stress*

$$K(\zeta) = \kappa \mathcal{H}(\zeta) + \mathbf{n} \cdot \mathbf{T}_e \mathbf{n} + p_e. \tag{1.6.8}$$

Hypotheses on $K(\zeta)$

We shall assume that $K(\zeta)$ is a *linear operator* as function of ζ.

For the sake of simplicity, for stability problems one may **linearize** the effect of surface tension as a result of the double mean curvature $\mathcal{H}$.

From the *mathematical point of view*, in our approach it is not difficult to take into consideration the strong nonlinearity of $\mathcal{H}$.

From the *physical point of view*, the effect of surface tension is important only when capillary phenomena are present.

To describe regular free boundaries, we will use models of structure describing linear elastic stresses, and we set

$$K'(\zeta) = \beta\Delta'\zeta - \alpha\Delta'^2\zeta + b\zeta, \tag{1.6.9}$$

where the positive constants α, β and b denote regularity coefficients. In subsequent chapters, for specific cases we will explain the physical meaning of α, β and b. At the boundary, one may distinguish the cases of elastic or viscoelastic walls and free boundaries as follows:

Viscoelastic membrane

$$K(\zeta) := -\partial_t^2\zeta + \gamma\partial_t\Delta'\zeta + K'(\zeta), \qquad \text{on } \Sigma, \tag{1.6.10}$$

where γ denotes the visco-elasticity coefficient, and α, β and b are pure elasticity coefficients. More specifically, $\alpha > 0$, $\beta \geq 0$, and $\gamma \geq 0$ are the non-dimensional rigidity, stretching, surface tension and friction coefficients.

Elastic membrane

$$K(\zeta) := -\partial_t^2\zeta + K'(\zeta), \qquad \text{on } \Sigma, \tag{1.6.11}$$

the meaning is the same as that explained above.

Free, regular boundary

$$K(\zeta) := H'(\zeta) = \beta\Delta'\zeta, \qquad \text{on } \Sigma, \tag{1.6.12}$$

where β denotes a surface tension.

If $T_e = -p_e\mathbf{I}$ we take into account the nonlinearity of the curvature which has been defined in (1.2.3).

Here either (1.6.10), (1.6.11), or (1.6.12) completely describe the motion of the fluid in the 'elastic' vessel, and a solution will be determined once the height, density and velocity fields are fixed at an initial time.

Remark 1.6.2 *The study of the elastic model can be developed by changing kinematic and dynamic boundary conditions; cf. [19].*

1.6.4   Boundary Conditions on Temperature

Concerning the value of temperature for fluid particles at the boundary, we can apply either the Dirichlet (1805-1859) or Neumann (1832-1925) conditions, depending on the kind of wall, which can be either perfect heat conductor or totally adiabatic, respectively. Furthermore, for generic walls, we use the mixed type Dirichlet and Neumann conditions.

For perfectly conducting walls S_1, we add the Dirichlet boundary condition

$$\Theta(x,t) = \Theta_1(x,t), \qquad (x,t) \in S_1 \times (0,\infty),$$

where Θ_1 is a given scalar function representing the temperature of the points of S_1, and Θ is the temperature of material points of $\partial\Omega$.

For perfectly adiabatic walls S_2 we add the Neumann boundary condition

$$\frac{\partial}{\partial n}\Theta(x,t) = \Theta_2(x,t), \qquad (x,t) \in S_2 \times (0,\infty),$$

where $\mathbf{n}$ is the exterior normal to S_2, Θ_2 is a given scalar function representing the temperature of the points of S_2, and Θ is the temperature of fluid particles at $\partial\Omega$.

Another boundary condition for the temperature at walls of mixed type S_3 is the Newton's cooling law:

$$\chi\mathbf{n} \cdot \nabla\Theta(x,t) + \nu(\Theta - \Theta_e)(x,t) = 0, \qquad (x,t) \in S_3 \times (0,\infty), \quad (1.6.13)$$

where $\mathbf{n}$ is the exterior normal, Θ is the temperature of fluid particles at S_3, Θ_e is a given scalar function representing the temperature of the particles of S_3, and ν is the global coefficient of heat exchange between fluid particles and particles of the ambient external to S_3. Equation (1.6.13) expresses the heat balance law at the boundary.

In subsequent chapters we consider only perfectly conducting walls, $S_1 = \partial\Omega$ and we add the Dirichlet boundary condition.

For unbounded domains, we add the condition at infinity for the temperature field

$$\lim_{|x|\to\infty} \Theta(x,t) = \Theta_\infty > 0.$$

Note that for the exterior case $\Omega = \mathbb{R}^3 - \mathcal{C}$ regarding heat conducting compressible fluids, in [112] there is research supporting the nonlinear stability of the rest state when the temperature of $\partial\mathcal{C}$ is higher than the temperature at infinity, provided the gravity is large enough.

1.6.5   Side Conditions on Density

In order to have physically meaningful solutions, we add the condition

$$\rho(x,t) \geq 0, \qquad (x,t) \in \Omega \times (0,\infty). \qquad (1.6.14)$$

Therefore, in the existence theorem we must look for regular solutions with ρ also satisfying condition (1.6.14).

(A) Type of boundary

(A-i) $\mathcal{S} \subseteq \partial\Omega$: rigid and impermeable.

When we cannot physically compute the density at the boundary, the type of domain is important.

(A-ii) $\mathcal{S}' \subseteq \partial\Omega$: rigid and porous.

In this case, [128], we distinguish whether the fluid in motion is inflowing or outflowing at the boundary $\mathcal{S}'$. To elaborate, at the side of boundary where fluid is inflowing the density is considered to be a known quantity and we may prescribe the density at points $\mathcal{S}_1' = \{x \in \mathcal{S} : \quad \mathbf{v} \cdot \mathbf{n} \leq 0\}$, whereas where fluid is incoming,

$$\rho = \rho(x,t), \qquad\qquad \text{at} \quad \mathcal{S}_1',$$

and the density is unknown.

At points of $\mathcal{S}_2' = \mathcal{S}' - \mathcal{S}_1'$ where fluid is outcoming we know the sign of $\mathbf{v} \cdot \mathbf{n}$ to be positive. No further conditions on the density are allowed.

(A-iii) $\partial\Omega_t$: deformable, *outside Ω_t there is a known body $\mathcal{C}(t)$*.

We recall that to (1.6.2) we have added the *dynamic condition* (1.6.5), which represents a condition also on density.

(A-iv) $\partial\Omega_t$: deformable, outside Ω_t there is a *unknown body $C(t)$*.

Please see the remarks for (**A-iii**).

We report a remark of the referee.

Remark 1.6.3 *The weak formulation of the continuity equation yields*

$$\int_\Omega \rho(x,t)dx = \int_\Omega \rho(x,0)dx + \int_0^t \int_{\partial\Omega} \rho\mathbf{v} \cdot \mathbf{n}\, dS\, dt'. \qquad (1.6.15)$$

Equation (1.6.15) states that the amount of total mass is ruled out by the momentum.

(i) If the momentum is prescribed at the whole boundary then total mass is also known

(ii) If the total momentum is zero, then we recover the conservation of mass in Ω.

In conclusion, the problem is a very challenging one to pose.

It is clear that uniqueness holds only if we make some additional assumptions regarding density, because we know that different amounts of fluid filling the same volume will have different behaviors. We cannot expect the same velocity field if we consider two different amounts of same fluid filling the same vessel under the same external forces, because one will be more dense than the other. Here we prove that, in a bounded domain, in the set of steady solutions, uniqueness holds if the total mass is prescribed, cf. [89]. We call such a condition a **side condition**.

We set here some **side conditions** for density. Side conditions will depend on the domain and on the physical problem one is dealing with. Following the work of the preceding subsection, we distinguish the cases **B-(i)** and **B-(ii)**.

B-(i) Ω bounded, with rigid boundary.
We assume that the total mass is a *given* positive quantity M. Therefore, to (1.6.14) we add the side condition for density

$$\int_\Omega \rho(x,t)dx = M.$$

Since $\rho \geq 0$ we deduce by default that density must be at least in $L^\infty(0, \infty; L^1(\Omega))$.

Remark 1.6.4 *It is clear that the side conditions on ρ must hold for both viscous and inviscid fluids.*

Remark 1.6.5 *For permeable walls $\mathcal{S}' = \partial\Omega$, there is a non-zero flux through the walls, $\mathbf{w} \cdot \mathbf{n} \neq 0$; applying the Gauss lemma to the continuity equation we find that*

$$\int_{\partial\Omega} \rho\mathbf{w} \cdot \mathbf{n}dS = 0, \tag{1.6.16}$$

where ρ is unknown. Hence, if the boundary $\partial\Omega$ is constituted by the union of more connected parts Σ_i, $i = 1, ...N$, naming by ϕ_i the flux of momentum outgoing from Σ_i, we ask

$$\sum_{i=1}^N \phi_i = 0 \tag{1.6.17}$$

We observe that, despite the case of incompressible fluids, (1.6.17) doesn't represent a compatibility condition (density is not given on $\partial\Omega$). (1.6.17) should be naturally satisfied by the solution of the model.

B-(ii) Ω exterior to the compact region $\mathcal{C}$.
In this case the fluid fills the region exterior to a compact material region $\mathcal{C}$ with bounded boundary $\partial\Omega$, or the whole space. For $\partial\Omega$ we prescribe the

above discussed boundary conditions, while at infinity we add side conditions. We distinguish between the following two cases:

(a) *Finite mass*

The fluid $\mathcal{F}$ moves around a rigid obstacle with fixed center of mass in $\mathcal{C}$. Imagine either $\mathcal{F}$ is the atmosphere of a star Ω that moves around a kernel $\mathcal{C}$ that has fixed center of mass G; or $\mathcal{F}$ may be a self gravitating star having fixed center of mass G.

For density we shall add one of the following side (limit) conditions.

(a) The body $\mathcal{C}$ may be rigidly fixed at the center of mass G, or it may be uniformly rotating around an axis passing through G. We assume finiteness for the total mass:

$$\int_\Omega \rho(x,t)dx = M, \qquad\qquad t \in (0,\infty).$$

(b) For density we assume

$$\lim_{|x|\to\infty} \rho(x,t) = 0, \qquad\qquad t \in (0,\infty).$$

For isothermal fluids, it has been proven that the rest state doesn't exist under small potential forces, as the vacuum is attractive; c.f. [99].

(b) *Infinite mass*

The fluid $\mathcal{F}$ flows around a moving rigid obstacle $\mathcal{C}$.

If the motion of $\mathcal{C}$ is a translation, we may imagine $\mathcal{F}$ as almost uniform air filling the region Ω around an airplane $\mathcal{C}$ that moves with velocity $\mathbf{v_G}$ of its center of mass G.

If $\mathcal{C}$ rotates uniformly around its axis passing through G, say z, we may imagine $\mathcal{F}$ as almost uniform air filling the region Ω around a rotating satellite.

For density we shall add the following side (limit) condition. The mass of $\mathcal{F}$ is no longer finite, and we shall assume that the density tends towards a uniform, in space and time, value ρ_∞

$$\lim_{|x|\to\infty} \rho(x,t) = \rho_\infty.$$

1.7   Three Model Problems

To explain the main ideas it is suitable to begin with a comparison between some mathematical aspects of incompressible and compressible Navier–Stokes equations in a domain with rigid boundaries.

Governing Equations

Steady Flows

(a) Incompressible Navier–Stokes equations are a system of elliptic type in the sense of Agmon–Douglis–Niremberg; cf. [3].
(b) Compressible Navier–Stokes equations do not subscribe to any usual classifications.

Unsteady Flows

(a) Incompressible Navier–Stokes equations are of parabolic type with spatially non local character.
(b) Compressible Navier–Stokes equations are of mixed parabolic-hyperbolic type, with nonlinearity (and degeneration) in the temporal leading term for the velocity.

In this section we provide the exact mathematical formulation for three models of fluids: incompressible, barotropic, polytropic.

1.7.1 Incompressible Fluids

Kinematic constraint

A fluid is said to be **incompressible** if along its motions it satisfies the kinematic constraint that all material volumes remain constant. This is represented by the condition

$$\nabla \cdot \mathbf{v}(x,t) = 0. \tag{1.7.1}$$

If the density at initial time $t = 0$ is assumed to be uniform in space, say $\rho(x,0) = 1$, $\forall x \in \Omega$, then we deduce the **Homogeneous Incompressible Model**

$$\partial_t \mathbf{v} + \nabla(\mathbf{v} \otimes \mathbf{v}) = -\nabla p + \nu \Delta \mathbf{v} + \mathbf{f}, \qquad (x,t) \in \Omega \times (0,T), \tag{1.7.2}$$
$$\nabla \cdot \mathbf{v} = 0,$$

with initial data

$$\mathbf{v}(x,0) = \mathbf{v}_0(x), \qquad x \in \Omega, \tag{1.7.3}$$

where $p = p(x,t)$ is a dynamic unknown, and $\nu = \mu/\rho > 0$ is the kinematic viscosity. System (1.7.2) is called the **Navier–Stokes system**. It constitutes a system of four equations in the four scalar unknown $\mathbf{v}$, p, functions of $(x,t) \in \Omega \times (0,T)$ and under suitable boundary conditions it is locally solvable.

To close system (1.7.2), (1.7.3), and to make the model problem solvable, we must control the boundary terms. The problem of fixing correct boundary conditions has constituted the subject of Sect. 1.6. We end the subsection recalling the energy equation.

Energy Equation. 1.7.1

Multiplying $(1.7.2)_2$ by $\mathbf{v}$ and integrating over Ω we obtain

$$\frac{d}{dt} \int_\Omega \frac{1}{2} |\mathbf{v}|^2 dx = -2\nu \int_\Omega D(\mathbf{v}) : D(\mathbf{v}) dx \tag{1.7.4}$$

$$+ \int_{\partial\Omega} \Big(-p(\rho)\mathbf{v} \cdot \mathbf{n} + 2\nu \mathbf{v} \cdot D(\mathbf{v})\mathbf{n} \Big) dS + \int_\Omega \rho \mathbf{f} \cdot \mathbf{v} dx.$$

Therefore, recalling the expression of the Cauchy stress tensor, we may claim that for incompressible fluids the following equation of mechanical energy holds:

$$\frac{d}{dt} \int_\Omega \frac{1}{2} |\mathbf{v}|^2 dx = -2\nu \mathcal{D}_\mathbf{v} + \int_{\partial\Omega} \mathbf{v} \cdot \mathbf{T} \cdot \mathbf{n} dS + \mathcal{P}_e, \tag{1.7.5}$$

where the mechanical dissipation $\mathcal{D}_\mathbf{v}$, the external mechanical power $\mathcal{P}_e$ are given by:

$$\mathcal{D}_\mathbf{v} = \int_\Omega D(\mathbf{v}) : D(\mathbf{v}) dx, \qquad \mathcal{P}_e = \int_\Omega \rho \mathbf{f} \cdot \mathbf{v} dx.$$

Equation (1.7.4) is useful to derive global a priori estimates on the solutions.

1.7.2   Barotropic Fluids

Thermodynamic constraint

The fluid $\mathcal{F}$ is said to be **barotropic** if in motion it satisfies the thermodynamic constraint that its pressure must be a function of density, $p = p(\rho)$. More specifically, the thermodynamics may modify the motion of $\mathcal{F}$ only modifying the pressure law $p = p(\rho)$. The phenomenological law $p = p(\rho)$ is said *constitutive equation* for the physical model called the **Barotropic Model**, governed by the system

$$\partial_t \rho + \nabla \cdot (\rho \mathbf{v}) = 0, \tag{1.7.6}$$

$$\partial_t (\rho \mathbf{v}) + \nabla(\rho \mathbf{v} \otimes \mathbf{v}) = -\nabla p(\rho) + (\lambda + \mu)\nabla(\nabla \cdot \mathbf{v}) + \mu \Delta \mathbf{v} + \rho \mathbf{f}, \qquad (x, t) \in \Omega \times (0, T),$$

With initial data

$$\rho(x,0) = \rho_0(x), \quad \mathbf{v}(x,0) = \mathbf{v}_0(x).$$

To close system (1.7.6) we must control the boundary terms. The problem of prescribing correct boundary conditions has constituted the subject of Sect. 1.6. Thus we fix one of those side conditions discussed in Sect. 1.6 of this chapter. System (1.7.6) is called the **Compressible Navier–Stokes system** or the **Poisson Stokes system**. It constitutes a system of four equations in the four scalar unknowns ρ, $\mathbf{v}$, and under suitable initial and boundary conditions it is locally solvable.

Energy Equation. 1.7.2

Multiplying $(1.7.6)_2$ by $\mathbf{v}$, integrating over Ω and using $(1.7.6)_1$ we obtain

$$\frac{d}{dt} \int_\Omega \rho\left(\frac{1}{2}|\mathbf{v}|^2 + \int^\rho \frac{p(s)}{s^2} ds\right) dx = -\int_\Omega \left(\lambda(\nabla \cdot \mathbf{v})^2 + 2\mu D(\mathbf{v}) : D(\mathbf{v})\right) dx$$

$$(1.7.7)$$

$$+ \int_{\partial\Omega} \left[\left(-p(\rho) + \lambda\nabla \cdot \mathbf{v}\right)\mathbf{v} \cdot \mathbf{n} + 2\mu\mathbf{v} \cdot D(\mathbf{v})\mathbf{n}\right] dS + \int_\Omega \rho\mathbf{f} \cdot \mathbf{v} dx,$$

Given as thermodynamic potential, the Helmholtz free energy function per unit of mass $\psi(\rho) = \int^\rho \frac{p(s)}{s^2} ds$, with ρ a scalar quantity, and with the symbol $\int^\rho \psi(s)\, ds = \Psi(\rho)$, we refer to the antiderivative of ψ; for example, a function Ψ such that $\frac{d\Psi}{d\rho} = \psi$. As mechanical energy, we refer to the sum of kinetic energy and Helmholtz free energy[9]; cf.(1.5.25). Therefore, recalling the expression of the Cauchy stress tensor, we may claim that for compressible fluids the following equation of mechanical energy holds:

$$\frac{d}{dt} \int_\Omega \rho\left(\frac{1}{2}|\mathbf{v}|^2 + \Psi(\rho)\right) dx = -2\mu\mathcal{D}_\mathbf{v} + \int_{\partial\Omega} \mathbf{v} \cdot \mathbf{T} \cdot \mathbf{n} dS + \mathcal{P}_e, \qquad (1.7.8)$$

where the mechanical dissipation $\mathcal{D}_\mathbf{v}$ and the external mechanical power $\mathcal{P}_e$ are given by:

$$2\mu\,\mathcal{D}_\mathbf{v} = \int_\Omega \left(\lambda(\nabla \cdot \mathbf{v})^2 + 2\mu D(\mathbf{v}) : D(\mathbf{v})\right) dx, \qquad \mathcal{P}_e = \int_\Omega \rho\mathbf{f} \cdot \mathbf{v} dx.$$

Equation (1.7.8) is useful to derive global a priori estimates on the solutions.

---

[9]In the energy equation there may appear another thermodynamical potential, called *enthalpy*, cf. (1.5.24). We shall use equivalent forms of the energy equation in subsequent chapters.

1.7.3   Polytropic Fluids

For a general fluid $\mathcal{F}$, thermodynamical processes modify the motion of $\mathcal{F}$ through a new independent thermodynamical variable and a new equation. If no chemical reactions are considered, we may choose only one thermodynamical variable; from now on we will use temperature $\theta = \theta(x,t)$ as the independent unknown function. In **Polytropic Models** the fluid motions are governed by the system

$$\frac{d\rho}{dt} + \rho\nabla \cdot \mathbf{v} = 0 \tag{1.7.9}$$

$$\rho\frac{d\mathbf{v}}{dt} = -\nabla p + (\lambda + \mu)\nabla(\nabla \cdot \mathbf{v}) + \mu\Delta\mathbf{v} + \rho\mathbf{f},$$

$$c_V\rho\frac{d\theta}{dt} = \nabla \cdot (\chi\nabla\theta) - p\nabla \cdot \mathbf{v} + \Xi + \rho r,$$

With initial data

$$\rho(x,0) = \rho_0(x), \qquad \mathbf{v}(x,0) = \mathbf{v}_0(x), \qquad \theta(x,0) = \theta_0(x).$$

Equation (1.7.9) represents a system of six scalar equations in the six unknown functions ρ, v_1, v_2, v_3, θ.

In order to close system (1.7.9) we must control boundary terms. The problem of prescribing correct boundary conditions has been addressed in Sect. 1.6.

Energy Equation. 1.7.3

For polytropic fluids the following balance of total energy holds:

$$\frac{d}{dt}\int_\Omega \rho\left(\frac{1}{2}|\mathbf{v}|^2 + c_V\rho\theta\right)dx$$

$$= \int_\Omega \rho\mathbf{f} \cdot \mathbf{v}dx + \int_\Omega \rho r\, dx + \int_{\partial\Omega}\left[\left(-p(\rho) + \lambda\nabla \cdot \mathbf{v}\right)\mathbf{v} \cdot \mathbf{n}\right. \tag{1.7.10}$$

$$\left. + 2\mu\mathbf{v} \cdot D(\mathbf{v})\mathbf{n} + \chi\nabla\theta \cdot \mathbf{n}\right]dS,$$

Recalling the expression of the stress tensor $\mathbf{T}$ and of the out-coming heat flux $\mathbf{q} = \chi\nabla\theta$, we may simplify the balance of energy as follows

$$\frac{d}{dt}\int_\Omega \rho\left(\frac{1}{2}|\mathbf{v}|^2 + c_V\rho\theta\right)dx = \mathcal{P}_e + \mathcal{Q}_e + \int_{\partial\Omega}\left(\mathbf{v} \cdot \mathbf{T} + \mathbf{q}\right) \cdot \mathbf{n}dS, \tag{1.7.11}$$

where the external thermal power $\mathcal{Q}_e$ and external mechanical power $\mathcal{P}_e$ are given by:

$$\mathcal{Q}_e = \int_\Omega \rho\, r\, dx, \qquad \mathcal{P}_e = \int_\Omega \rho \mathbf{f} \cdot \mathbf{v} dx.$$

Equation (1.7.11) is useful to derive global *a priori* estimates of the solutions.

1.7.4   Bibliographical Notes

The problem of the existence and uniqueness of steady and non-steady solutions to incompressible Navier–Stokes equations has been studied extensively. Concerning existence of regular solutions and uniqueness to steady Navier-Stokes equations there is a very large literature that we leave out; cf. [36, 63]. The existence of regular solutions and the uniqueness to unsteady Navier-Stokes equations have been deeply studied in the past century, and the interest is still incredible. We give here only two references as examples; cf. [63, 129]. It is not our present purpose, hence the absence of references on such a wide subject; we recognize the omission. The field has, however, recently produced new research and new answers, and where relevant, we cite this research. For compressible fluids we are at the sunrise of their mathematical study.

(I) Regarding the well-posedness of **steady motions** of barotropic fluids in a bounded rigid domain with a given mass, several existence theorems of regular and weak solutions have recently been proven for equations governing steady viscous barotropic and polytropic flows. For *regular small data* (non potential force) there have been proof, uniqueness cf.[89, 92, 93], and existence theorems for regular solutions since 1981 under assumptions on viscosity coefficients, cf. [90, 91, 95], and without restrictions on the coefficients, cf. [13, 146].[10] Yet for the existence of steady flows in domains with regular rigid boundaries, we cite [8, 12, 25, 52, 56, 57, 77, 78, 82–86, 95, 100–102, 111, 114, 120, 121, 146]. Regarding the existence of steady flows in domains with corners, see [78]. The existence of steady barotropic flows has been proven in both cases $\mathbf{v}_\infty = 0$ and $\mathbf{v}_\infty \neq 0$; cf. [38, 85, 86]. A new scheme to treat numerical solutions is given in [7]. Recently Frehse [206] et al. have proved an existence theorem of steady solutions of compressible fluid in two dimensions for large data.

(II) Regarding the well-posedness of **unsteady motions** of barotropic and polytropic fluids, for small initial data, there have been existence and uniqueness theorems of *regular flows* close to the rest state published since 1962; we recall some [69–71, 73, 130, 141, 145, 148]. For information about the existence of unsteady regular flows with numerical algorithms, see [42–44, 51, 122, 125, 128, 130].

---

[10]Concerning barotropic fluids, only a potential inviscid flow can be classified as an elliptic system; cf. [40].

Existence theorems have also been proven for *weak solutions for large data*; see P.L. Lions [66], Fereisl and Novotny [26–28] in the isentropic case, and [120–122] for the isothermal case. In these theorems there is no proof of uniqueness.

In order to prove stability theorems, strong regularity is required on solutions of the leading equations.[11] Our method of proof assumes existence of regular solutions, nevertheless it does not require any smallness assumptions for initial data. Since the known existence theorems of regular unsteady solutions assume smallness for initial data, we claim that our results remain valid for a regularity class of solutions larger than that till now known. The existence theorems of weak solutions do not assume smallness in their initial data, but unfortunately this class is too weak to prove our theorems on asymptotic stability. However, we posit that stable results can be proven for at least two dimensions, and we leave it as a very challenging open problem.

The stability of the rest state of polytropic fluids, when $\mathbf{v}_\infty = 0$ and with nonhomogeneous boundary data on temperature has been proven in [112], see also [99].

Domains with non compact boundaries are not studied here. Examples of such domains are given by pipes with bounded and unbounded cross sections; cf. [57, 77, 111].

The one dimensional case is entirely omitted even though in this case the existence class of solutions is known.

---

[11]Indeed for two dimensions it would be enough to assume that weak solutions have bounded density in space and time. Such an hypothesis on the density would be sufficient to prove the stability theorem.

Chapter 2
Topics in Stability

Pura potenza tenne la parte ima;
nel mezzo strinse potenza con atto
tal vime, che giammai non si divima.
34, XXIX, Paradiso A. Dante

The flows that occur in Nature must not only obey the equations of fluid dynamics, but also be stable. L.D. Landau & E.M. Lifschitz (1959)

2.1 Introduction

In this chapter we introduce some definitions and qualitative methods useful in the study of nonlinear stability with respect to the initial data of a basic fluid motion. The aim of the chapter is to recall the energy and Dirichlet methods used to study distinctive properties of nonlinear stability for incompressible fluids (parabolic equations) and for elastic bodies (hyperbolic equations), respectively, and to give an overview of the results obtained in the book.

The central part of this chapter is a *generalization of the Dirichlet method*, achieved using the free work equation, that we call the *modified energy method*. Specifically, we will study the asymptotic behavior in time of perturbations to a basic state, introducing an auxiliary equation called *the free work equation*. To give a preview of the two main technical tools we will use some simple examples.

It is worth emphasizing that our method is intended to be naive and straightforward, and does not require complicated analysis.

We will first present a short survey of the *energy and Dirichlet methods* used to study nonlinear stability of parabolic and hyperbolic systems. We

M. Padula, *Asymptotic Stability of Steady Compressible Fluids*,
Lecture Notes in Mathematics 2024, DOI 10.1007/978-3-642-21137-9_2,
© Springer-Verlag Berlin Heidelberg 2011

recall that in mechanics the energy and Dirichlet methods are used for incompressible and elastic media, using Eulerian and Lagrangian coordinate systems, respectively.

Next we will introduce two key tools which will be used in the proofs of nonlinear stability and the asymptotic decay to steady compressible flows, which are central to the thesis of this book.

The first tool concerns an *extension of the Dirichlet method* in the wake of the work by Arnold, cf. [4]. In order not to obscure the main idea, we will explain the Dirichlet method by studying the stability of steady flows particular to inviscid fluids both incompressible and compressible.

The second tool is represented by the *free work equation* (FWE), and appears useful for systems of mixed parabolic-hyperbolic type. The FWE allows the transfer of asymptotic behavior in time characteristic of the parabolic part to the hyperbolic one.

For pedagogical reasons, in this introductive chapter we limit ourselves to simple examples; generalizations will be considered in the remaining chapters.

Section 2.2 Basic definitions of nonlinear stability of steady fluids motions will be reviewed. The classical "Energy Method" will be employed to study the nonlinear stability of a steady viscous incompressible flow.

Section 2.3 The Lagrange–Dirichlet method is outlined, and four applications are explained. The first two applications are known stability theorems for the rest state of both inviscid, incompressible fluids and elastic continua; cf. [2, 4]. The second two applications concern the stability of a basic flow for both inviscid and viscous, barotropic gases.

Section 2.4 Nonlinear stability and instability theorems to be proved in the next three chapters are listed.

2.2 Nonlinear Stability

Let us begin with some abstract settings. If we set by Y a scalar Banach space, then $X = [Y]^n$ will denote the vector Banach space given by the Cartesian product of Y n times. In this section we introduce some definitions of stability.

Given a steady flow S_b, the physical concept of the *stability* of S_b is linked to the concept of *observability*. Assume, as a qualitative definition of **stability**, the following proposition:

> *The stability of a given motion is its capacity to 'hold' (to be observed) in the presence of the small perturbations present in any physical system.*

This definition allows us to introduce the correct definition of stability of a given steady solution S_b to the equations of motion that will be referred to as *stability of the basic motion* S_b. At time $t = 0$ we perturb the basic motion S_b, and call $\widetilde{S}_0 = \widetilde{S}(0)$ the perturbed initial data that produce the perturbed motion $\widetilde{S}(t)$. Correspondingly, the **perturbation** $\widetilde{S}_0 - S_b$ at initial time produces the evolution in time of the perturbation $\widetilde{S}(t) - S_b$. The stability question ask:

Is it possible to control $\widetilde{S}(t) - S_b$, in a given spatial norm $\|.\|_X$, for $t \in (0, \infty)$, provided $\widetilde{S}_0 - S_b$ is sufficiently small in the same norm $\|.\|_X$?

Definition 2.2.1 Stability *The motion S_b is said **stable in the fixed norm**[1] $\|.\|_X$, **with respect to initial data** if and only if for all numbers $\epsilon > 0$, there exists $\delta > 0$ such that, for all initial perturbations $\widetilde{S}_0 - S_b$ having norm in X less than δ, i.e. $\|\widetilde{S}_0 - S_b\|_X < \delta$, the corresponding perturbations $\widetilde{S}(t) - S_b$ in the norm $\|.\|_X$, remain less that ϵ, i.e. $\|\widetilde{S}(t) - S_b\|_X < \epsilon$ for all time t.*

Basic flows that verify Definition 2.2.5 are sometimes called *stable in the mean*, but this notation is not generally accepted.

Definition 2.2.2 Asymptotic Stability *If perturbations come back to zero as time goes to infinity,*

$$\lim_{t \to \infty} \|\widetilde{S}(t) - S_b\|_X = 0,$$

*we say that the basic motion is **asymptotically stable**.*

Definition 2.2.3 Instability *A motion $S_b(t)$ is said to be **unstable** in the $X-$norm if it is not stable; that is, if there exists $\epsilon > 0$, a sequence of initial data $\{S_i(0)\}$ approaching S_b, and a sequence of times t_i, such that $\|S_i(t) - S_b\|_X \geq \epsilon$ for any $i \in \mathbb{N}$.*

Under nonlinear stability hypotheses, if we can physically control that at initial time the norm in X of $\widetilde{S}_0 - S_b$ is less than δ, then the basic flow S_b will also be experimentally observable in the class of perturbed flows having initial data sufficiently close to S_b.

2.2.1 Abstract Settings

We denote by f a C^1 vector function defined in the vector Banach space X, with values in X, $f : V \in X \to f(V) \in X$. We study the *abstract* **autonomous evolution problem**

$$\frac{dV}{dt} = f(V);$$

$$V(0) = V_0.$$

(2.2.1)

---

[1]The distance between two motions may be calculated by a measure, instead that by a norm. This will be not considered here.

Usually, (2.2.1) represents an evolution law of a physical quantity of the
system S. In our case, $V(t, V_0)$ represents the motion of S, corresponding
to initial state V_0. The notation $V(t) = V(t; V_0)$ may also be used. We
assume that system (2.2.1) has existence and uniqueness theorems of global
in time regular solutions in correspondence of large initial data. In general,
the problem of finding explicit solutions to (2.2.1) is very difficult, if not
impossible, to solve. As such, it is worth applying qualitative information,
such as the uniqueness or the asymptotic behavior of a solution in time, in
a given norm $\|.\|_X$. We are trying to frame the concept of time control for a
solution (stability) according to a rigorous mathematical formulation. Such
arguments were developed in the second half of the nineteenth century, mainly
by the French mathematician H. Poincare' (1854–1912), and by the Russian
mathematician A.M. Liapunov (1857–1918). In this section we will study the
direct method developed by Lyapunov. In referring to a *direct method*, we
mean to describe a mechanism that would allow for the direct deduction of
certain types of qualitative data once external data are given, and without
integrating the equations.

Definition 2.2.4 *A value $V_b \in X$ is said to be a **critical point** of* (2.2.1)
if $f(V_b) = 0$.

Note that the existence of critical points infers knowledge of steady
solutions $V(t; V_b) = V_b$.

If there exists a critical point for (2.2.1), then by a simple subtraction we
deduce that the function $W(t, W_0)) = V(t; V_0) - V_b \in X$ with $W_0 = V_0 - V_b$
solves the problem

$$\frac{dW}{dt} = g(W) \qquad W(0) = W_0, \tag{2.2.2}$$

where

$$g(W) = f(V_b + W) - f(V_b).$$

In our case we are directly studying the difference of motions, hence S_b is the
critical point of (2.2.2), and verifies $W_b(t) = W(t; 0) = 0$.

Definition 2.2.5 *Nonlinear Stability* *The zero solution $W_b(t) = 0$ to*
(2.2.2) *is said to be **nonlinearly stable** in the $X-$norm if*

$$\forall \epsilon > 0 \quad \exists \delta(\epsilon) > 0 : \ \|W_0\|_X < \delta \quad \Rightarrow \quad \|W(t; W_0)\|_X < \epsilon, \qquad \forall t \in (0, \infty). \tag{2.2.3}$$

A solution $W_b(t) = 0$ to (2.2.2) *is said to be **unstable** in the $X-$norm if it
is not stable; that is, if there exists $\epsilon > 0$, a sequence of initial data $\{W_i\}$
approaching zero, and a sequence of times t_i, such that $\|W(t_i, W_i)\|_X \geq \epsilon$ for
any $i \in \mathbb{N}$.*

*The difference between continuous dependence and stability lies within the
times intervals $(0, T)$ and $(0, \infty)$, where control occurs.*

Definition 2.2.6 *Nonlinear   Exponential   Stability   The   solution* $W_b(t, 0) = 0$ *to (2.2.2) is said to be* **nonlinearly unconditionally stable** *in the* $X-$*norm if there is control of perturbations in terms of initial data in the* $X-$*norm, however large are the perturbations at initial time in the* $X-$*norm.*

The solution $W_b(t, 0) = 0$ *to (2.2.2) is said to be* **nonlinearly exponentially stable** *in the* $X-$*norm if it is stable and, for any initial datum* W_0, *it holds*

$$\lim_{t \to \infty} \|W(t; W_0)\|_X = 0, \qquad \|W(t; W_0)\|_X < c \exp^{-\beta t}, \qquad resp.,$$

with c, β *suitable constants,* β *is the time decay constant.*

Notice that *stability Definitions 2.2.6 are not intrinsic properties of the critical point* V_b, *but rather depend on the norm* $\|.\|_X(t)$ *and on the radius* δ *of the ball in the space* X, *on the difference* W_0 *between the basic motion* V_b, *and the initial data* $V(0)$; *specifically, it is a local statement.*

The term '*unconditional*' means without the condition of smallness for initial data, however this adjective is not generally accepted.

Definition 2.2.7 *The rest state is said to be* **unstable** *if it is not nonlinearly stable.*

For the linearized problem associated with (2.2.2), all definitions are simplified.

Definition 2.2.8 *The rest state* S_b *is said to be* **linearly stable** *in the norm* $\| \cdot \|_X(t)$ *if it is stable in the system obtained by linearizing around zero the term* $g(W)$ *at right hand side of (2.2.2). Namely, setting*

$$\frac{dW'}{dt} = f'(V_b) W',$$

if there exists a constant $\beta > 0$ *such that*

$$\|W'(t; W_0)\|_X < \|W_0\|_X \exp^{-\beta t}, \qquad \forall t > 0.$$

If $\beta = 0$ *then we have* **marginal stability**.[2]

Definition 2.2.9 *The rest state* S_b *is said to be* **unstable** *in the norm* $\| \cdot \|_X(t)$ *if there exists a constant* $\beta > 0$ *such that*

$$\|W(t; W_0)\|_X \geq \|W_0\|_X e^{\beta t}, \qquad \forall t > 0.$$

---

[2]Definitions for linear stability are given in the complex plane, employing the eigenvalues of the linearized operator. Thus the marginal stability involves only the real part of the eigenvalues. However, boundedness can be proven.

Notice that the Definitions 2.2.8 and 4.3.4 are independent of the size of the initial data; namely they are global statements.

Once more we wish to outline that *the **difference between linear and nonlinear stability** with respect to initial data in classical fluid-dynamics is due to the **size** of the distance between the initial data $S(0)$ and S_b, and to the **decay rate** of perturbations.*

*To questions of linear and nonlinear stability we require regularity on steady and unsteady flows, and such regularity **do depend** on external forces.*

Below we recall the **Linearization Principle.** The *linearization principle* refers to a theorem proving that stability properties of the exact steady solution S_b to the nonlinear system, are deduced from those of the system linearized around S_b, provided that the initial data are sufficiently close to S_b. Therefore, if a *linearization principle* holds for the rest state S_b, any linearly stable or unstable S_b is also nonlinearly stable or unstable, respectively, for a class of suitable small initial data.

If a linearization principle holds, then any linearly stable state S_b is also nonlinearly stable. This means that solutions corresponding to initial data in a sufficiently small neighborhood of 0 remain close to 0 for all time. In reality we may prescribe large initial data, so let us study what will happen in this circumstance.

2.2.2   Initial Data Control

In reality, we may prescribe initial data far from the stable state S_b, and we may wish to study what happens under these circumstances. *Previous definitions of nonlinear stability say nothing about the control of solutions with finite initial data.* Indeed, in nonlinear phenomena with large initial data, a solution $S(t)$ may lose its control from initial data, even though S_b is nonlinearly stable *(for small initial perturbations)*. This situation occurs frequently, and it constitutes the real discrepancy between linear and nonlinear stability. To this day it appears that there are no rigorous definitions for this phenomenon, thus we introduce here two new definitions:

Definition 2.2.10 *A perturbation $W(t; W_0)$ to the rest state S_b is said to be **controlled by initial data** in the range $\mathcal{I}_{2a}/\mathcal{I}_a$ if, and only if, for all initial data W_0 satisfying*

$$a \; < \; \|W_0\|_Y \; < 2a, \tag{2.2.4}$$

the solution $W(t; W_0)$ is bounded for all time; that is, there exists a suitable constant $\alpha = \alpha(a) > 0$ such that

$$\|W(t; W_0)\|_X(t) \leq \alpha, \qquad \forall t > 0. \tag{2.2.5}$$

Definition 2.2.11 *The rest state is said to **lose the control of the initial data** if there exists a positive number a and initial data W_0 satisfying (2.2.4), such that the corresponding solution $W(t; W_0)$ is not controlled by the initial data; that is, whenever given $\alpha > 0$, there exists $T > 0$ such that the solution $W(t; W_0)$ to problem (2.2.2) with initial data satisfying (2.2.4) satisfies the inequality*

$$\|W(t; W_0)\|_X(T) \geq \alpha. \qquad (2.2.6)$$

Definitions 2.2.10, 2.2.11 are meaningful only for nonlinear systems, because the definition of linear stability is valid for all initial data.

2.2.3 Lyapunov Method

We begin with the abstract settings (2.2.1). Let V_b denote a critical point of (2.2.1), namely

$$f(V_b) = 0. \qquad (2.2.7)$$

Stability studies the evolution in time of the disturbance

$$W(t; V_0 - V_b) = V(t; V_0) - V_b, \qquad W(0; V_0 - V_b) = V_0 - V_b,$$

in some prescribed norm $|.|_X$.

Definition 2.2.12 *Let $W \in X$ be a solution to (2.2.2). A smooth function*

$$\mathcal{F} : W \in X \longrightarrow \mathcal{F}(W) \in R,$$

*is said to be a **Lyapunov functional** for the null solution $W_b = 0$ in the abstract space X if:*

(1) It is positive definite in the neighborhood of the origin $\mathcal{I}$, i.e.

$$\mathcal{F}(0) = 0, \qquad \mathcal{F}(W) > 0, \qquad W \neq 0, \qquad \forall W \in \mathcal{I};$$

(2) It is continuous in the neighborhood of 0 of radius R, i.e.

$$\forall R > \epsilon > 0, \quad \exists \delta > 0 : \quad \|W\|_X < \delta \longrightarrow \quad \mathcal{F}(W) < \epsilon;$$

(3) It is decreasing along the solution to (2.2.2), i.e.

$$\frac{d\mathcal{F}(W(t))}{dt} \leq 0, \qquad \forall t > 0. \qquad (2.2.8)$$

Theorem 2.2.1 *Lyapunov Theorem* If there exists a Lyapunov functional for system (2.2.2), then the zero solution $W_b(t; 0) = 0$ to (2.2.2) is stable in the X norm.

Theorem 2.2.2 *If there exists a Lyapunov functional for system (2.2.1), then the stationary solution $V(t; V_b) = V_b$ to (2.2.1) is stable.*

Remark 2.2.1 *For stable motions of fixed space X, there exist infinite Lyapunov functionals. One problem lies in the construction of the most appropriate Lyapunov functional $\mathcal{F}$.*

As such, we will now limit ourselves to describing the energy method which proposes **one choice** of a Lyapunov functional. Note that other generalized energy methods have been proposed to construct more refined Lyapunov functionals; cf. [37,53].

The stability result given below provides nonlinear stability results in the class of regular motions.

2.2.4 Energy Method

The energy method takes as a starting point the initial value problem described in (2.2.2), the "perturbations system." It deduces Lyapunov functions by operating on system (2.2.2). A typical operation of this method is the multiplication of $(2.2.2)_1$ by a function of W in X, the latter of which is usually a Hilbert space $X = H$ where (2.2.2) is defined.[3]

The energy method is customarily used to study the nonlinear stability of incompressible viscous flows, governed by "generalized" parabolic systems, see Orr (1842–1912) [87], Reynolds (1842–1912) [124], etc. Notice that the 'energy method' generally doesn't use physical energy.

Let $\mathbf{v}, p$ be a solution to the **Navier–Stokes unsteady equations** in a fixed domain Ω that satisfies

$$\partial_t \mathbf{v} + \mathbf{v} \cdot \nabla \mathbf{v} - \nu \Delta \mathbf{v} = -\nabla p + \mathbf{f}$$
$$\nabla \cdot \mathbf{v} = 0, \qquad (x, t) \in \Omega \times (0, \infty),$$
$$\mathbf{v}(x, 0) = \mathbf{v}_0(x), \qquad x \in \Omega, \tag{2.2.9}$$
$$\mathbf{v}|_{\partial\Omega} = \mathbf{v}_\Sigma,$$

with $\mathbf{f}$ being external force, $\mathbf{v}_\Sigma$ being boundary velocity, and $\mathbf{v}_0$ the initial data.

---

[3]If $W \in L^p$, we could multiply $(2.2.2)$ by $|W|^{p-2}W$ and integrate over Ω to get

$$\frac{d}{dt} \int_\Omega \frac{W^p}{p} dx = \int_\Omega W^{p-2} W\, g(W) dx.$$

In the energy method one usually takes $p = 2$.

In this case, a critical point of (2.2.9) represents a solution to the **Navier–Stokes steady equations** given by

$$\mathbf{v}_b \cdot \nabla \mathbf{v}_b - \nu \Delta \mathbf{v}_b = -\nabla p_b + \mathbf{f}$$

$$\nabla \cdot \mathbf{v}_b = 0, \qquad x \in \Omega, \qquad (2.2.10)$$

$$\mathbf{v}_b|_{\partial\Omega} = \mathbf{v}_\Sigma,$$

with the same external force and the same boundary velocity.

Here we wish to study the **stability with respect initial data** $\mathbf{v}_0 = \mathbf{v}_b + \mathbf{u}_0$. Then, the perturbation $\mathbf{u} = \mathbf{v} - \mathbf{v}_b$ satisfies the equation

$$\partial_t \mathbf{u} + \mathbf{v} \cdot \nabla \mathbf{u} - \nu \Delta \mathbf{u} = -\nabla(p - p_b) - \mathbf{u} \cdot \nabla \mathbf{v}_b$$

$$\nabla \cdot \mathbf{u} = 0, \qquad (x, t) \in \Omega \times (0, \infty),$$

$$\mathbf{u}(x, 0) = \mathbf{u}_0(x), \qquad x \in \Omega, \qquad (2.2.11)$$

$$\mathbf{u}|_{\partial\Omega} = 0.$$

Multiplying $(2.2.11)_1$ by $\mathbf{u}$ and integrating over Ω, we deduce the following **energy equation**

$$\frac{d}{dt} \int_\Omega \frac{|\mathbf{u}|^2}{2} dx = -\nu \int_\Omega |\nabla \mathbf{u}|^2 dx - \int_\Omega \mathbf{u} \cdot \nabla \mathbf{v}_b \mathbf{u} \, dx, \qquad t \in (0, \infty). \quad (2.2.12)$$

From (2.2.12) one can easily prove a continuous dependence theorem for suitable regularity classes of solutions. In general nothing can be said about the stability of the stationary solution $\mathbf{v}_b$ of (2.2.11), except that unsteady perturbations $\mathbf{u}(x, t)$ do depend on the size of $\mathbf{v}_b$. Furthermore, it is clear that a candidate Lyapunov functional $\mathcal{F}(\mathbf{u})$ coincides with the spatial L^2–norm of $\mathbf{u}$. $\|\mathbf{u}\|_{L^2}$ will become a Lyapunov functional for system (2.2.11) if

$$-\nu \int_\Omega |\nabla \mathbf{u}|^2 dx - \int_\Omega \mathbf{u} \cdot \nabla \mathbf{v}_b \mathbf{u} \, dx \leq 0. \qquad (2.2.13)$$

Drawback $\nu \neq 0$ To prove stability one must prove that the right hand side of (2.2.12) is less than zero. To this end, we notice that the second integral at l.h.s. of (2.2.13) $A := -\int_\Omega \mathbf{u} \cdot \nabla \mathbf{v}_b \cdot \mathbf{u} \, dx$, has no definite sign. Hence one requires the dissipative term $D := -\nu \int_\Omega |\nabla \mathbf{u}|^2 dx$ be larger than A. This in turn requires that $\nu \neq 0$, thus the fluid must be viscous. Under this assumption, in order to control the term A, we require that the basic motion $\mathbf{v}_b$ and the domain be such that there exists a constant $c = c(\Omega, \mathbf{v}_b)$ such that

$$\left| \int_\Omega \mathbf{u} \cdot \nabla \mathbf{v}_b \cdot \mathbf{u} \, dx \right| \leq c \int_\Omega |\nabla \mathbf{u}|^2 dx, \qquad \forall \mathbf{u} \in J_1(\Omega).$$

Finally, to prove stability it is enough to assume that $c \leq \nu$.

2.3   Lagrange–Dirichlet Method

The Lagrange–Dirichlet method considers the abstract initial value problem (2.2.1), and constructs Lyapunov functionals by using direct physical balance laws (in particular conservation laws).

The Lagrange–Dirichlet method is customarily used to study nonlinear stability of the rest state of elastic and thermodynamic systems; cf. [24]. For the stability study one computes the first and second variation of the total energy. More generally, the Lagrange–Dirichlet Method states that those V_b which are the minima of total energy are stable steady states.

Nonlinear stability has been investigated by Dirichlet using general conservation laws.

2.3.1   Hyperbolic First Order Systems

Arnold [4] has extended the Dirichlet method to study the nonlinear stability of potentially non-smooth steady solutions of two-dimensional Euler equations; for incompressible fluids in symmetric bounded domains, see also [127]. He used as a Lyapunov functional a linear combination of conservation laws, first integrals, due to the symmetry of the problem. Using as prototype a functional $\mathcal{E}$, the basic idea was to look for conditions ensuring the vanishing of the first variation of $\mathcal{E}$ and the positivity of its second variation. We will give an outline of the stability theorem proved by **Arnold** for a class of steady solutions to **Euler equations**.

Consider a two-dimensional domain Ω, bounded by two smooth, fixed, closed, non-intersecting curves C_0, C_1, or the internal and external boundaries, respectively. Let us denote by (x, y) the independent space variables in the plane π containing Ω, and by $\mathbf{k}$ a direction orthogonal to π.

For the local angular velocity we set

$$\omega = curl\mathbf{v} \cdot \mathbf{k} = \partial_x v_y - \partial_y v_x,$$

where (v_x, v_y) denotes the components of $\mathbf{v}$ along π.

Let the velocity $\mathbf{v}_b$, p_b and the pressure be a **steady solution of Euler equations** to the Boundary Value Problem (BVP)

$$\mathbf{v}_b \cdot \nabla \mathbf{v}_b = -\nabla p_b,$$

$$\nabla \cdot \mathbf{v}_b = 0, \qquad\qquad \text{in } \Omega;$$

$$\mathbf{v}_b \cdot \mathbf{n}|_{C_i} = v_i, \qquad\qquad\qquad (2.3.1)$$

$$\int_{C_i} v_i ds = 0, \qquad\qquad i = 0, 1.$$

Denoted by ψ_b the stream function $\mathbf{v}_b = \nabla \psi_b \times \mathbf{k}$, it is well known, cf. [65], Sect. 10 in Chap. 1, that any function of $\Delta \psi_b = 0$ is constant through the paths of fluid particles. Furthermore, by the motion equation $(2.3.1)_1$ we also know that

$$\mathbf{v}_b \cdot \nabla \omega_b = \nabla \psi_b \times \mathbf{k} \cdot \nabla \Delta \psi_b = 0. \qquad (2.3.2)$$

The parallelism between $\nabla \psi_b \times \nabla \Delta \psi_b$ and $\mathbf{k}$, combined with (2.3.2) implies

$$\nabla \psi_b \times \nabla \Delta \psi_b = 0.$$

The parallelism between the gradients of ψ_b, and of $\Delta \psi_b$, if $\nabla \Delta \psi_b \neq 0$, infers a functional dependence between these two functions, expressed by

$$\psi_b = \aleph(\Delta \psi_b). \qquad (2.3.3)$$

Therefore, all basic flows must satisfy (2.3.3).

Let $\mathbf{v}, p$ be solutions to the **incompressible Euler unsteady equations**, namely let $\mathbf{v}, p$ solve the Initial Boundary Value Problem (IBVP)

$$\partial_t \mathbf{v} + \mathbf{v} \cdot \nabla \mathbf{v} = -\nabla p;$$
$$\nabla \cdot \mathbf{v} = 0;$$
$$\mathbf{v}(x, 0) = \mathbf{v}_0; \qquad (2.3.4)$$
$$\mathbf{v} \cdot \mathbf{n}|_{C_i} = v_i, \qquad i = 0, 1.$$

Let dl be the infinitesimal element of the line tangent to oriented curves C_i, $i = 1, 2$. It is not difficult to show that the **modified energy** functional $\mathcal{E}(\mathbf{v})$ remains constant along the motion

$$\mathcal{E}(\mathbf{v}) = \frac{1}{2} \int_\Omega v^2 dx dy + \int_\Omega \Phi(\omega) dx dy + \sum_{i=0}^{2} a_i \int_{C_i} \mathbf{v} \cdot dl, \qquad (2.3.5)$$

where $\Phi : R \longrightarrow R$ is any smooth function with a_i real numbers.

Set $\mathbf{u} = \mathbf{v} - \mathbf{v}_b$. Then, as **possible Lyapunov functional** we may choose

$$\mathcal{F}(\mathbf{u}) = \mathcal{E}(\mathbf{v}) - \mathcal{E}(\mathbf{v}_b). \qquad (2.3.6)$$

Choice of Φ. For functionals Φ in (2.3.5), we take any function having a derivative coinciding with

$$\Phi'(\cdot) = \aleph((\cdot)).$$

It's worth checking under which conditions $\mathcal{F}$ becomes a Lyapunov functional.

Let J and J_1 denote the subspaces of $L^2(\Omega)$ and $W^{1,2}(\Omega)$, respectively, of solenoidal vector fields. Conditions on the first and second variations of $\mathcal{F}$ are given by (2.3.6),

$$\delta\mathcal{F}(\psi_b)[\varphi] = 0, \qquad \forall\varphi \in J,$$

$$\delta^2\mathcal{F}(\psi_b)[\varphi]^2 > 0, \qquad \forall\varphi \in J_1, \qquad (2.3.7)$$

and ensure that $\mathcal{F}$ is a right Lyapunov functional.

Conditions (2.3.7) are conditions on the basic motion ψ_b. Therefore, we are lead to select the stable steady flows $\mathbf{v}_b$ in the set of functions that satisfy (2.3.7).

The second variation introduces the norm, called the **natural norm** in which the perturbation is controlled. In this case, the natural norm with which to study the problem of stability is

$$\delta^2\mathcal{E}(\mathbf{v}_b)[\mathbf{u}]^2 = \int_\Omega u^2 dxdy + \int_\Omega \aleph(\bar{\omega})\omega^2 dxdy,$$

with $\bar{\omega}$ between ω_b and ω. Finally, the hypotheses on $\aleph$ ensuring $\delta^2\mathcal{E}$ is positive definitely ensure that $\mathcal{F}$ is a good Lyapunov function.

2.3.2 Second Order ODE

We wish to explore the **Dirichlet method** by studying the solutions $x = x(t, x_0, \dot{x}_0)$ of an ordinary second order system

$$\frac{d^2}{dt^2}x = f(x) - h\dot{x}, \qquad (2.3.8)$$

$$x(0) = x_0, \qquad \frac{d}{dt}x(0) = \dot{x}_0.$$

In (2.3.8) h is a positive constant, and sometime we have set $\frac{dx}{dt} = \dot{x}$. Let x_b be a critical point of f that satisfies $f(x_b) = 0$. This leads to the knowledge of an equilibrium solution $x(t, x_b, 0) = x_b$ for (2.3.8).

The **energy equation** is written as

$$\frac{d}{dt}\mathcal{E}(x) = -h\dot{x}^2, \qquad (2.3.9)$$

with

$$\mathcal{E}(x) = \frac{1}{2}\dot{x}^2 - F(x), \qquad \frac{d}{dx}F(x) =: F'(x) = f(x). \qquad (2.3.10)$$

Therefore, for $y = x - x_b$, $\dot{y} = \dot{x}$ we see that

$$\mathcal{F}(y, \dot{y}) = \mathcal{E}(x_b + y) - \mathcal{E}(x_b)$$

becomes a Lyapunov functional once $(x_b, 0)$ is a minimum for $\mathcal{E}$. A sufficient condition then is that the Hessian $\delta^2 \mathcal{F}(x_b, 0)$ defines a positive definite quadratic form. Since the Hessian of $\mathcal{F}$ is in diagonal form, it is enough to compute the two eigenvalues $\partial_{\dot{x}}^2 F(x_b, 0)$, $\partial_x^2 F(x_b, 0)$. The derivative $\partial_{\dot{x}}^2 F(x_b, 0)$ is given by $1/2$, and is always positive. The derivative $\partial_x^2 F(x_b, 0)$ is $-f'(x_b)$, hence we must assume that $-f'(x_b) > 0$, that is $f'(x_b) < 0$.

To give an **asymptotic result**, we first observe that $\dot{x} = \dot{y}$. Thus we multiply (2.3.8) times $y = x - x_b$, recalling that x_b is a critical point. So we find the **free work equation**

$$\frac{d}{dt}\left(y\dot{y} + \frac{hy^2}{2}\right) - \dot{y}^2 = \left(f(x_b + y) - f(x_b)\right)y = f'(x_b + \overline{y})y^2, \qquad (2.3.11)$$

where $\overline{y}$ is a point between y and 0. The hypothesis $f'(x_b) < 0$, together with the regularity of solutions, implies that there exists a neighborhood of x_b in which $f'(x) < 0$. Notice that by applying the Taylor expansion to the variable x with initial point x_b up to the second order, and recalling that x_b is a critical point, if $F'(x_b) = f(x_b) = 0$, we obtain

$$\mathcal{E}(x) - \mathcal{E}(x_b) = \frac{1}{2}\dot{y}^2 - \left(F(x) - F(x_b)\right) = \frac{1}{2}\dot{y}^2 - \frac{1}{2}f'(\overline{x})y^2. \qquad (2.3.12)$$

Multiplying (2.3.11) by an arbitrary positive constant ϵ and adding (2.3.9), one obtains

$$\frac{d}{dt}\mathbb{E}(t) = -(h - \epsilon)\dot{y}^2 + \epsilon f'(x_b + \overline{y})y^2$$

where we have introduced the **modified energy**

$$\mathbb{E} = \frac{1}{2}\dot{y}^2 - \frac{1}{2}f'(x_b + \overline{y})y^2 + \epsilon\, y\dot{y} + \epsilon\,\frac{h}{2}y^2.$$

It is trivial to check that there exists a b such that

$$\mathbb{E}(t) \le \frac{1}{b}(y^2 + \dot{y}^2). \qquad (2.3.13)$$

Furthermore we observe that, for $f'(\overline{x}) < 0$, it is $h - \sqrt{h^2 - 4f'(\overline{x})} < 0$, hence condition

$$(h - \sqrt{h^2 - 4f'(\overline{x})}) < 2\epsilon < (h + \sqrt{h^2 - 4f'(\overline{x})}),$$

is satisfied for

$$0 < 2\epsilon < (h + \sqrt{h^2 - 4f'(\overline{x})}),$$

namely for ϵ sufficiently small. Under this assumption the modified energy $\mathbb{E}$ is always a positive definite quadratic form in the variables y, $\dot{y}$. Moreover, it holds

$$\frac{d}{dt}\mathbb{E}(t) = -a^2(\dot{y}^2 + y^2), \tag{2.3.14}$$

where

$$a^2 = min\{(h - \epsilon), -\epsilon \, f'(x_b + \overline{y})\}.$$

Hence (2.3.13) in (2.3.14) yields the following modified energy differential inequality

$$\frac{d}{dt}\mathbb{E}(t) \le -a^2 b\mathbb{E}(t), \tag{2.3.15}$$

which integrated infers

$$\mathbb{E}(t) \le \mathbb{E}(0)\exp^{-a^2bt}.$$

If $f'(x_b) < 0$, then nonlinear stability holds and asymptotic decay has been proved.

Analogous calculations have been developed for elastic media; cf. [2, 24, 106, 108]. In particular, we note that the method is applicable to hyperbolic systems.

Question:
Is the Dirichlet method applicable to coupled hyperbolic-parabolic equations? In particular, where the rest state of a compressible fluid is asymptotically stable? In the following chapters we shall give a partial answer to this question. Below we apply the Dirichlet method to compressible fluids, studying two model problems.

2.3.3 Stability of Barotropic Inviscid Fluids $v_b \ne 0$

Here we give an example of natural norms in which the perturbation is controlled by initial data, for all times, for zero viscosity and non-zero basic velocity.

We recall that the nonsteady flows of a barotropic inviscid fluid in a bounded domain are governed by the **compressible Euler equations**. The unknown velocity $\mathbf{v}$ and density ρ are governed by the following initial boundary value problem in a bounded fixed domain Ω:

$$\partial_t \rho + \nabla \cdot (\rho \mathbf{v}) = 0, \tag{2.3.16}$$

$$\rho(\partial_t \mathbf{v} + \mathbf{v} \cdot \nabla \mathbf{v}) = -\nabla p(\rho) + \rho \nabla U, \qquad (x, t) \in \Omega \times (0, T),$$

$$\mathbf{v}(x, 0) = \mathbf{v}_0(x), \qquad\qquad \rho(x, 0) = \rho_0(x), x \in \Omega,$$

$$\int_\Omega \rho(x, t)dx = M, \qquad\qquad \mathbf{v} \cdot \mathbf{n}|_{\partial\Omega} = 0.$$

If the pressure $p(\rho)$ is a convex function of ρ, system (2.3.16) is strictly hyperbolic.

Set $x \in \Omega$, $x \equiv (x, y, z)$, and denote by $\mathbf{v}_b$, ρ_b the steady solution to the system (2.3.16). We give sufficient conditions for nonlinear stability of steady potential flows, with respect to three dimensional perturbations. We work in Eulerian coordinates.

*We study nonlinear stability of the steady flow of a **barotropic inviscid** fluid, governed by the **compressible Euler equations**.* Begin by recalling the Boundary Value Problem governing steady flows

$$\nabla \cdot (\rho_b \mathbf{v}_b) = 0,$$

$$(\rho \mathbf{v}_b \cdot \nabla)\, \mathbf{v}_b = -\nabla p(\rho_b) + \rho_b \mathbf{f}, \qquad x \in \Omega,$$

$$\mathbf{v}_b \cdot \mathbf{n}\Big|_{\partial \Omega} = 0, \qquad \int_\Omega \rho_b(x)\, dx = M, \qquad \rho_b \geq 0,$$

$$(2.3.17)$$

where $\mathbf{n}$ is the normal to the boundary, and $p = p(\rho_b)$. As we know, the rest state $S_b = \{\mathbf{v} = 0, \quad \rho_b = \rho_b(x)\}$ exists only when forces $\mathbf{f}$ are positional and derived from a uniform potential $\mathbf{f} = \nabla U$. Indeed in this case, the rest state is the exact solution to (2.3.17), with ρ_b implicitly given by

$$\int^{\rho_b} \frac{p'(s)}{s}\, ds = U + c, \qquad \int_\Omega \rho_b\, dx = M. \qquad (2.3.18)$$

Notice that the constant c is given by the condition that the total mass is prescribed (2.3.17)$_4$, and therefore (2.3.18) may furnish a complex value for the density. In order to have real positive solutions ρ_b (densities), we are led to assume the following:

Hypotheses on the basic flow $(\rho_b, \mathbf{v}_b)$

  (i) The flow $(\rho_b, \mathbf{v}_b)$ satisfies the boundary problem (2.3.17).
 (ii) The velocity $\mathbf{v}_b$ is *potential*.
(iii) The momentum $\rho_b \mathbf{v}_b$ is *potential*

$$\rho_b \mathbf{v}_b = \nabla \chi.$$

Assumptions (i) and (ii) allow us to write (for regular flows) **Bernoulli (1700-1782) equation**

$$\nabla\left(\frac{1}{2}\mathbf{v}_b^2 + \Phi(\rho_b) - U\right) = 0, \qquad x \in \Omega$$

$$\Phi(\rho) = \int^\rho \frac{p'(s)}{s}\, ds, \qquad p'(\rho) = \frac{dp}{d\rho},$$

$$(2.3.19)$$

where Φ is the enthalpy; cf. (1.5.24). We define

$$\mathcal{E}(\mathbf{v}_b, \rho_b) = \int_\Omega \rho_b \Big(\frac{1}{2}\mathbf{v}_b^2 + \Phi(\rho_b) - U\Big)dx.$$

We rewrite (1.7.7) for inviscid fluids and find

$$\frac{d}{dt}\int_\Omega \rho\Big(\frac{1}{2}|\mathbf{v}|^2 + \int^\rho \frac{p(s)}{s^2}ds\Big)dx = -\int_{\partial\Omega} p(\rho)\mathbf{v}\cdot\mathbf{n}dS + \int_\Omega \rho\mathbf{v}\cdot\nabla U dx.$$

$$(2.3.20)$$

In this case, the total energy is given by

$$E(\mathbf{v}, \rho) = \int_\Omega \{\frac{1}{2}\rho\mathbf{v}^2 + \rho\Psi(\rho) - \rho U(x)\}dx,$$

$$(2.3.21)$$

$$\Psi(\rho) = \int^\rho \frac{p(s)}{s^2}ds,$$

where Ψ is the Helmholtz free energy; cf. (1.5.25). Now, set

$$\mathbf{u} = \mathbf{v} - \mathbf{v}_b, \qquad\qquad \sigma = \rho - \rho_b,$$

Solutions $\mathbf{v}_b + \mathbf{u}$, $\rho_b + \sigma$ to (2.3.16) satisfy the side condition

$$\int_\Omega \sigma dx = 0.$$

The Lyapunov functional is

$$\mathcal{F}(\mathbf{u}, \sigma) = E(\mathbf{v}_b + \mathbf{u}, \rho_b + \sigma) - \mathcal{E}(\mathbf{v}_b, \rho_b)$$

$$= \int_\Omega \{\frac{1}{2}(\rho_b + \sigma)(\mathbf{v}_b + \mathbf{u})^2 + (\rho_b + \sigma)(\Psi(\rho_b + \sigma) - U(x))\}dx \qquad (2.3.22)$$

$$- \int_\Omega \{\frac{1}{2}\rho_b\mathbf{v}_b^2 + \rho_b(\Phi(\rho_b) - U(x))\}dx.$$

We notice that the time derivative of $\mathcal{E}(\mathbf{v}_b, \rho_b)$ is zero, thus from (2.3.19) we get

$$\frac{d}{dt}\Big(E(\mathbf{v}, \rho) - \mathcal{E}(\mathbf{v}_b, \rho_b)\Big) = \frac{d}{dt}\int_\Omega \{\frac{1}{2}\rho(\mathbf{v}^2 - \mathbf{v}_b^2) + \rho(\Psi(\rho) - \Phi(\rho_b))\}dx = 0.$$

$$(2.3.23)$$

We have now proven that $\mathcal{F}(\mathbf{u}, \sigma)$ is a good Lyapunov functional.

Applying the **Taylor (1685-1731) polynomial** formula, with initial point $(\mathbf{v}_b, \rho_b)$, we get:

$$\mathcal{F}(\mathbf{u}, \sigma) = E(\mathbf{v}_b + \mathbf{u}, \rho_b + \sigma) - \mathcal{E}(\mathbf{v}_b, \rho_b)$$

$$= E(\mathbf{v}_b + \mathbf{u}, \rho_b + u) - E(\mathbf{v}_b, \rho_b) + E(\mathbf{v}_b, \rho_b) - \mathcal{E}(\mathbf{v}_b, \rho_b) \qquad (2.3.24)$$

$$= \delta E(\mathbf{v}_b, \rho_b)[\mathbf{u}, \sigma] + \frac{1}{2}\delta^2 E(\bar{\mathbf{v}}, \bar{\rho})[\mathbf{u}, \sigma]^2 + \int_\Omega \rho_b(\Psi(\rho_b) - \Phi(\rho_b))dx,$$

where $\bar{\mathbf{v}}, \bar{\rho}$ is a point between $(\mathbf{v}_b, \rho_b)$ and $(\mathbf{v}, \rho)$.

Since the antiderivatives are defined up to a constant, the definitions of Ψ, and Φ yield

$$\int_\Omega \rho_b(\Psi(\rho_b) - \Phi(\rho_b))dx = \int_\Omega \rho_b \int^{\rho_b} \frac{d}{ds}\left(\frac{p(s)}{s}\right) ds\, dx \qquad (2.3.25)$$

$$= \int_\Omega \rho_b\left(\frac{p(\rho_b)}{\rho_b} + c\right) dx \qquad = \int_\Omega p(\rho_b)dx + cM.$$

The left hand side of (2.3.25) vanishes for c suitable

$$c = -\frac{\int_\Omega p(\rho_b)dx}{M}.$$

If such a choice is made for c, the function $\mathcal{F}$ reduces to

$$\mathcal{F}(\mathbf{u}, \sigma) = \delta E(\mathbf{v}_b, \rho_b)[\mathbf{u}, \sigma] + \frac{1}{2}\delta^2 E(\bar{\mathbf{v}}, \bar{\rho})[\mathbf{u}, \sigma]^2. \qquad (2.3.26)$$

Let'a now compute the first and second order variations of E taking as our initial point $(\mathbf{v}_b, \rho_b)$. We get

$$\delta E(\mathbf{v}_b, \rho_b)[\mathbf{u}, \sigma] = \int_\Omega \frac{\partial}{\partial \rho}\left[\rho(\mathbf{v}^2 - \mathbf{v}_b^2) + \rho\Psi(\rho) - \rho\Phi(\rho_b)\right]_{\rho_b, \mathbf{v}_b} \sigma dx + \int_\Omega \rho_b \mathbf{v}_b \cdot \mathbf{u} dx,$$

$$\frac{1}{2}\delta^2 E(\bar{\mathbf{v}}, \bar{\rho})[\mathbf{u}, \sigma]^2 dx = \int_\Omega \left\{\frac{1}{2}\rho\mathbf{u}^2 + \frac{1}{2}\frac{\partial^2(\rho\Psi(\rho))}{\partial\rho^2}\Big|_{(\bar{u}, \bar{\rho})} \sigma^2 + \bar{\mathbf{v}}_b \cdot \mathbf{u}\sigma\right\} dx.$$

$$(2.3.27)$$

Concerning the first integral in the first variation of E from the basic state, we notice that,

$$\frac{\partial}{\partial \rho}\left[\rho(\mathbf{v}^2 - \mathbf{v}_b^2) + \rho\Psi(\rho)) - \rho\Phi(\rho_b)\right]_{(\mathbf{v}_b, \rho_b)} = \rho_b\Psi'(\rho_b) + (\Psi(\rho_b) - \Phi(\rho_b))$$

$$= \frac{p(\rho_b)}{\rho_b} + \int^{\rho_b}\frac{p(s)}{s^2}ds - \int^{\rho_b}\frac{p'(s)}{s}ds = \frac{p(\rho_b)}{\rho_b} - \int^{\rho_b}\frac{d}{ds}\left(\frac{p(s)}{s}\right)ds = const = C.$$

$$(2.3.28)$$

Finally, $\int_\Omega C\sigma \, dx = 0$ because $\int_\Omega \sigma dx = 0$, thus the first integral in δE is zero.

Concerning the second integral in the first variation of E, using the **Helmholtz decomposition** for $\mathbf{u}$, $\mathbf{u} = \mathbf{w} + \nabla\xi$, with $\mathbf{w}$ solenoidal and with zero normal component at the boundary, recalling the second hypothesis (ii) together with the boundary conditions and $(2.3.17)_1$, we get

$$\int_\Omega \rho_b \mathbf{v}_b \cdot \mathbf{u} \, dx = \int_\Omega \rho_b \mathbf{v}_b \cdot (\mathbf{w} + \nabla\xi) \, dx = \int_\Omega \nabla\chi \cdot \mathbf{w} \, dx + \int_\Omega \rho_b \mathbf{v}_b \cdot \nabla\xi \, dx = 0.$$

Thus, the first variation of E is zero at $(\mathbf{v}_b, \rho_b)$. Now, we need to calculate the second order variation of E given by $(2.3.27)_2$. We recall that the point $(\mathbf{v}_b, \rho_b)$ is a minimum for the Helmhotz free energy, and therefore it holds that

$$\frac{\partial^2(\rho\Psi(\rho))}{\partial\rho^2}\bigg|_{(\mathbf{v}_b,\rho_b)} > 0,$$

which infers by continuity that it remains positive for all values of $I_R(\mathbf{v}_b, \rho_b)$, for R suitably small, and we can deduce

$$\frac{\partial^2(\rho\Psi(\rho))}{\partial\rho^2}\bigg|_{(\overline{u},\overline{\rho})} > 0, \qquad (\overline{u}, \overline{\rho}) \in I_R(\mathbf{v}_b, \rho_b).$$

If $\mathbf{v}_b$ is not too large as compared to basic density[4]

$$esssup_\Omega |\overline{\mathbf{v}}_b| < essinf_{\Omega \times (0,infty)} \rho(x,t) \frac{\partial^2\rho\Psi}{\partial\rho^2}\bigg|_{\overline{\rho}}$$

$(2.3.27)_2$ is equivalent to the L^2 norm of perturbations $\mathbf{u}$ and σ. Finally, integrating $(2.3.23)$ in time yields

$$a\left(\|\mathbf{u}\|_{L^2}^2 + \|\sigma\|_{L^2}^2\right) \leq \int_\Omega \left\{ \frac{1}{2}\rho\mathbf{u}^2 + \frac{1}{2}\frac{\partial^2(\rho\Psi(\rho))}{\partial\rho^2}\bigg|_{(\overline{u},\overline{\rho})} \sigma^2 + \overline{\mathbf{v}}_b \cdot \mathbf{u}\sigma \right\} dx$$

$$= \int_\Omega \left\{ \frac{1}{2}\rho_0\mathbf{u}_0^2 + \frac{1}{2}\frac{\partial^2(\rho_0\Psi(\rho_0))}{\partial\rho_0^2}\bigg|_{(\overline{u_0},\overline{\rho_0})} \sigma_0^2 + \overline{\mathbf{v}}_b \cdot \mathbf{u}_0\sigma_0 \right\} dx, \qquad (2.3.29)$$

which delivers control of perturbations $\mathbf{u}$ and σ in the L^2 norm, for all times $t > 0$.

---

[4]For isothermal flows it is enough to assume

$$esssup_\Omega |\overline{\mathbf{v}}_b| < k,$$

where $k = R_*\theta_b$ and R_* is the universal gas constant.

2.3.4   *Isothermal Viscous Fluids* $\mathbf{v}_b = 0$

In this subsection we use the energy method in an unorthodox manner to find the behavior in time of the difference between the energies $E(t)$ of the non steady motion and E_b of the rest state. We observe that $E(t) - E_b$ dominates the spatial L^2-norm of the difference of two solutions $(\rho, \mathbf{v})$, $(\rho_b, \mathbf{v}_b)$. Next, to get the decay to zero in time, we apply the free work equation, which furnishes dissipative terms for the perturbation in the density. This will be achieved by using suitable test functions whose existence is ensured by Lemma 3.7.5.

Let us consider the system

$$\partial_t \rho + \mathbf{u} \cdot \nabla \rho = -\rho \nabla \cdot \mathbf{u},$$

$$\rho \partial_t \mathbf{u} + \rho \mathbf{u} \cdot \nabla \mathbf{u} - \mu \Delta \mathbf{u} - (\lambda + \mu) \nabla \nabla \cdot \mathbf{u} = -k \nabla \rho, \qquad (x, t) \in \Omega \times (0, T),$$

$$\mathbf{u}(x, 0) = \mathbf{u}_0(x), \qquad \rho(x, 0) = \rho_0(x), \qquad x \in \Omega,$$

$$\mathbf{u}|_{\partial \Omega} = 0, \qquad \int_\Omega \rho \, dx = M, \qquad (2.3.30)$$

with $k = R_* \theta_b$ a positive constant. System (2.3.30) describes isothermal fluids moving in the absence of external force. It is easy to verify that $\mathbf{v}_b = 0$, $\rho_b = M/|\Omega|$ is a solution to (2.3.30). Furthermore, it is also standard to verify that the **energy equation** holds

$$\frac{d}{dt} \left\{ \int_\Omega \rho \frac{u^2}{2} + k\rho \ln \rho \right\} dx + \mu \, D_u(t) = 0, \qquad (2.3.31)$$

with

$$\mu \, D_u(t) = \int_\Omega \left[(\lambda + \mu)(\nabla \cdot \mathbf{u})^2 \, dx + \mu |\nabla \, \mathbf{u}|^2 \right] dx.$$

Notice that (2.3.31) can be equally rewritten in the following form

$$\frac{d}{dt} \mathcal{F}(\mathbf{u}, \sigma) + \mu \, D_u(t) = 0. \qquad (2.3.32)$$

where $\mathcal{F}$ is the energy of perturbations $\mathbf{u}$, σ:

$$\mathcal{F}(\mathbf{u}, \sigma) = \int_\Omega \left\{ \rho \frac{u^2}{2} + k\rho \ln \rho - k\rho \ln \rho_b - k(\rho - \rho_b) \right\} dx = \int_\Omega \left(\rho \frac{u^2}{2} + k \frac{\sigma^2}{2\bar{\rho}} \right) dx,$$

with $\bar{\rho}$ between ρ and ρ_b. It follows from the definition and from (2.3.32) that $\mathcal{F}$ is a Lyapunov functional, hence the rest state is stable.

Remark 2.3.1 *The stability in the mean continues to hold for inviscid fluids, for example if $\mu = 0$. In fact, $\mathcal{F}$ is a Lyapunov functional despite the viscosity term.*

For f and g, two vector functions in dual spaces, we use the notation (f, g) to denote the integral over Ω of the scalar product between these two functions.

Let $\mu > 0$; we wish to analyze the asymptotic behavior of a perturbation in time. To this end we construct a dissipative term for σ.

We recall that in (2.3.2) to get asymptotic decay we have multiplied the equation (2.3.8), that coincides with the equation of perturbation, times the free displacement $x(t) - x_b$ obtaining the free work equation (2.3.11). Here, in order to provide a dissipative term for the perturbation $\sigma = \rho - \rho_b$, we also construct a free work equation where now the free displacement is given by a suitable test function $\mathbf{V}$.

Let us multiply (2.3.30) by an auxiliary function $\mathbf{V}$, having a dimension of displacement "**free displacement**," and integrate over Ω. We obtain the **free work equation**

$$\frac{d}{dt}(\rho\mathbf{u}, \mathbf{V}) + k(\nabla\sigma, \mathbf{V}) = \mathcal{I}, \qquad (2.3.33)$$

with

$$\mathcal{I} = (\rho\mathbf{u}, \partial_t\mathbf{V} + \mathbf{u}\cdot\nabla\mathbf{V}) - (\lambda + \mu)\left(\nabla\cdot\mathbf{u}, \nabla\cdot\mathbf{V}\right) - \mu\left(\nabla\,\mathbf{u}, \nabla\,\mathbf{V}\right).$$

We call equation (2.3.33) the **free work equation** because it equates the time derivative of $(\rho\mathbf{u}, \mathbf{V})$ to an appropriate work. Notice that the *displacement* $\mathbf{V}$ is a free vector field *to be suitably chosen*.

Given σ as smooth function $\sigma = \sigma(x, t)$, now we can choose the displacement $\mathbf{V}$ as a solution to the boundary value problem; cf. Lemma 3.7.5 of Chap. 3,

$$\nabla\cdot\mathbf{V} = \sigma, \qquad (x, t) \in \Omega \times (0, \infty),$$

$$\mathbf{V}|_{\partial\Omega} = 0. \qquad (2.3.34)$$

We note that, since $\int_\Omega \sigma\,dx = 0$, the compatibility condition is satisfied.

Furthermore, for solutions to (2.3.34) there exist constants c' $\bar{c}$, c_* such that the following estimates hold, cf. (3.7.4) Lemma 3.7.5 of Chap. 3,

$$\|\mathbf{V}\|_{L^2} \leq c'\|\sigma\|_{L^2},$$

$$\|\nabla\mathbf{V}\|_{L^2} \leq \bar{c}\|\sigma\|_{L^2}, \qquad (2.3.35)$$

$$\|\partial_t\mathbf{V}\|_{L^2} \leq c_*\|\nabla\mathbf{u}\|_{L^2}.$$

Notice that (2.3.34) furnishes the **free work equation**

$$-\frac{d}{dt}(\rho\mathbf{u}, \mathbf{V}) + k\|\sigma\|_{L^2}^2 = -\mathcal{I}. \qquad (2.3.36)$$

Equation (2.3.36) appears in the form of the time derivative of a term, having the dimensions of the integral in time of an energy, plus a dissipative term for σ that equals the functional $\mathcal{I}$ having the dimension of a work.

If we multiply (2.3.33) by $-\nu$, where ν is an arbitrary positive constant having the dimension of inverse of time 1/sec, and add it to the energy equation (2.3.31), we can deduce, as an alternative Lyapunov functional $\mathcal{F}$, the **modified energy** $\mathbb{E}$

$$\mathbb{E} = \left\{ \int_\Omega \frac{u^2}{2} + k\frac{\sigma^2}{2\rho} - \nu\,\rho\mathbf{u}\cdot\mathbf{V} \right\}dx. \tag{2.3.37}$$

that satisfies the **modified energy equation**

$$\frac{d\mathbb{E}}{dt} = -\nu\|\nabla\mathbf{u}\|_{L^2}^2 dx - \nu k\|\sigma\|_{L^2}^2 - \nu\mathcal{I} =: -\mathcal{D} - \nu\mathcal{I}. \tag{2.3.38}$$

Employing (2.3.35), it is trivial to verify that the modified energy function (2.3.37) becomes equivalent to a norm of perturbations for ν sufficiently small

$$m_1(\|\mathbf{u}\|_{L^2}^2 dx + \|\sigma\|_{L^2}^2) \le \mathbb{E} \le m_2(\|\mathbf{u}\|_{L^2}^2 dx + \|\sigma\|_{L^2}^2).$$

We now use the Dirichlet generalized method, and analyze $\mathcal{I}$, which for small ν as well as for regular solutions, constitutes a positive quadratic form in the L^2 norm of perturbations $\mathbf{u}$, σ. From the estimates enjoyed by $\mathbf{V}$, it follows that

$$|\mathcal{I}| \le c(\|\nabla\mathbf{u}\|_{L^2}\|\sigma\|_{L^2} + \|\nabla\mathbf{u}\|_{L^2}^2),$$

where $c = c(\mathbf{u}, \sigma, \rho_b)$. Hence the term

$$\mathcal{D} - \nu\mathcal{I} = \int_\Omega \left(\mu|\nabla\mathbf{u}|^2 + (\lambda + \mu)(\nabla\cdot\mathbf{u})^2 \right)dx + \nu k\|\sigma\|_{L^2}^2 - \nu\mathcal{I}$$

is a positive definite quadratic form in the L^2 norm in the perturbations $\mathbf{u}$, σ. It follows that for ν small enough, using the Poincaré inequality, it is possible to prove that there exists a constant $\beta > 0$ such that

$$\mathcal{D} \ge \beta\mathbb{E}.$$

Substituting this information in (2.3.38) we deduce the differential inequality for the modified energy

$$\frac{d\mathbb{E}}{dt} + \beta\mathbb{E} \le 0,$$

that implies exponential decay to zero of the L^2 norm of the perturbations $\mathbf{u}$, σ. The decay constant depends upon the value of ν thus in general is very small.

Open problem *To extend the method to study the stability of a general steady flow.*

2.4 Main Theorems

Here we list the main theorems proved in next three chapters:

(a) The first set of theorems will be proven in Chap. 3. They concern the
steady flows S_b of *barotropic viscous fluids* filling a domain Ω, with a **rigid
boundary**. Once the basic flow S_b is given, there are proven theorems of
uniqueness of S_b in the class of *steady motions*, stability and asymptotic
stability of S_b in a suitable regularity class of solutions.

(b) The second set of theorems will be proven in Chap. 4. They concern
the *rest state* of a horizontal layer of *isothermal viscous fluids* filling a
periodicity cell Ω, which has as bottom a rigid flat surface, and above
a **free boundary**. There are proven uniqueness theorems for the rest
state S_b in the class of steady motions, stability, and asymptotic stability
theorems of S_b in a suitable regularity class of solutions. Next, for fluids
having a free boundary below the rigid flat surface, we introduce the
concept of initial data control, thus proving one instability theorem, and
one theorem relative the loss of initial data control.

(c) The third set of theorems will be proven in Chap. 5. They concern the
rest state of a horizontal layer of *polytropic viscous fluids* in a periodicity
cell Ω between **rigid** flat horizontal **boundaries**, heated from below.

There are proven uniqueness theorems of the rest state S_b in the class
of steady motions, stability and asymptotic stability theorems of S_b, in a
suitable regularity class of solutions.

Remark 2.4.1 *In cases (b) and (c), in order to simplify the problem, we
have considered the infinite horizontal plane as union of rectangular cells
periodic in two horizontal directions.*

All the proofs employ the "free work equation" FWE, which is one essential
tool of the book.

2.4.1 Case (a) Barotropic Fluid, Rigid Boundary

We study the following initial boundary value problem IBVP

$$
\begin{aligned}
\partial_t \rho + \nabla \cdot (\rho \mathbf{v}) &= 0, \\
\rho \partial_t \mathbf{v} + (\rho \mathbf{v} \cdot \nabla)\mathbf{v} &= -\nabla p + (\lambda + \mu)\nabla \nabla \cdot \mathbf{v} + \mu \Delta \mathbf{v} + \rho \mathbf{f}, && (x,t) \in \Omega \times (0,T), \\
\mathbf{v}(x,0) &= \mathbf{v}_0(x), && \rho(x,0) = \rho_0(x), && x \in \Omega, \\
\mathbf{v}|_{\partial\Omega}(x,t) &= \mathbf{w}(x,t), && x \in \partial\Omega, \\
\int_\Omega \rho &= M, && \rho \geq 0,
\end{aligned}
$$

$$(2.4.1)$$

where $\mathbf{w}$ is the velocity of the point of boundary, M the total mass of the fluid, and we have assumed

$$p = p(\rho), \qquad \mu \geq 0, \qquad (2\mu + 3\lambda) \geq 0.$$

Associated to (2.4.1), we consider the Boundary Value Problem

$$\nabla \cdot (\rho_b \mathbf{v}_b) = 0,$$

$$\rho_b \mathbf{v}_b \cdot \nabla) \mathbf{v}_b = -\nabla p(\rho_b) + (\lambda + \mu)\nabla\nabla \cdot \mathbf{v}_b + \mu \Delta \mathbf{v}_b + \rho_b \mathbf{f}, \qquad x \in \Omega, \qquad (2.4.2)$$

$$\mathbf{v}_b\Big|_{\partial\Omega} = \mathbf{w}(x), \qquad \int_\Omega \rho_b \, dx = M, \qquad \rho_b \geq 0,$$

with $p = p(\rho)$ a smooth function.

(a_1) Rest State

As we know, the rest state exists only when external forces derive from a uniform potential $\mathbf{f} = \nabla U$, and $\mathbf{w} = 0$. Indeed, in this case, the rest state $\mathbf{v} = 0$, $\rho_b = \rho_b(x)$ is the exact solution to (2.4.2), with ρ_b implicitly given by

$$\int^{\rho_b} \frac{p'(s)}{s} ds = U + c, \qquad \int_\Omega \rho_b \, dx = M. \qquad (2.4.3)$$

For the steady case, in order to have real positive solutions ρ_b (densities), we are lead to assume

Hypothesis R *The force is such that there exists a positive real solution to* (2.4.3).

Our uniqueness result will hold for **large potential forces** satisfying only **Hypothesis R**.

If $p'(\rho) > 0$, we may introduce the **Orlicz (1903-1990) space** $L_\phi(\Omega)$ with the following convex function

$$\phi(x) = (\rho - \rho_b) \int_{\rho_b(x)}^{\rho(x)} \frac{p'(s)}{s} ds,$$

in $L_1(\Omega)$.

Theorem 2.4.1 *Uniqueness of the Rest State in the Class of Steady Solutions. Let* $\mathbf{f} = \nabla U \in H^{-1}(\Omega)$ *satisfies Hypothesis R and* $\mathbf{w} = 0$, *then the rest state* $\mathbf{v}_b = 0$, $\rho_b = \rho_b(x)$, *with* ρ_b *implicitly given by* (2.4.3), *is unique in the class of solutions* $\mathbf{v}$, ρ, *to* (2.4.2) *where* $\mathbf{v} \in H_0^1(\Omega)$, *and* $\rho - \rho_b$ *belongs to* $L_\phi(\Omega)$.

The proof of this theorem is achieved by absurdum procedure, with the help of the FWE.

In order to state our stability theorems we introduce the class of generalized unsteady solutions $\big(\mathbf{u}(\mathbf{x},t),\rho(\mathbf{x},t)\big)$ in $\mathcal{W}$

$$\mathcal{W} = L^\infty\left(0,\infty; L^3(\Omega)\right) \cap L^2\left(0,\infty; W^{1,2}(\Omega)\right) \times L^\infty\left(0,\infty; C^0(\bar{\Omega})\right). \quad (2.4.4)$$

Roughly speaking, we prove that in bounded domains Ω the L^2-norm of any $'regular'$ perturbation $(\mathbf{u}, \sigma)$ to the unique basic rest state S_b decays to zero as $t \to \infty$, at an exponential rate.

We begin by deriving an energy equation that represents the energy conservation law of unsteady motions $\mathbf{v}(x,t) = \mathbf{u}(x,t)$ and $\rho(x,t) = \rho_b(x) + \sigma(x,t)$ Dirichlet method. In doing this we shall pay attention on the perturbation terms $\mathbf{u}(x,t)$ and $\sigma(x,t)$.

Theorem 2.4.2 *Energy Equation.* *Let* $\mathbf{f} = \nabla U \in L^\infty(0,\infty; H^{-1}(\Omega))$, *and let* $\mathbf{u} = \mathbf{v}$, $\rho = \rho_b + \sigma$ *solve the Initial Boundary Value Problem* (2.4.1) *with* $(\mathbf{u},\rho) \in \mathcal{W}$. *Then, the energy equation holds*

$$\frac{d}{dt}\Big[E_u + E_\sigma\Big] + \mu\, D_u(t) = 0, \quad (2.4.5)$$

$$E_u(t) = \frac{1}{2}\int_\Omega \rho \mathbf{u}^2\, dx;$$

$$E_\sigma(t) = \int_\Omega \rho\bigg(\int^\rho \frac{p(s)}{s^2}ds - U\bigg)\, dx = \frac{1}{2}\int_\Omega \frac{p'(\bar{\rho})}{\bar{\rho}}\sigma^2\, dx;$$

$$\mu\, D_u(t) = \int_\Omega \Big[(\lambda+\mu)(\nabla\cdot\mathbf{u})^2\, dx + \mu|\nabla\ \mathbf{u}|^2\Big]\, dx.$$

We give now the asymptotic result.

Theorem 2.4.3 Nonlinear Exponential Stability. *Let* $\mathbf{f} = \nabla U \in L^\infty(0,\infty; H^{-1}(\Omega))$, *then the rest state* $\mathbf{v}_b = 0$, $\rho_b = \rho_b(x)$, *with* ρ_b *given implicitly by* (2.4.3), *is **exponentially stable** in the class of motions* $\mathbf{u} = \mathbf{v}$, $\rho = \rho_b + \sigma$; *solutions to Initial Boundary Value Problem* (2.4.1) *with* $(\mathbf{u},\sigma) \in \mathcal{W}$.

(a_2) Non-potential Forces

Let us define the regularity class where uniqueness is proved

$$\mathcal{V} = \{(\mathbf{v},\rho) \in L^3(\Omega) \cap W^{1,2}(\Omega) \times (L^\infty(\Omega) \cap W^{1,\infty}(\Omega)) :$$
$$\inf p'(\rho_1) =: m_1 > 0, \quad \inf \rho > 0\},$$
$$\mathcal{V}_b = \{(\mathbf{v}_b,\rho_b) \in (L^3(\Omega) \cap W^{1,2}(\Omega)) \times (L^\infty(\Omega) \cap W^{1,\infty}(\Omega));$$
$$\mathbf{v}_b \cdot \nabla\mathbf{v}_b \in L^3(\Omega); \quad \inf p'(\rho_b) =: m_b > 0\}.$$

Notice that $\mathcal{V}_b \subseteq \mathcal{V}$. We remark that *it is reasonable to make assumptions on* $\mathbf{v}_b$ *because it is a given known motion, in this case regularity properties need only be verified, not proved.*

Theorem 2.4.4 Uniqueness of Steady Flows. *Let $f \in L^\infty(0, \infty; L^3(\Omega))$, and let $(v_b, \rho_b) \in \mathcal{V}_b$ be a given solution to the Boundary Value Problem (2.4.2). Then if $\|f\|_3$ and constants c_0, c_1, c_2, C_1, C_2, defined by (3.3.16), (3.3.24), satisfy (3.3.18), (3.3.27), (3.3.31), (v_b, ρ_b) is the unique solution to Boundary Value Problem (2.4.2) in the regularity class of solutions with $(v, \rho) \in \mathcal{V}$.*

In order to prove stability theorems within the presence of small non-potential forces we now define the following **regularity classes:**
Perturbed unsteady flows

$$\mathcal{W} = \{(\mathbf{v}, \rho) \in L^\infty(0, \infty; L^3(\Omega)) \cap L^2(0, \infty; W^{1,2}(\Omega)) \times L^\infty(0, \infty; W^{1,\infty}(\Omega));$$

$$\inf p'(\rho) > 0; \ \inf \rho > 0; \ \partial_t \rho \in L^\infty(0, \infty; L^\infty(\Omega)),$$

$$\sqrt{\rho} \mathbf{v} \in L^\infty(0, \infty; L^2(\Omega))\}.$$

Basic steady flows

$$\mathcal{V}_b = \{\mathbf{v}_b, \rho_b \in L^3(\Omega) \cap W^{1,2}(\Omega)) \times W^{1,\infty}(\Omega);$$

$$\mathbf{v}_b \cdot \nabla \mathbf{v}_b \in L^3(\Omega); \quad \inf_x p'(\rho_b) =: m_1 > 0\}.$$

Assume there exists a steady solution $\mathbf{v}_b$, ρ_b to (2.4.2) in the regularity class $\mathcal{V}_b$, then it is derived an energy inequality, which provides an energy stability result. Next, employing the FWE, it is proved the exponential decay of suitable norms of perturbations to zero.

Theorem 2.4.5 *Energy Equation.* *Let $f \in L^\infty(0, \infty; L^3(\Omega))$, and $(\mathbf{v}_b, \rho_b) \in \mathcal{V}_b$. Let $\mathbf{v} = \mathbf{v}_b + \mathbf{u}$, $\rho = \rho_b + \sigma$, solve the Initial Boundary Value Problem (2.4.1) with $(\mathbf{v}, \rho) \in \mathcal{W}$. Then, the following energy equation holds true*

$$\frac{d}{dt}\left[E_u + E_\sigma\right] + \mu D_u(t) = \mathcal{I}_1, \tag{2.4.6}$$

$$E_u(t) = \frac{1}{2}\int_\Omega \rho \mathbf{u}^2 \, dx, \qquad E_\sigma(t) = \int_\Omega \frac{p'(\bar{\rho})}{\rho}\frac{\sigma^2}{2} \, dx,$$

$$\mu D_u(t) = \int_\Omega \left[(\lambda + \mu)(\nabla \cdot \mathbf{u})^2 \, dx + \mu|\nabla \mathbf{u}|^2\right] dx,$$

$$\mathcal{I}_1(t) = \int_\Omega \mathbf{u} \cdot \mathbf{b} \, dx - \int_\Omega \frac{p'(\bar{\rho})}{\rho}\sigma \mathbf{u} \cdot \nabla \rho \, dx$$

$$- \int_\Omega \left[\frac{p'(\bar{\rho})}{\rho}\nabla \cdot \mathbf{v}_b + \frac{\rho_b}{2}\left(\partial_t + \mathbf{v}_b \nabla\right) \cdot \frac{p'(\bar{\rho})}{\rho_b \rho}\right]\sigma^2 \, dx,$$

$$\mathbf{b} = -\rho_b \mathbf{u} \cdot \nabla \mathbf{v}_b + \sigma(\mathbf{f} - \mathbf{v} \cdot \nabla \mathbf{v}_b).$$

Theorem 2.4.6 *Nonlinear Exponential Stability.* Let $f \in L^\infty(0, \infty;$ $L^3(\Omega))$, and let $(v_b, \rho_b) \in V_b$ be a given solution to the Boundary Value Problem (2.4.2) with $\mathbf{w} = 0$. If $\|f\|_3$ and constants c_0, c_1, c_2, C_1, C_2, C_3, defined by (3.3.16), (3.3.24), satisfy (3.3.18), (3.3.27) and (3.3.31), then (v_b, ρ_b) is exponentially stable with respect to motions in the regularity class $\mathcal{W}$.

(a_3) Non Zero Boundary Data

Let us suppose that $\mathbf{f} \in L^\infty(0, \infty; L^3(\Omega))$, $\mathbf{w} \in L^\infty(0, \infty; H^{1/2}(\partial\Omega))$, and let $\mathbf{w} \neq 0$.

Theorem 2.4.7 *Uniqueness in the Class of Steady Flow* Let $f \in L^\infty(0, \infty; L^3(\Omega))$, $\mathbf{w} \in L^\infty(0, \infty; H^{1/2}(\partial\Omega))$, and let $(v_b, \rho_b) \in \mathcal{S}_b$ be a given solution to Boundary Value Problem (2.4.2). Then if $\|f\|_3$ and constants c_0, c_1, c_2, C_1, C_2, C_3, defined by (3.3.16), (3.3.24), satisfy (3.3.18), (3.3.27), (3.3.31) thus (v_b, ρ_b) is the unique solution to Boundary Value Problem (2.4.2) in the regularity class of solutions with $(v, \rho) \in \mathcal{V}$.

Theorem 2.4.8 *Nonlinear Exponential Stability* Let $f \in L^\infty(0, \infty;$ $L^3(\Omega))$, $\mathbf{w} \in L^\infty(0, \infty; H^{1/2}(\partial\Omega))$, and let $(v_b, \rho_b) \in V_b$ a given solution to (2.4.2). Let $\|f\|_3$ and constants c_0, c_1, c_2, C_1, C_2, C_3, defined by (3.3.16), (3.3.24), satisfy (3.3.18), (3.3.27) and (3.3.31), then (v_b, ρ_b) is asymptotically stable with regards to motions in the regularity class W.

(a_4) Domains Exterior to a Fixed Bounded Body $\mathcal{C}$

Consider the Initial Boundary Value Problem IBVP

$$\partial_t \rho + \nabla \cdot (\rho \mathbf{v}) = 0,$$

$$\partial_t(\rho \mathbf{v}) + \nabla(\rho \, \mathbf{v} \otimes \mathbf{v}) = -\nabla p(\rho) + \lambda \nabla \nabla \cdot \mathbf{v} + 2\mu \nabla \cdot D(\mathbf{v}) + \rho \mathbf{f}, \qquad (x, t) \in \Omega \times (0, T),$$

$$\mathbf{v}|_{\partial\mathcal{C}} = 0, \qquad \lim_{x \to \infty} \mathbf{v} = 0 \qquad \lim_{x \to \infty} \rho = \rho_\infty,$$

$$(2.4.7)$$

with $p = p(\rho)$.

Consider the Boundary Value Problem BVP

$$\nabla \cdot (\rho \mathbf{v}) = 0,$$

$$\nabla(\rho \, \mathbf{v} \otimes \mathbf{v}) = -\nabla p + \lambda \nabla \nabla \cdot \mathbf{v} + 2\mu \nabla \cdot D(\mathbf{v}) + \rho \mathbf{f}, \qquad x \in \Omega, \qquad (2.4.8)$$

$$\mathbf{v}|_{\partial\mathcal{C}} = 0, \qquad \lim_{x \to \infty} \mathbf{v} = 0 \qquad \lim_{x \to \infty} \rho = \rho_\infty,$$

with $p = p(\rho)$.

Theorem 2.4.9 *Uniqueness of the Rest State in the Class of Steady Solutions. Let $f \in L^\infty(0, \infty; L^3(\Omega))$, and let $(v_b, \rho_b) \in \mathcal{V}_b$ be a given solution to (2.4.8). Assume $\|f\|_3$ and constants c_0, c_1, c_2, C_1, C_2, C_3, defined by (3.3.16), (3.3.24), satisfy (3.3.18), (3.3.27) and (3.3.31), then the solution $(v_b, \rho_b) \in \mathcal{V}_b$ to (2.4.8) is unique in the regularity class of steady solutions in $\mathcal{V}$.*

Notice that for compressible fluids moving in exterior domains, despite the incompressible case, it is still possible to prove exponential stability. Of course, this requires severe hypotheses of regularity on the density $\rho \in L^{3/2}$, that in turn intuitively require that the mass of gas is not too large. We give the proof for isothermal fluids; different state equations for pressure is an open problem.

Theorem 2.4.10 *Stability Let $f \in L^\infty(0, \infty; L^3(\Omega))$, and let $(v_b, \rho_b) \in \mathcal{V}_b$ be a given solution to (2.4.8). If $\|f\|_3$ and constants c_0, c_1, c_2, C_1, C_2, C_3, defined by (3.3.16), (3.3.24), satisfy (3.3.18), (3.3.27) and (3.3.31), then the solution $(v_b, \rho_b) \in \mathcal{V}_b$ to (2.4.8) is stable in the regularity class of unsteady solutions $\mathcal{W}$.*

Theorem 2.4.11 *Nonlinear Exponential Stability Let the gas be perfect, i.e. let $p(\rho) = k\rho$. Let $f \in L^\infty(0, \infty; L^3(\Omega))$, and let $(v_b, \rho_b) \in \mathcal{V}_b$ be a given solution to (2.4.8). Assume $\|f\|_3$ and constants c_0, c_1, c_2, C_1, C_2, C_3, defined by (3.3.16), (3.3.24), satisfy (3.3.18), (3.3.27) and (3.3.31), then the solution $(v_b, \rho_b) \in \mathcal{V}_b$ to (2.4.8) is asymptotically stable in the regularity class of unsteady solutions $\mathcal{W}$.*

With the term "**phase change**" we mean a barotropic fluid whose pressure, for some intervals of density ρ, becomes a decreasing function of ρ.

(a_5) Instability of Rest State in a Phase Change

Hypothesis of instability (HI):
Before dealing with instability questions, let us observe that for given pressure, it is possible to construct more density fields such that the pressure has constant value; cf. footnote in Sect. 3.6.
   Let us assume that:

(a) there exists more than one equilibrium configuration, i.e. the condition $p(\rho) = c$, when c is fixed, is satisfied by several values of ρ, say ρ_i, $i = 1, \ldots, N$, corresponding to the same given mass M and different volumes V_i.

(b) in at least one equilibrium configuration, say ρ_1, it holds

$$\frac{\partial^2 \rho \Psi}{\partial \rho^2}\bigg|_{\rho_1} < 0. \qquad (2.4.9)$$

In this case, we shall prove the following theorem:

Theorem 2.4.12 Instability *Assume the second order derivative of the Helmholtz free energy per unit of volume $\rho\Psi(\rho)$ satisfies the hypothesis of instability (2.4.9), then the equilibrium position S_1 at ρ_1 is unstable.*

2.4.2   Case (b) Isothermal Fluid, Deformable Boundary

For problems with deformable boundaries we prove a uniqueness theorem of steady fluid motions occurring in a rectangular section of the horizontal layer Σ, having the rigid bottom below and a free upper surface. The problem is described in Cartesian coordinates by $z = \hat{\zeta}(x)$; $\mathbf{k}$ is upward oriented, directed toward the free surface. We suppose that the domain occupied by the fluid is given by the cartesian representation $\Omega_t = \{\mathbf{x} = (x', z) : x' \in \Sigma, 0 < z < \zeta(x', t)\}$, remarking that such a representation is possible only when we exclude the formation of reversal flows. For the sake of simplicity we assume periodicity conditions at lateral walls. We use the notation $\nabla' = (\partial_1, \partial_2), \mathrm{div}'u' = \partial_1 u_1 + \partial_2 u_2$.

Given the system

$$\rho_t + \nabla \cdot (\rho\mathbf{v}) = 0$$

$$\rho(\mathbf{v}_t + \mathbf{v} \cdot \nabla\mathbf{v}) = -k\nabla\rho + \nabla \cdot \mu\mathbf{S}(\mathbf{v}) + \rho\nabla U, \qquad (x,t) \in \Omega \times (0,T),$$

$$\zeta_t(x',t) = \mathbf{v} \cdot \tilde{\mathbf{n}}(x', \zeta(x',t), t),$$

$$-k\rho\mathbf{n} + \mu\mathbf{S}(\mathbf{v}) = \alpha\mathcal{H}(\zeta)\mathbf{n} - p_e\mathbf{n}, \quad \text{on } \Gamma_t,$$

$$\mathbf{v}(x', 0, t) = 0, \tag{2.4.10}$$

$$\mathbf{v}(x', z, 0) = \mathbf{v}_0(x', z),$$

$$\rho(x', z, 0) = \rho_0(x', z), \quad \zeta(x', 0) = \zeta_0(x'),$$

$$M = \int_{\Omega_b} \rho_b dx = \int_{\Omega_0} \rho_0 dx = \int_{\Omega_t} \rho(x,t)dx,$$

$$\mu\mathbf{S}(\mathbf{v}) = 2\mu\mathbf{D} + \lambda\nabla \cdot \mathbf{v}\mathbf{I}.$$

Here, $\mathcal{H}(\zeta)$ is the **double mean curvature** to Γ_t, and it holds

$$\mathcal{H}(\zeta) = \mathrm{div}'\left(\frac{\nabla'\zeta}{\sqrt{1 + |\nabla'\zeta|^2}}\right),$$

and $\mathbf{n}$ denotes the exterior unit normal vector at the point of free surface Γ_t,

$$\mathbf{n} = \frac{1}{\mathcal{G}}(-\partial_1\zeta, -\partial_2\zeta, 1) = \frac{1}{\mathcal{G}}(-\nabla'\zeta, 1), \qquad \mathcal{G} = \sqrt{1 + |\nabla'\zeta|^2}, \quad \tilde{\mathbf{n}} = \mathcal{G}\mathbf{n}.$$

We assume here that the initial density ρ_0, and height ζ_0, are everywhere positive. We state the following

Initial Free Boundary Value Problem

Given a periodicity cell Σ, external potential forces with potential U, and uniform external pressure p_e, initial data $(\mathbf{v}_0,\ \rho_0,\ \zeta_0)$, and total mass M, find the triple of functions $(\mathbf{v}(x', z, t),\ \rho(x', z, t),\ \zeta(x', t))$ defined in $\Omega_t,\ t \in (0, \infty)$ for a solution to system (2.4.10).

To (2.4.10) we associate the steady system

$$\nabla \cdot (\rho \mathbf{u}) = 0$$

$$\rho(\mathbf{u} \cdot \nabla)\mathbf{u} - \mu\Delta\mathbf{u} - (\lambda + \mu)\nabla\nabla \cdot \mathbf{u} = -\nabla p + \rho\nabla U + \rho\mathbf{f},$$

$$\mathbf{u} \cdot \mathbf{n} = 0,$$

$$\mu S(\mathbf{u})\mathbf{n} - p\mathbf{n} = (-p_e + \kappa\mathcal{H}(\zeta))\mathbf{n}, \qquad (2.4.11)$$

$$\mathbf{u}(x', 0) = 0,$$

$$\int_{\Omega_\zeta} \rho(x)dx = M.$$

We also study the boundary value problem below.

Free Boundary Value Problem

Given a periodicity cell Σ, external potential forces with potential U, uniform external pressure p_e, and total mass M, find the triple of functions $(\mathbf{v}(x', z),\ \rho(x', z),\ \zeta(x'))$ defined in $\left(\Omega_\zeta^2 \times \Sigma\right)$ for a solution to system (2.4.11).

(b_1) Uniqueness of Rest State

The first theorem is concerned with the uniqueness of the rest state of an isothermal fluid under gravity action, in the class of steady weak solutions $(\mathbf{v}_b(x', z), \rho_b(x', z), \zeta_b(x')) \in \mathcal{V}_b$

$$\mathcal{V}_b = L^2_\sharp(\Omega) \times W^{1,2}_\sharp(\Omega) \times W^{1,1}_\sharp(\Sigma) \cap L^\infty_\sharp(\Sigma),$$

where $\sharp$ means periodicity in the horizontal directions.

Assume that the rigid side of a layer is below the fluid, and that the gravity force $U = -gz$, with z upward oriented, is acting.

In the rectangle $\Omega_b = \Sigma \times (0, h)$ there exists at least the rest state S_b, with

$$S_b = \left\{ \mathbf{v}_b = 0, \rho_b = \rho_* \exp\left(-\frac{gh}{k}\right), h = \frac{k}{g}\ln\left(1 + \frac{Mg}{p_e|\Sigma|}\right)\right\},$$

and ρ_* given in (4.4); cf. [6].

We now state the uniqueness theorem of S_b in the class of three-dimensional steady regular solutions to the boundary value problem (2.4.10), corresponding to the same force g, the same total mass M, the same periodicity Σ, and to the same external pressure p_e.

In order to present the main result we introduce the following regularity classes for steady $(\mathbf{u}(x', z), \rho(x', z), \zeta(x'))$, and unsteady $(\mathbf{u}(x', z, t), \rho(x', z, t), \zeta(x', t))$ solutions

$$\mathcal{V} = W_\sharp^{1,2}(\Omega) \times C_\sharp^0(\Omega) \times W_\sharp^{1,1}(\Sigma) \cap L_\sharp^\infty(\Sigma).$$

$$\mathcal{W} = L^2(0, \infty; W_\sharp^{1,2}(\Omega)) \times C_\sharp^0(\Omega \times (0, \infty)) \times L^\infty(0, \infty; W_\sharp^{1,1}(\Sigma)).$$

We prove the following uniqueness theorem.

Theorem 2.4.13 *Uniqueness of the Rest State in the Class of Steady Solutions.* *The rest state S_b is the unique solution to the system (2.4.11) in the class of steady solutions $(\mathbf{u}, \rho, \zeta)$ to the system (2.4.11) belonging to $\mathcal{V}$, corresponding to the same external data.*

(b_2) Stability of Rest State

We begin by deriving an energy equation that represents the difference between the energy equations of the non-steady motion and of the rest, respectively. Such a method is also known as **Dirichlet method**.

Theorem 2.4.14 *Energy Equation.* *Let $\mathbf{u}$, $\rho = \rho_b + \sigma$, $\zeta = h + \eta$ solve (2.4.10) with $(\mathbf{u}, \rho, \zeta) \in \mathcal{W}$. Then, the following energy equation holds*

$$\frac{d}{dt}\Big[E_u + E_\sigma + E_\zeta\Big] + \mu\, D_u(t) = 0, \tag{2.4.12}$$

$$E_u(t) = \frac{1}{2}\int_{\Omega_t} \rho \mathbf{u}^2\, dx,$$

$$E_\sigma(t) = k\int_{\Omega_t} \Big\{\rho(\ln\rho - \ln\rho_b) - (\rho - \rho_b)\Big\}\, dx,$$

$$E_\eta = \kappa\int_\Sigma \Big(\sqrt{1 + |\nabla'\eta|^2} - 1\Big)\, dx' + kM$$

$$\qquad - k\int_\Sigma\int_h^\zeta \Big(\rho_b(z) - \rho_b(h)\Big)\, dz dx',$$

$$\mu\, D_u(t) = \frac{\mu}{2}\int_{\Omega_t} |S(\mathbf{u})|^2\, dx.$$

It remains to prove exponential decay to the rest state for the L^2 norm of solutions to the full system (2.4.10), under the action of **large potential forces without smallness conditions on initial data**. To this end, we use again the FWE.

Theorem 2.4.15 *Nonlinear Exponential Stability.* *Assume that there exist solutions to (2.4.10)* $\mathbf{u}(x', z, t)$, $\rho(x', z, t)$, $\zeta(x', t)$ *in the regularity class* $\mathcal{W}$, *corresponding to initial data*

$$\mathbf{u}_0(x', z), \quad \rho_0(x', z) = \rho_b(x', z) + \sigma_0(x', z), \quad \zeta_0(x') = h + \eta_0(x').$$

Then, for any data $(\mathbf{u}_0, \rho_0, \eta_0)$ *in* $L^2(\Omega_0) \times C^0(\Omega_0) \times W^{1,\infty}(\Sigma)$, *the rest state* S_b *is exponentially stable in the energy norm in the class of solutions in* $\mathcal{W}$.

(b_3) Instability

Before giving our result, we introduce the non-dimensional characteristic number

$$Gr := \frac{g_0 \rho_*}{\kappa}, \tag{2.4.13}$$

with $\rho_* = g_0 M/(k|\Sigma|)$.

Theorem 2.4.16 *Assume that for all* $\delta > 0$, *there exists at least one initial data* $\mathbf{u}_0$, σ_0, η_0 *with* $\|(\mathbf{u}_0, \sigma_0, \eta_0)\|_Y < \delta$, *such that the initial energy* E_0 *is negative,*

$$0 < -E_0 = k \int_\Sigma \int_h^\zeta \Big(\rho_b(z) - \rho_b(h) \Big) dz dx' \tag{2.4.14}$$

$$- kM - \kappa \int_\Sigma \left(\sqrt{1 + |\nabla' \eta_0|^2} - 1 \right) dx' - E_{u_0} - E_{\sigma_0},$$

then the rest state S_b *is nonlinearly unstable. More precisely, there exists an* $\epsilon > 0$ *such that for all* $\delta > 0$ *there exists an initial value* $(\mathbf{u}_0, \rho_0 = \rho_b + \sigma_0, \zeta_0 = h + \eta_0) \in W^{1,2}(\Omega_0) \times W^{1,\infty}(\Sigma)$ *less than* δ, *and there exists* $T > 0$ *such that the solution* $(\mathbf{u}, \rho = \rho_b + \sigma, \zeta = h + \eta)$ *of the problem (2.4.10) satisfies the inequality*

$$\|(\mathbf{u}, \sigma, \eta)\|_X(T) \geq \epsilon. \tag{2.4.15}$$

(b_4) Loss of Initial Data Control

To deal with full nonlinear instability problems, we give the following

Definition 2.4.1 *The rest state is said to **lose the control from initial data** if there exists a positive large number a and there exists a perturbation* $(\mathbf{u}_0, \sigma_0, \eta_0)$ *to initial data satisfying*

$$a \; < \; \|(\mathbf{u}_0, \sigma_0, \eta_0)\|_Y \; < 2a, \tag{2.4.16}$$

such that the corresponding perturbation $(\mathbf{u}(x,t), \sigma(x,t), \eta(x,t))$ *is not controlled by the initial data. That is, given* $\alpha > 0$*, there exists* $T > 0$ *such that the perturbation* $(\mathbf{u}, \sigma, \eta)$ *with initial data satisfying* (2.4.16) *satisfies the inequality*

$$\|(\mathbf{u}, \sigma, \eta)\|_X(T) \; := \; \|\mathbf{u}(T)\|_{W^{1,2}(\Omega_T)} + \|\sigma(T)\|_{L^2(\Omega_T)} + \|\eta(T)\|_{W^{1,\infty}(\Sigma)} \geq \alpha.$$
$$\tag{2.4.17}$$

Next we construct a solution $(\mathbf{u}(x,t), \sigma(x,t), \eta(x',t))$ that though linearly stable, is not controlled by initial data when the data set is larger than a computable constant A.

Theorem 2.4.17 *We assume that the linear stability hypothesis:*

$$Gr \; < \; 1 \tag{2.4.18}$$

holds; cf. see Sect. 4.3.1. We may construct initial values $(\mathbf{u}_0, \sigma_0, \eta_0) \in W^{1,2}(\Omega_0) \times C_0(\Omega_0) \times W^{1,\infty}(\Sigma)$ *sufficiently small such that the value of initial energy* E_0 *is negative.*

Following upon Theorem 2.4.16 we may prove:

Theorem 2.4.18 *Loss of Initial Data Control of Solution* *There exists a positive large number a, such that the solution* $(\mathbf{u}(x,t), \sigma(x,t), \eta(x,t))$*, corresponding to initial data* $(\mathbf{u}_0, \sigma_0, \eta_0)$ *satisfying* (2.4.16)*, is not controlled by the initial data. That is, however fixed* $\alpha > 0$*, there exists initial data* $(\mathbf{u}_0^\alpha, \sigma^\alpha, \eta_0^\alpha)$ *satisfying* (2.4.16)*, and an instant* $T^\alpha > 0$ *such that the perturbation* $(\mathbf{u}^\alpha, \sigma^\alpha, \eta^\alpha)$ *corresponding to problem* (2.4.10) *satisfies the inequality*

$$\|(\mathbf{u}^\alpha, \sigma^\alpha, \eta^\alpha)\|_X(T^\alpha) \geq \alpha. \tag{2.4.19}$$

We conjecture that all proofs and considerations of this subsection continue to hold in the case of an infinite layer. We leave it as an open problem.

2.4.3 Case (c) Polytropic Fluid, Rigid Boundary

Let us consider the initial boundary value problem

$$\rho_t + \nabla \cdot (\rho \mathbf{u}) = 0,$$

$$\rho(\mathbf{u}_t + \mathbf{u} \cdot \nabla)\mathbf{u}) = \nabla \cdot \mathbf{T} - \rho g \mathbf{k}, \qquad (x,t) \in \Omega \times (0,T),$$

$$\rho c_v(\Theta_t + \mathbf{u} \cdot \nabla\Theta) = \chi\Delta\Theta - R_*\rho\Theta\nabla \cdot \mathbf{u} + 2\mu\mathbf{S}(\mathbf{u})^2 + \lambda(\nabla \cdot \mathbf{u})^2, \qquad (x,t) \in \Omega \times (0,T),$$

$$\mathbf{u}(x',0,t) = \mathbf{u}(x',h,t) = 0, \qquad \Theta(x',0,t) = \Theta_h + \beta h, \quad \Theta(x',h) = \Theta_h,$$

$$\mathbf{u}(x',z,0) = \mathbf{u}_0(x',z), \qquad \Theta(x',z,0) = \Theta_0(x',z), \qquad (x',z) \in \Sigma \times (0,h),$$

$$\int_\Omega \rho = M.$$

$$(2.4.20)$$

where $\mathbf{T} = -p\mathbf{I} + 2\mu D(\mathbf{u}) + \lambda\nabla \cdot \mathbf{u}\mathbf{I}$ is the stress tensor, $p = R_*\rho\Theta$ is the pressure, R_* the universal gas constant, μ is the shear viscosity, λ the bulk viscosity, $D(\mathbf{u}) = (\nabla\mathbf{u} + \nabla^T\mathbf{u})/2$ is the rate-of-strain tensor, c_v is the specific heat at constant volume, and χ is the coefficient of thermal conductivity.

We also give the system

$$\nabla \cdot (\rho\mathbf{u}) = 0,$$

$$\rho(\mathbf{u} \cdot \nabla)\mathbf{u} - \mu\Delta\mathbf{u} - (\lambda + \mu)\nabla\nabla \cdot \mathbf{u} = -\nabla p - \rho g \nabla z + \rho\mathbf{f}, \qquad x \in \Omega,$$

$$\rho c_v(\Theta_t + \mathbf{u} \cdot \nabla\Theta) = \chi\Delta\Theta - R_*\rho\Theta\nabla \cdot \mathbf{u} + 2\mu\mathbf{S}(\mathbf{u})^2 + \lambda(\nabla \cdot \mathbf{u})^2, \quad x \in \Omega,$$

$$\mathbf{u}(x',0,t) = \mathbf{u}(x',h,t) = 0, \qquad\qquad (2.4.21)$$

$$\Theta(x',h) = \Theta_h, \quad \Theta(x',0,t) = \Theta_h + \beta h, \quad (x',z) \in \Sigma \times (0,h),$$

$$\int_\Omega \rho = M.$$

In order to present the results we introduce the following regularity classes

$$\mathcal{V} = \left\{ (\rho(x',z), \mathbf{u}(x',z), \Theta(x')) \in C^0_\sharp(\Omega) \times W^{1,2}_\sharp(\Omega) \times W^{1,2}_\sharp(\Omega) \right\}.$$

$$\mathcal{W} = \left\{ \rho(x',z,t), \mathbf{u}(x',z,t), \Theta(x',z,t) \right.$$

$$\left. \in C^0_\sharp(\Omega \times (0,\infty)) \times L^2(0,\infty; W^{1,2}_\sharp(\Omega)) \times L^2(0,\infty; W^{1,2}_\sharp(\Omega)) \right\}.$$

A non-linear stability result for heat conducting fluids in exterior domain has been proven in cf. [111].

Theorem 2.4.19 *Uniqueness in the Class of Steady Flow* The rest state S_b is the unique steady solution to system $(2.4.21)$ in the class of motions $\mathcal{V}$, corresponding to the same data, provided temperature gradient is sufficiently small.

Theorem 2.4.20 *Nonlinear Exponential Stability* Let the conditions $(5.3.25)$ be verified. Then the rest state S_b is asymptotically stable for solutions to system $(2.4.20)$ in the class $\mathcal{W}$, corresponding to the same data.

2.5 Bibliographical Notes

Stability results, compressible fluids : The stability of compressible fluids is a challenging problem related to the study of the structure of linearized problems, which in turn depends upon the basic flow. In the case of parallel fluid flows, it has been studied in cf. [61].

Concerning the existence of regular solutions close to the rest state, there have been existence theorems since 1981 under assumptions of viscosity coefficients, cf. [90, 91, 95], and without restrictions on the coefficients, cf. [13, 56, 69–72, 130, 146].

Since 1986, several stability results have been proven; cf. [50, 94, 147], we quote also [72, 96, 97, 103, 104, 140]. Where bounded and unbounded domains with rigid, compact boundaries and strictly positive, bounded densities are considered. In these papers, stability of steady flows is proven for barotropic gases, either under large potential forces, or under small non-potential ones. Concerning unbounded domains, we cite [102] for isothermal fluids with density bounded from below, and for isothermal fluids with nonnegative density, uniqueness theorems have been proved; cf.[93, 99].

The method of the free work inequality appears to be useful as a new hint in the study of nonlinear asymptotic stability; a similar method has been introduced in the theory of elasticity by Abeyaratne and Knowles, cf. [2]. Concerning the stability and instability results proved with the free work inequality, we quote the papers by the author in cf. [104–109]. The stability proofs are formal, if initial data are large.

Chapter 3
Barotropic Fluids with Rigid Boundary

La vista mia nell'ampio e nell'altezza
non si smarriva, ma tutto prendeva
il quanto e il quale di quell'allegrezza.
118, XXX, Paradiso, A. Dante

The size of braveness is in a honest reasoning.

3.1 Introduction

In this chapter concerning several regular basic steady flows S_b we reach three goals: (a) the uniqueness of S_b in suitable regularity classes of steady flows; (b) the nonlinear asymptotic stability of S_b in the class of barotropic gases filling different domains Ω; (c) instability of S_b if $p'(\rho) < 0$.

Specifically at points (a) and (b) as domains we consider bounded domains which are rigid, impermeable, or porous, and domains exterior to a fixed, rigid bounded obstacle $\mathcal{C}$, as forces we consider potential and non potential forces. In exterior domains, summability of the perturbation σ to the density is required in order to prove uniqueness and stability.

To prove uniqueness in the class of steady flows, and stability theorems of rest state, it is sufficient to require only the non-negativeness of density.

To prove an asymptotic result it is suitable to distinguish between isothermal and isentropic gases. For isentropic fluids to prove an asymptotic result with exponential decay rate, both strict positivity and summability of density ρ_b are needed. These requests on density are incompatible in exterior domains, in particular the exponential decay rate is a very strong decay rate. We recall that for incompressible fluids the decay rate is polynomial like. Therefore the problem of decay of solutions in exterior domains remains an open problem, when ρ_b is not summable. We guess that the decay in time

M. Padula, *Asymptotic Stability of Steady Compressible Fluids*,
Lecture Notes in Mathematics 2024, DOI 10.1007/978-3-642-21137-9_3,
© Springer-Verlag Berlin Heidelberg 2011

has a polynomial rate. For isothermal fluids only summability of density ρ_b is requested. Actually in case of isothermal gases $p = k\rho$, also called perfect gases, the requirement of positive infimum on the density ρ is no longer needed. In this case, the exponential decay of energy can be proven in the class of non-negative densities.

Between *open questions* to solve, we quote the following two questions that depend upon the boundedness of Ω. Find: *(a) the most general conditions on supremum and infimum of density; (b) the largest regularity class of perturbations, **where** both stability, **and** asymptotic stability theorems hold.*

Assume the fluid is barotropic. The results proven in this chapter are the following:

Section 3.2 The uniqueness of the rest state, in a weak regularity class of steady flows, the nonlinear exponential stability of the rest state for barotropic fluids under large potential forces and large initial perturbations are proved; it represents a generalization of [104].

Section 3.3 The nonlinear exponential stability of steady state of barotropic fluids under small non-potential forces, and large initial perturbations is proved. The result is achieved assuming smallness of basic flow, for perturbations that are almost incompressible, with small divergence of velocity and gradient of density, let us call H_1 such hypotheses. It represents a generalization of cf. [103].

Section 3.4 The nonlinear exponential stability of the steady state of barotropic fluids with non-homogeneous boundary data and large initial perturbations is proved. The result is achieved assuming the same smallness assumptions H_1 for the basic flow, and on the perturbations as well. It is a new result.

Section 3.5 The nonlinear asymptotic stability of steady state barotropic fluids in exterior domains with large initial perturbations is proved. The result is achieved assuming the same smallness assumptions H_1 for the basic flow, and on the perturbations as well. The summability assumptions play a crucial role. It is a new result.

Section 3.6 The instability of the rest state for real fluids in the presence of a change of phase is proved. It represents a generalization of results of cf. [109].

Section 3.7 The existence of test functions, name them "**free work test functions**," as requested in the proofs of previous sections, is proven in several auxiliary Lemmas. Where the construction of free work test functions is known, the proofs are omitted.

Remark 3.1.1 *Through the chapter computations are carried out using regular functions, by applying a density argument, one may also obtain the final inequality for generalized solutions.*

3.2   Ω Bounded, Potential Forces

Regarding the existence of the rest state for compressible fluids subjected to potential forces, we quote two drawbacks in the logical application of the current theory of compressible fluids:

(a) The problem of existence of a **rest state** for *isentropic fluids*, where the polytropy index $\gamma > 1$, in a bounded rigid domain with a given mass, and subjected to potential force, is ill-posed if the volume and the forces are suitably large; cf. [12, 102].

(b) The problem of existence of a **rest state** for *isothermal fluids*, where the polytropy index $\gamma = 1$, in an exterior domain, with a given mass, and subjected to potential force, is ill-posed only if the external force is suitably small; cf. [99, 102].

In this section we propose an extension to general barotropic fluids of the result given in [104]. For bounded domains we consider the boundary value problem (BVP)

$$\nabla \cdot (\rho \mathbf{v}) = 0,$$

$$(\rho \mathbf{v} \cdot \nabla)\mathbf{v} = -\nabla p(\rho) + (\lambda + \mu)\nabla\nabla \cdot \mathbf{v} + \mu\Delta\mathbf{v} + \rho\nabla U, \qquad x \in \Omega, \tag{3.2.1}$$

$$\mathbf{v}\Big|_{\partial\Omega} = 0, \qquad \int_\Omega \rho\, dx = M, \quad \rho \geq 0,$$

with $p = p(\rho)$ a smooth function. As known, the rest state exists only when forces derive from a uniform potential U. Indeed, in this case, the rest state $\mathbf{v}_b = 0$, $\rho_b = \rho_b(x)$ is the exact solution to (3.2.1), with ρ_b implicitly given by equating the thermodynamic potential of *enthalpy* per unit of mass $\int^{\rho_b} \frac{p'(s)}{s} ds$, cf. (1.5.24), to the mechanical potential U,

$$\int^{\rho_b} \frac{p'(s)}{s} ds = U + c, \qquad \int_\Omega \rho_b\, dx = M. \tag{3.2.2}$$

Notice that the constant c is given by the condition that the total mass is known $(3.2.2)_2$, and therefore it may furnish a complex value for the density. In order to have real positive solutions ρ_b (densities), we are led to assume

Hypothesis R *The force* $\mathbf{f}$ *is such that there exists a positive real solution to* (3.2.2)*, that we call by* $(0, \rho_b)$.

3.2.1 Uniqueness of the Rest State

With the rest state as baseline.

Let us define the regularity class where uniqueness is proved

$$\mathcal{V} = \{(\mathbf{v}, \rho) \in W^{1,3}(\Omega) \times W^{1,\infty}(\Omega); \quad \inf p'(\rho) > 0; \quad \inf \rho > 0\},$$

$$\mathcal{V}_b = \{\rho_b \in W^{1,\infty}(\Omega); \qquad\qquad \inf p'(\rho_b) > 0\}.$$

Remark 3.2.1 *Notice that it is reasonable to make assumptions on $\mathcal{V}_b$ because it represents the regularity class of a given flow. However, it is not a weak hypothesis to assume the non-steady flow to belong to regularity class $\mathcal{V}$; indeed, such regularity has to be mathematically proven for large data.*

Let us give the thermodynamic potential enthalpy per unit of mass $\Phi = \int^\rho \frac{p'(s)}{s} ds$. Here we prove a uniqueness theorem that holds for **large potential forces** satisfying only Hypothesis R. We introduce the Orlicz space $L_\Phi(\rho_b)$ corresponding to the following convex function

$$\varphi(x) = (\rho - \rho_b) \int_{\rho_b(x)}^{\rho(x)} \frac{p'(s)}{s} ds = (\rho - \rho_b)(\Phi(\rho) - \Phi(\rho_b)).$$

Theorem 3.2.1 - Uniqueness of the Rest State. *Let $\rho f = \rho \nabla U \in H^{-1}(\Omega)$, if U satisfies Hypothesis R, then the rest state $\mathbf{v} = 0$, $\rho_b = \rho_b(x)$, with ρ_b given by (3.2.2), is unique in the class of steady solutions $\mathbf{v}$, ρ to (3.2.1), where $\mathbf{v} \in H_0^1(\Omega)$ and $\rho - \rho_b$ belongs to $L_\phi(\rho_b)$.*

Proof. Assume by absurdum there exists another steady solution $\mathbf{v}$, ρ to the boundary value problem (3.2.1) corresponding to the same force ∇U and the same mass M. We observe that the perturbation is $\mathbf{u} = \mathbf{v}$, $\sigma = \rho - \rho_b$. We first multiply (3.2.1)$_2$ by $\mathbf{u}$ and integrate over Ω to get the energy equation

$$(\lambda + \mu) \int_\Omega (\nabla \cdot \mathbf{u})^2 \, dx + \mu \int_\Omega |\nabla \mathbf{u}|^2 \, dx = - \int_\Omega \mathbf{u} \cdot \nabla p(\rho) \, dx + \int_\Omega \rho \mathbf{u} \cdot \nabla U \, dx.$$
$$(3.2.3)$$

We notice that, from (3.2.1)$_1$ and boundary conditions, integrating by parts, it yields

$$- \int_\Omega \mathbf{u} \cdot \nabla p(\rho) \, dx = - \int_\Omega \rho \mathbf{u} \frac{p'(\rho)}{\rho} \cdot \nabla \rho \, dx = - \int_\Omega \rho \mathbf{u} \cdot \nabla \int^\rho \frac{p'(s)}{s} ds \, dx = 0,$$
$$(3.2.4)$$

$$\int_\Omega \rho \mathbf{u} \cdot \nabla U \, dx = 0. \qquad (3.2.5)$$

Hence substituting in (3.2.3) the identities (3.2.4), (3.2.5) we deduce

$$(\lambda + \mu) \int_\Omega (\nabla \cdot \mathbf{u})^2 \, dx + \mu \int_\Omega |\nabla \mathbf{u}|^2 \, dx = 0. \qquad (3.2.6)$$

This together with the boundary conditions implies $\mathbf{u} = \mathbf{0}$.

In order to obtain uniqueness we must prove $\sigma = 0$, to this end we employ the free work equation.

Free Work Equation FWE

We need now an estimate for $\|\sigma\|_{L^2}$, to this end we construct the FWE. Notice that we may use Lemma 3.7.1 because the total mass is fixed and it holds

$$\int_\Omega \sigma \, dx = 0.$$

Let us observe that it holds

$$0 = -\nabla p(\rho) + \rho \frac{\nabla p(\rho_b)}{\rho_b} = \rho \nabla \left(\frac{\nabla p(\rho)}{\rho} - \frac{\nabla p(\rho_b)}{\rho_b} \right) = \rho \nabla \left(\int_{\rho_b}^{\rho} \frac{p'(s)}{s} \right). \quad (3.2.7)$$

Multiplying (3.2.7) by $\mathbf{V}/\rho$ given in Lemma 3.7.1, and integrating over Ω, we obtain the free work equation

$$0 = -\int_\Omega \mathbf{V} \cdot \nabla \left(\int_{\rho_b}^{\rho} \frac{p'(s)}{s} \right) dx = \int_\Omega \nabla \cdot \mathbf{V} \left(\int_{\rho_b}^{\rho} \frac{p'(s)}{s} ds \right) dx$$

$$= \int_\Omega \sigma \left(\int_{\rho_b}^{\rho} \frac{p'(s)}{s} ds \right) dx = \|\sigma\|_{L_\phi}^2. \quad (3.2.8)$$

Where L_ϕ is the Orlicz space generated by the convex function $\phi = (\rho - \rho_b) \int_{\rho_b}^{\rho} \frac{p'(s)}{s}$, the integrand is a positive function in σ, thus we obtain $\sigma = 0$ and the Theorem 3.2.1 is completely proved.

3.2.2 Nonlinear Stability

This subsection, deals with the evolution in time of unsteady compressible flows obtained perturbing at initial time the rest state in correspondence with the same potential U and the same mass M. We shall give the control in time of a spatial norm of perturbations with large initial data. This problem has been solved for isothermal fluids in [104].

For bounded domains we consider the initial boundary value problem (IBVP)

$$\rho_t + \mathbf{v} \cdot \nabla \rho = -\rho \nabla \cdot \mathbf{v},$$

$$\rho \mathbf{v}_t + \rho(\mathbf{v} \cdot \nabla) \, \mathbf{v} = -\nabla p + (\lambda + \mu) \nabla \nabla \cdot \mathbf{v} + \mu \Delta \mathbf{v} + \rho \mathbf{f},$$

$$\mathbf{v}(x,0) = \mathbf{v}_0(x), \qquad \rho(x,0) = \rho_0(x), \quad\quad (3.2.9)$$

$$\mathbf{v}\Big|_{\partial\Omega} = 0, \qquad \int_\Omega \rho_0 = M, \qquad \rho \geq 0,$$

with

$$p = p(\rho) \qquad x \in \Omega,\ t \in (0, T).$$

By conservation of mass we deduce

$$\int_\Omega \rho = M. \qquad (3.2.10)$$

It is worth reminding that in Theorem 3.2.1, uniqueness of the rest state has been proved in correspondence to large potential forces in rigid domains for a class of suitably "weak" steady solutions. Also note that in our stability proofs we shall make use of a class of suitably weak solutions; this will be achieved by modifying the method of the free work equation *FWE* previously used. Such a method provides dissipative terms for the variables satisfying hyperbolic equations.

In order to state our theorem we introduce the class $\mathcal{W}$ of weak solutions $(\mathbf{u}(x,t), \rho = \rho_b + \sigma(x,t))$ through the formula

$$\mathcal{W} = \Big\{ (\mathbf{u}, \sigma) : \quad \rho \in L^\infty \left(0, \infty; C^0(\bar\Omega)\right) \int_\Omega \sigma\, dx = 0; \qquad (3.2.11)$$

$$\mathbf{u} \in L^\infty \left(0, \infty;\, L^3(\Omega)\right) \cap L^2 \left(0, \infty;\, W^{1,2}(\Omega)\right) \Big\}.$$

In the bounded domain Ω, we prove that the L^2-norm of any suitably "*regular*" perturbation to the (unique) basic rest state S_b decays to zero as $t \to \infty$ at an exponential rate. To achieve this goal, first we use the Dirichlet method to find the behavior in time of the difference between the energy $E(t)$ of non-steady motion and that E_b of the rest state. For barotropic fluids with $p'(\rho) > 0$, we observe that $E(t) - E_b$ is positive, and larger than a constant times the spatial L^2-norm of perturbations. Next to get the decay to zero in time, we use the FWE that furnishes dissipative terms for the perturbation related to the basic density. To construct the FWE it is crucial to have suitable test functions constructed in Lemma 3.7.3.

Here we don't study the existence problem, hence we shall tacitly assume that

Hypothesis E *In the correspondence of regular data, there exists a global in time regular solution to (3.2.9) $(\mathbf{v}, \rho) \in \mathcal{W}$.*

Statement E has been proved for rigid domains, if the initial data belong to a neighborhood of $(0, \rho_b)\ \mathcal{U}_\epsilon(0, \rho_b) \subseteq X$, of radius ϵ, and ϵ is sufficiently small; cf. [70–72, 130, 145, 146].

Our stability result will hold for **large potential forces** satisfying only Hypothesis R introduced in the previous chapter.

We begin by deriving an energy equation, say the balance law for the sum of kinetic, Helmholtz free, and potential energies of unsteady motions $\mathbf{v}(x,t) = \mathbf{u}(x,t)$, $\rho(x,t) = \rho_b(x) + \sigma(x,t)$. In case of inviscid fluids we shall

deduce a conservation law for the total energy. In doing this we pay attention
to perturbation terms $\mathbf{u}(x,t), \sigma(x,t)$.

Before stating the energy theorem, we notice that for the difference
between the **Helmholtz free energy** and the enthalpy per unit of volume,
the following identity holds

$$\rho\left(\int^\rho \frac{p(s)}{s^2}ds - \int^{\rho_b} \frac{p'(s)}{s}ds\right) = \frac{1}{2}\frac{p'(\bar\rho)}{\bar\rho}\sigma^2. \tag{3.2.12}$$

Moreover, it holds the identity

$$\rho\frac{d^2(\rho\Psi)}{d\rho^2} = p'(\rho). \tag{3.2.13}$$

Theorem 3.2.2 *Energy Equation of Perturbation. Let* $\rho\mathbf{f} = \rho\nabla U \in L^\infty$
$(0,T; H^{-1}(\Omega))$, *and* $\mathbf{u} = \mathbf{v}$, $\rho = \rho_b + \sigma$, *solve* (3.2.9) *with* $(\mathbf{u},\sigma) \in \mathcal{W}$.
Then, the energy equation holds

$$\frac{d}{dt}\Big[E_u + E_\sigma\Big] + \mu\, D_u(t) = 0, \tag{3.2.14}$$

$$E_u = \frac{1}{2}\int_\Omega \rho\mathbf{u}^2\, dx;$$

$$E_\sigma = \frac{1}{2}\int_\Omega \frac{p'(\bar\rho)}{\bar\rho}\sigma^2\, dx;$$

$$\mu\, D_u = \int_\Omega \Big[(\lambda+\mu)(\nabla\cdot\mathbf{u})^2\, dx + \mu|\nabla\ \mathbf{u}|^2\Big]\, dx.$$

Proof. Let $\mathbf{v}$, ρ be a solution to the IBV problem (3.2.9) corresponding to the
same potential force $\mathbf{f} = \nabla U$ and the same mass M as the basic rest state.
We observe that the perturbation is $\mathbf{u} = \mathbf{v}$, $\sigma = \rho - \rho_b$. We first multiply
$(3.2.9)_2$ by $\mathbf{u}$ and integrate over Ω to get the equation

$$\frac{1}{2}\frac{d}{dt}\int_\Omega \rho\mathbf{u}^2\, dx + (\lambda+\mu)\int_\Omega (\nabla\cdot\mathbf{u})^2\, dx + \mu\int_\Omega |\nabla\ \mathbf{u}|^2\, dx$$

$$= -\int_\Omega \mathbf{u}\cdot\nabla p(\rho)\, dx + \int_\Omega \rho\mathbf{u}\cdot\nabla U\, dx. \tag{3.2.15}$$

Notice that multiplying $(3.2.9)_1$ by $p(\rho)/\rho$ and integrating over Ω, using
boundary conditions, it yields

$$\frac{d}{dt}\int_\Omega \rho\int^\rho \frac{p(s)}{s^2})ds\, dx = -\int_\Omega p(\rho)\nabla\cdot\mathbf{u}\, dx. \tag{3.2.16}$$

Also, since U is not an explicit function of time, by the Reynolds transport theorem it follows the identity

$$\int_\Omega \rho \mathbf{u} \cdot \nabla U \, dx = \frac{d}{dt} \int_\Omega \rho U \, dx. \tag{3.2.17}$$

Hence, substituting into (3.2.15) the identities (3.2.16), (3.2.17) we deduce the **energy equation**

$$\frac{1}{2} \frac{d}{dt} \int_\Omega \rho \mathbf{u}^2 \, dx + \frac{d}{dt} \int_\Omega \rho \Big[\int^\rho \frac{p(s)}{s^2} ds - U \Big] \, dx + \int_\Omega \Big[(\lambda + \mu)(\nabla \cdot \mathbf{u})^2 + \mu |\nabla \mathbf{u}|^2 \Big] \, dx = 0. \tag{3.2.18}$$

Observe that from the $(3.2.2)_1$, and (3.2.10) with hypothesis R, applying the Taylor polynomial formula of second grade with initial point ρ_b, we get:

$$\int_\Omega \rho \Big[\int^\rho \frac{p(s)}{s^2} ds - U \Big] \, dx = \int_\Omega \rho \Big[\int^\rho \frac{p(s)}{s^2} ds - \int^{\rho_b} \frac{p'(s)}{s} ds + c \Big] \, dx$$

$$= \int_\Omega \rho_b \Big[\int^{\rho_b} \frac{p(s)}{s^2} ds - \int^{\rho_b} \frac{p'(s)}{s} ds + c \Big] \, dx$$

$$+ \int_\Omega \Big[\int^\rho \frac{p(s)}{s^2} ds - \int^{\rho_b} \frac{p'(s)}{s} ds + \frac{p(\rho)}{\rho} \Big]_{\rho_b} \sigma \, dx$$

$$+ \frac{1}{2} \int_\Omega \frac{p'(\bar\rho)}{\bar\rho} \sigma^2 \, dx \tag{3.2.19}$$

where $\bar\rho$ is a point between ρ_b and ρ. Now the first integral term at right hand side of (3.2.19) has the integrand

$$\rho_b \Big[\int^{\rho_b} \frac{p(s)}{s^2} ds - \int^{\rho_b} \frac{p'(s)}{s} ds \Big] = -\rho_b \Big[\int^{\rho_b} \frac{d}{ds} \Big(\frac{p(s)}{s} \Big) ds \Big] = -p(\rho_b) + c\rho_b, \tag{3.2.20}$$

which is constant in time. Therefore, the integral of this term over a constant domain has a zero time derivative.

Next we analyze the integrand in the second term of (3.2.19), creating a relation between the Helmholtz free energy (1.5.25) and the enthalpy (1.5.24); cf. see Definition 1.5.7[1],

---

[1]Notice that it holds

$$\Big[\int^\rho \frac{p(s)}{s^2} ds - \int^{\rho_b} \frac{p'(s)}{s} ds + \frac{p(\rho)}{\rho} \Big] = \int_{\rho_b}^\rho \frac{p(s)}{s^2} ds + \frac{p(\rho)}{\rho} - \frac{p(\rho_b)}{\rho_b} =: \Phi(\rho)$$

and

$$\Phi'(\rho) = \frac{p'(\rho)}{\rho}.$$

$$\int^{\rho_b} \frac{p(s)}{s^2} ds - \int^{\rho_b} \frac{p'(s)}{s} ds + \frac{p(\rho_b)}{\rho_b} = -\frac{p(\rho_b)}{\rho_b} + \frac{p(\rho_b)}{\rho_b} = 0. \qquad (3.2.21)$$

Finally we obtain

$$\frac{d}{dt} \int_\Omega \rho \left[\int^\rho \frac{p(s)}{s^2} ds - U \right] dx = \frac{1}{2} \frac{d}{dt} \int_\Omega \frac{p'(\bar\rho)}{\bar\rho} \sigma^2 \, dx. \qquad (3.2.22)$$

Substituting (3.2.22) into (3.2.18) we obtain the wanted energy of perturbation equation

$$\frac{1}{2} \frac{d}{dt} \int_\Omega \left\{ \rho \mathbf{u}^2 + \frac{p'(\bar\rho)}{\bar\rho} \sigma^2 \right\} dx + \int_\Omega \left[(\lambda + \mu)(\nabla \cdot \mathbf{u})^2 + \mu |\nabla \mathbf{u}|^2 \right] dx = 0.$$
$$(3.2.23)$$

Integrating in time (3.2.23), and recalling that $p'(\rho) > 0$, we deduce at once the a priori estimate for the weighted L^2 norms of perturbations,

$$\frac{1}{2} \int_\Omega \left\{ \rho \mathbf{u}^2 + \frac{p'(\bar\rho)}{\bar\rho} \sigma^2 \right\} dx + \int_0^t \int_\Omega \left[(\lambda + \mu)(\nabla \cdot \mathbf{u})^2 + \mu |\nabla \mathbf{u}|^2 \right] dx$$
$$(3.2.24)$$
$$= \frac{1}{2} \int_\Omega \left\{ \rho_0 \mathbf{u}_0^2 + \frac{p'(\bar\rho_0)}{\bar\rho_0} \sigma_0^2 \right\} dx.$$

Equation (3.2.24) provides a control for weighted norms of σ, and $\mathbf{u}$ in $L^2(\Omega)$, and for the norm of $\mathbf{u}$ in $L^2(0, \infty; W^{1,2}(\Omega))$. $\qquad \square$

Remark 3.2.2 *On E_σ. Reminding that $\frac{p'(\rho)}{\rho}$ has a minimum*

$$m := \min_\rho \frac{p'(\rho)}{\rho} \qquad (3.2.25)$$

we find

$$\frac{p'(\bar\rho)}{\bar\rho} \geq m, \qquad (3.2.26)$$

which yields

$$E_\sigma = \frac{1}{2} \int_\Omega \frac{p'(\bar\rho)}{\bar\rho} \sigma^2 \, dx \geq m \|\sigma\|_{L^2}^2,$$

and the term $E_u + E_\sigma$ represents a Lyapunov functional.

It is worth nothing that (3.2.24) furnishes a stability result, say a control of L^2-norm of perturbations for all time, also for non negative densities, provided $p'(\rho) > 0$, e.g. for perfect gases.

In the general situation of non negative density $\rho \geq 0$, the energy of perturbations is no more equivalent to the L^2 norms of σ and $\mathbf{u}$, but to the L^2

norms of **u** *and* σ *weighted by the functions* ρ, *and* $\frac{p'(\bar{\rho})}{\bar{\rho}}$, *respectively, called with Arnold (1937-2010) [4], the* **natural norms**.

For isentropic gases with $\gamma > 2$ *the strict positiveness of density is needed to prove the asymptotic result.*

Remark 3.2.3 *From equation (3.2.24) stability follows also for non viscous fluids* $\mu = \lambda = 0$.

3.2.3 Nonlinear Exponential Stability

We now denote by c_e any embedding constant.

In order to obtain exponential decay we must find a dissipative term for σ. To this end we use the method of free work equation.

Using (3.2.7) we may rewrite $(3.2.9)_2$ as

$$\rho \mathbf{u}_t + \rho(\mathbf{u} \cdot \nabla)\,\mathbf{u} = -\rho \nabla \Big(\int_{\rho_b}^{\rho} \frac{p'(s)}{s} \Big) + (\lambda + \mu)\nabla \nabla \cdot \mathbf{u} + \mu \Delta \mathbf{u}. \qquad (3.2.27)$$

We will provide now the dissipative term for $\|\sigma\|_{L^2}$ by using the free work equation. We notice that the perturbation to density σ satisfies all hypotheses requested on the given function σ in Lemma 3.7.3. Actually, because the total mass, and the volume are fixed, it holds

$$\int_{\Omega} \sigma\,dx = 0, \qquad \forall\,t > 0. \qquad (3.2.28)$$

We are now in the position to prove exponential decay to the rest state of the L^2 norm of perturbations $(\mathbf{v} = \mathbf{u},\,\sigma)$, solutions to the initial boundary value problem $(3.2.9)_{1,3-7}$, (3.2.27), under the action of **large potential forces without smallness conditions on initial data**. A previous result can be found in [72], where also existence has been proved, see also [99].

Theorem 3.2.3 Modified Energy Equation *Let* $\rho \nabla U$ *belong to the Bochner space* $L^{\infty}(0, T; H^{-1}(\Omega))$, *and let* $(\mathbf{u}, \sigma) \in \mathcal{W}$ *solve the initial boundary value problem* $(3.2.9)_{1,3-7}$, *(3.2.27), corresponding to large data. Then the following modified energy equation holds*

$$\frac{d}{dt}\mathbb{E} + \mathcal{D} = \nu \mathcal{I}, \qquad (3.2.29)$$

where

$$\mathbb{E} = E_u + E_\sigma - \nu(\rho \mathbf{u}, \mathbf{V});$$

$$\mathcal{D} = \mu\,D_u + \nu\,D_\sigma,$$

$$D_\sigma = \int_{\Omega} \sigma \Big(\int_{\rho_b}^{\rho} \frac{p'(s)}{s}\,ds \Big)\,dx,$$

$$\mathcal{I} = \Big(\partial_t \mathbf{V}, \rho \mathbf{u}\Big) + \Big(\rho \mathbf{u} \cdot \nabla \mathbf{V}, \mathbf{u}\Big) + \mu \Big(\nabla \mathbf{u}, \nabla \mathbf{V}\Big) + (\lambda + \mu)\Big(\nabla \cdot \mathbf{u}, \nabla \cdot \mathbf{V}\Big), \quad (3.2.30)$$

where D_u has been defined in Theorem 3.2.2, ν is an arbitrary dimensional (one over time) positive constant, and $\mathbf{V}$ is the vector field constructed in Lemma 3.7.10.

Proof. Let us multiply (3.2.27) by $\mathbf{V}$, given in Lemma 3.7.3 below. Integrating over Ω by parts we obtain the free work equation

$$\frac{d}{dt} \int_\Omega \rho \mathbf{u} \cdot \mathbf{V}\, dx - \int_\Omega \rho \mathbf{u} \cdot \Big(\mathbf{V}_t + (\mathbf{u} \cdot \nabla)\, \mathbf{V}\Big) dx$$

$$= \int_\Omega \sigma \Big(\int_{\rho_b}^{\rho} \frac{p'(s)}{s}\, ds\Big)\, dx - \int_\Omega \Big((\lambda + \mu)\nabla \cdot \mathbf{u} \nabla \cdot \mathbf{V} + \mu \nabla \mathbf{u} \cdot \nabla \mathbf{V}\Big)\, dx.$$

$$(3.2.31)$$

The function $\sigma \int_{\rho_b}^{\rho} \frac{p'(s)}{s}\, ds$ is convex in σ, hence the first integrand at the right hand side in (3.2.31) is a positive function in σ; thus, we obtain a dissipative term for σ in the norm of the Orlicz space L_ϕ. We observe that $\sigma \int_{\rho_b}^{\rho} \frac{p'(s)}{s}\, ds = \sigma(\Phi(\rho) - \Phi(\rho_b))$, where Φ is the enthalpy. Thus it holds

$$\phi = \sigma \int_{\rho_b}^{\rho} \frac{p'(s)}{s}\, ds = \frac{p'(\widetilde{\rho})}{\widetilde{\rho}}\sigma^2 = \frac{p'(\rho_b + \alpha \sigma)}{\rho_b + \alpha \sigma}\sigma^2, \quad (3.2.32)$$

where $0 < \alpha < 1$. Also note that $\overline{\rho}$ is the same point between ρ_b, ρ, defined in (3.2.19). Equivalently, using (3.2.8) we rewrite (3.2.31) as the **free work equation**

$$-\frac{d}{dt} \int_\Omega \rho \mathbf{u} \cdot \mathbf{V}\, dx + \int_\Omega \frac{p'(\widetilde{\rho})}{\widetilde{\rho}}\sigma^2\, dx = \mathcal{I}, \quad (3.2.33)$$

where $\mathcal{I}$ is given in (3.2.30).

Adding (3.2.33) multiplied by an arbitrary positive constant ν to (3.2.23), we get the **modified energy equation** for $\mathbb{E}$

$$\frac{1}{2}\frac{d}{dt} \int_\Omega \Big\{ \rho \mathbf{u}^2 + \frac{p'(\overline{\rho})}{\overline{\rho}}\sigma^2 - \nu \rho \mathbf{u} \cdot \mathbf{V} \Big\} dx + \int_\Omega \Big[(\lambda + \mu)(\nabla \cdot \mathbf{u})^2$$

$$(3.2.34)$$

$$+ \mu |\nabla\, \mathbf{u}|^2 + \nu \frac{p'(\widetilde{\rho})}{\widetilde{\rho}}\sigma^2 \Big]\, dx = \nu \mathcal{I},$$

and the proof of theorem is completed. □

Remark 3.2.4 *Let us remark that we have supposed $(p(\rho)/\rho)$ to be an increasing function of ρ; this hypothesis is certainly satisfied by isentropic gases $\gamma > 1$, while for isothermal gases it is $\gamma = 1$, and $(p(\rho)/\rho)$ is constant in ρ. Thus in both cases, either isentropic or isothermal gases, the hypothesis* *of $(p(\rho)/\rho)$ not decreasing in ρ_b is satisfied. Integrating by parts in D_σ, it yields*

$$D_\sigma = \int_\Omega \Big(\int^\rho \frac{p'(s)}{s} ds - \int^{\rho_b} \frac{p'(s)}{s} ds \Big) \sigma \, dx = E_\sigma + \int_\Omega \sigma \Big(\frac{p(\rho)}{\rho} - \frac{p(\rho_b)}{\rho_b} \Big) \, dx \geq E_\sigma.$$

$$(3.2.35)$$

From the previous calculation, we infer that D_σ defines a norm for σ in an Orlicz space, which is included a priori in the Orlicz space defined by E_σ. Furthermore, under our regularity assumptions on ρ we deduce

$$m\|\sigma\|_{L^2} \leq D_\sigma \leq c(\rho)\|\sigma\|_{L^2},$$

with m given in (3.2.25). Consequently, the total dissipation $\mathcal{D}$ defines a norm for $(\mathbf{u}, \sigma)$ in $W^{1,2}(\Omega) \times L^2(\Omega)$. However, the modified energy functional is not always positive definite, this depends on the value of the parameter ν.

We will now explore the asymptotic result.

Theorem 3.2.4 Nonlinear Exponential Stability. *Let $\rho \nabla U(x) \in H^{-1}(\Omega)$, then the rest state $\mathbf{v} = 0$, $\rho_b = \rho_b(x)$, with ρ_b given implicitly by (3.2.2) when it exists, is **exponentially stable** in the class of motions $(\mathbf{u}, \rho) \in \mathcal{W}$.*

Proof. We begin by noting that in (3.2.34) there is a dissipative term in σ; as we shall see, this term is not sufficient to prove asymptotic stability. In order to do this, we need the stronger assumption

$$\sigma \int_{\rho_b}^\rho \frac{p'(s)}{s} ds \geq m\sigma^2, \qquad (3.2.36)$$

where m is given in (3.2.25). With this last position, (3.2.31) furnishes the **free work inequality**

$$-\frac{d}{dt} \int_\Omega \rho \mathbf{u} \cdot \mathbf{V} \, dx + m \int_\Omega |\sigma|^2 \, dx \leq \mathcal{I}, \qquad (3.2.37)$$

with $\mathcal{I}$ given in (3.2.30), satisfying

$$\mathcal{I} = -\int_\Omega \rho \mathbf{u} \cdot \Big(\mathbf{V}_t + (\mathbf{u} \cdot \nabla)\mathbf{V} \Big) \, dx + \int_\Omega \Big((\lambda + \mu) \nabla \cdot \mathbf{u} \nabla \cdot \mathbf{V} + \mu \nabla \mathbf{u} \cdot \nabla \mathbf{V} \Big) \, dx$$

$$\leq \Big[\|\sqrt{\rho}\mathbf{u}\|_{L^2} \|\sqrt{\rho}\mathbf{V}_t\|_{L^2} + \|\sqrt{\rho}\mathbf{u}\|_{L^4} \|\sqrt{\rho}\nabla \mathbf{V}\|_{L^2} + (\lambda + 2\mu)\|\nabla \mathbf{u}\|_{L^2}\|\nabla \mathbf{V}\|_{L^2} \Big]$$

$$\leq c\|\nabla \mathbf{u}\|_{L^2}\|\sigma\|_{L^2} + c\|\nabla \mathbf{u}\|_{L^2}^2,$$

$$(3.2.38)$$

where the constant c is a function of c_e, $\mathbf{u}$ and ρ.

Adding to (3.2.23) the free work inequality (3.2.37), multiplied by an arbitrary positive number ν, we deduce the inequality for the **modified energy** $\mathbb{E}$

$$\frac{d}{dt}\mathbb{E} + \mathcal{D} \leq \nu\mathcal{I}, \qquad (3.2.39)$$

with the modified energy functional given by

$$\mathbb{E} = \frac{1}{2}\int_\Omega \left\{ \rho\mathbf{u}^2 + \frac{p'(\bar\rho)}{\bar\rho}\sigma^2 - \nu\rho\mathbf{u}\cdot\mathbf{V} \right\} dx, \qquad (3.2.40)$$

$$\mathcal{D} = \int_\Omega \left[(\lambda+\mu)(\nabla\cdot\mathbf{u})^2 + \mu|\nabla\ \mathbf{u}|^2 + m\nu\sigma^2 \right] dx \qquad (3.2.41)$$

$$= (\mu+\lambda)\|\nabla\cdot\ \mathbf{u}\|_{L^2}^2 + \mu\|\nabla\mathbf{u}\|_{L^2}^2 + m\nu\|\sigma\|_{L^2}^2.$$

Notice that

$$\mathbb{E} \geq \left\{ \|\sqrt{\rho}\mathbf{u}\|_{L^2}^2 + m\|\sigma\|_{L^2}^2 - \nu\|\sqrt{\rho}\mathbf{u}\|_{L^2}\|\sqrt{\rho}\mathbf{V}\|_{L^2} \right\}, \qquad (3.2.42)$$

and $\mathbb{E}$ is positive definite when ν is small enough. If ν is sufficiently small, it results that

$$a\,\|(\mathbf{u},\sigma)\|_{L^2}^2 := a\left(\|\mathbf{u}\|_{L^2}^2 + \|\sigma\|_{L^2}^2 \right) \leq \mathbb{E}(t) \leq A\left(\|\mathbf{u}\|_{L^2}^2 + \|\sigma\|_{L^2}^2 \right) = A\|(\mathbf{u},\sigma)\|_{L^2}^2, \qquad (3.2.43)$$

and

$$\mathcal{D} \geq b(\nu)\left(\|\nabla\mathbf{u}\|_{L^2}^2 + \|\sigma\|_{L^2}^2 \right) \geq \frac{b(\nu)}{A}\mathbb{E}(t), \qquad (3.2.44)$$

with $b(\nu) = \min\{c_P\mu, m\nu\}$, and c_P the Poincare' constant. Inequality (3.2.39), with (3.2.38) and (3.2.43), provides the differential modified energy inequality

$$\frac{d}{dt}\mathbb{E}(t) + \left(\mu\|\nabla\mathbf{u}\|_{L^2}^2 + m\nu\|\sigma\|_{L^2}^2 \right) \leq \nu\,c\|\nabla\mathbf{u}\|_{L^2}\|\sigma\|_{L^2} + c\nu\,\|\nabla\mathbf{u}\|_{L^2}^2. \qquad (3.2.45)$$

Hence, for ν sufficiently small

$$\nu \leq \min\left\{ \frac{\mu}{4c}, \frac{m}{2c^2\mu} \right\}$$

it holds

$$\left(\mu\|\nabla\mathbf{u}\|_{L^2}^2 + m\nu\|\sigma\|_{L^2}^2 \right) - \nu\,c\|\nabla\mathbf{u}\|_{L^2}\|\sigma\|_{L^2} - \nu\,c\|\nabla\mathbf{u}\|_{L^2}^2$$

$$\geq \left((\mu-c\nu)\|\nabla\mathbf{u}\|_{L^2}^2 + m\nu\|\sigma\|_{L^2}^2 \right) - c^2\mu\nu^2\|\sigma\|_{L^2}^2 - \frac{\mu}{4}\|\nabla\mathbf{u}\|_{L^2}^2$$

$$\geq \frac{1}{2}\left(\mu\|\nabla\mathbf{u}\|_{L^2}^2 + m\nu\|\sigma\|_{L^2}^2 \right),$$

which in turn yields

$$\frac{d}{dt}\mathbb{E}(t) + \frac{b(\nu)}{2A}\mathbb{E}(t) \leq 0. \tag{3.2.46}$$

The application of Gronwall's (1877-1932) Lemma yields

$$\mathbb{E}(t) \leq \mathbb{E}(0)\ \exp^{-\frac{b}{2A}t}.$$

Finally, the inequality (3.2.43) implies

$$\|(\mathbf{u},\sigma)\|_{L^2}^2 \leq \frac{A}{a}\|(\mathbf{u}_0,\sigma_0)\|_{L^2}^2\ e^{-\frac{b}{a}t},$$

and resultingly, the exponential decay of perturbations is completely proved.

Remark 3.2.5 *It arises from the proof that, for two dimensional domains, the regularity hypotheses on* $\mathbf{u}$ *can be weakened.*

Remark 3.2.6 *Global existence theorems of weak solutions, with large initial data, for isentropic fluids, have been proven; cf. [26–28, 66]. Unfortunately, we have not been able to work all estimates for the term* $\mathcal{I}$ *by using only this assumption. Provided one is able to work all estimates employing only the hypothesis that* $\rho - \rho_b$ *belongs to the Orlicz space* L_ϕ, *one would derive also an existence theorem in the same class of stability, and the stability theorem would be no more formal. We leave it as a very challenging open problem.*

Remark 3.2.7 *Notice that Theorem 3.2.2 appears to hold without any conditions on initial data. However, smallness of initial data is being applied as a condition because existence theorems of global regular flows have been proven only under the assumption of smallness; cf. Sect. 3.7.3.*

3.3 Ω Bounded, Non-Potential Forces

This section represents an extension to the results given in [103] for general barotropic fluids.

3.3.1   *Uniqueness* $\boldsymbol{f} \neq \nabla U$

With the steady state as baseline.

In order to prove stability of steady solutions to system (3.2.1) for small non potential forces, we now define the following regularity classes: Perturbed steady flows

$$\mathcal{V} = \{(\mathbf{v},\rho):\quad \mathbf{v} \in W^{1,2}(\Omega);\ \inf p'(\rho) > m_1 > 0;\ \inf \rho > m > 0;\ \rho \in W^{1,\infty}\Omega)\}.$$

Basic steady flows

$$\mathcal{V}_b = \{(\mathbf{v}_b, \rho_b) : \quad \mathbf{v}_b \in W^{1,\infty}(\Omega); \rho_b \in W^{1,\infty}(\Omega) : \quad \inf_x \rho_b > 0\}$$

Notice that $\mathcal{V}_b \subseteq \mathcal{V}$.

Consider the following boundary value problem: find four scalar fields ρ_b, $\mathbf{v}_b$, defined in the domain Ω solutions to the following boundary value problem

$$\nabla \cdot (\rho_b \mathbf{v}_b) = 0, \qquad (3.3.1)$$

$$\rho_b \mathbf{v}_b \cdot \nabla \mathbf{v}_b = \mu \Delta \mathbf{v}_b + (\lambda + \mu)\nabla\nabla \cdot \mathbf{v}_b - \nabla p(\rho_b) + \rho_b \mathbf{f}$$

$$\mathbf{v}_b|_{\partial\Omega} = 0, \qquad \int_\Omega \rho_b = M.$$

Assuming that for regular forces there exists a solution $\mathbf{v}_b$, ρ_b in the regularity class $\mathcal{V}_b$, we first derive an energy inequality. Then, through the free work equation, we prove uniqueness of the basic steady flow S_b in a class of 'regular' steady flows, provided that S_b is 'not too large' in a sense to be clarified.

Theorem 3.3.1 Uniqueness of Steady Flow. *Let* $\mathbf{f} \in L^3(\Omega)$, *and let be* $(\mathbf{v}_b, \rho_b) \in \mathcal{V}_b$ *be a given solution to* (3.3.1). *If* $\|\mathbf{f}\|_3$ *and constants* c_0, c_1, c_2, C_1, C_2, *defined by* (3.3.16)$_1$ *and* (3.3.24), *satisfy* (3.3.18), (3.3.50) *and* (3.3.51), *then* $(\mathbf{v}_b, \rho_b)$ *is the unique solution to* (3.3.1) *in the regularity class* $\mathcal{V}$.

Proof. Given the steady solution $(\mathbf{v}_b, \rho_b)$ to (3.3.1) in $\mathcal{V}_b$, we assume by absurdum that $(\mathbf{v}, \rho)$ in $\mathcal{S}$ is another steady solution to system (3.3.1) corresponding to the same force, the same mass, and the same boundary data. Let $\mathbf{u} = \mathbf{v} - \mathbf{v_b}$, and $\sigma = \rho - \rho_b$ be the difference between the two solutions. The pair $(\mathbf{u}, \sigma)$ will satisfy the perturbation system

$$\nabla \cdot (\rho \, \mathbf{u}) + \nabla \cdot (\sigma \, \mathbf{v}_b) = 0, \qquad x \in \Omega,$$
$$\rho \, \mathbf{v} \cdot \nabla \, \mathbf{u} + (\rho_b \, \mathbf{u} + \sigma \, \mathbf{v}) \cdot \nabla \, \mathbf{v}_b = -\nabla \Big(p(\rho) - p(\rho_b)\Big) + (\lambda + \mu)\nabla\nabla \cdot \mathbf{u}$$
$$+ \mu\Delta \, \mathbf{u} + \sigma \, \mathbf{f},$$

$$(3.3.2)$$

or more simply,

$$\nabla \cdot (\rho \, \mathbf{u}) + \nabla \cdot (\sigma \, \mathbf{v}_b) = 0, \qquad x \in \Omega,$$
$$\rho \, \mathbf{v} \cdot \nabla \, \mathbf{u} = -\nabla \Big(p(\rho) - p(\rho_b)\Big) + (\lambda + \mu)\nabla\nabla \cdot \mathbf{u} + \mu\Delta \, \mathbf{u} + \mathbf{b}, \qquad (3.3.3)$$

with

$$\mathbf{b} = -\rho_b \mathbf{u} \cdot \nabla \, \mathbf{v}_b + \sigma(\, \mathbf{f} - \mathbf{v} \cdot \nabla \, \mathbf{v}_b). \qquad (3.3.4)$$

Let us multiply $(3.3.3)_2$ by $\mathbf{u}$ and integrate over Ω to have

$$\int_\Omega \rho \mathbf{v} \cdot \nabla \left(\frac{|\mathbf{u}|^2}{2} | \right) dx = -\int_\Omega \mathbf{u} \cdot \nabla (p'(\bar\rho)\sigma) dx + \int_\Omega \left[(\lambda+\mu)\mathbf{u} \cdot \nabla\nabla \cdot \mathbf{u} + \mu\mathbf{u} \cdot \Delta\mathbf{u} \right] dx + \int_\Omega \mathbf{u} \cdot \mathbf{b} dx,$$
$$(3.3.5)$$

where $\bar\rho$ is the value of ρ between ρ, and ρ_b. Integrating by parts in (3.3.5), and taking into account that the boundary terms

$$B_1 := -\int_{\partial\Omega} \frac{\rho}{2} |\mathbf{u}|^2 \mathbf{v} \cdot \mathbf{n} \, dS + \int_{\partial\Omega} \left[(\lambda+\mu)\mathbf{u} \cdot \mathbf{n}\nabla \cdot \mathbf{u} + \mu\mathbf{n} \cdot \nabla \ \mathbf{u} \cdot \mathbf{u} \right] dS, \ (3.3.6)$$

vanish, we get

$$\int_\Omega \left[(\lambda+\mu)(\nabla \cdot \mathbf{u})^2 + \mu |\nabla \ \mathbf{u}|^2 \right] dx = -\int_\Omega \mathbf{u} \cdot \nabla (p'(\bar\rho)\sigma) \, dx + \int_\Omega \mathbf{u} \cdot \mathbf{b} \, dx. \ (3.3.7)$$

We rewrite $(3.3.3)_1$ as follows

$$0 = -\rho\nabla \cdot \ \mathbf{u} - \ \mathbf{u} \cdot \nabla\rho - \nabla \cdot (\sigma\mathbf{v}_b). \qquad (3.3.8)$$

Multiplying (3.3.8) by $(p'(\bar\rho)/\rho)\sigma$, and integrating over Ω, we have

$$0 = -\int_\Omega p'(\bar\rho)\sigma\nabla \cdot \ \mathbf{u} \, dx - \int_\Omega \frac{p'(\bar\rho)}{\rho}\sigma\mathbf{u} \cdot \nabla\rho \, dx - \int_\Omega \frac{p'(\bar\rho)}{\rho}\sigma^2\nabla \cdot \mathbf{v}_b \, dx - \int_\Omega \frac{p'(\bar\rho)}{\rho}\mathbf{v}_b \cdot \nabla \left(\frac{\sigma^2}{2} \right) dx,$$

which when integrated by parts yields

$$0 = -\int_\Omega p'(\bar\rho)\sigma\nabla \cdot \mathbf{u} \, dx - \int_\Omega \frac{p'(\bar\rho)}{\rho}\sigma\mathbf{u} \cdot \nabla\rho \, dx$$
$$-\int_\Omega \frac{p'(\bar\rho)}{\rho}\sigma^2\nabla \cdot \mathbf{v}_b \, dx + \int_\Omega \nabla \cdot \left(\frac{p'(\bar\rho)}{\rho}\mathbf{v}_b \right) \frac{\sigma^2}{2} \, dx - B_2,$$
$$(3.3.9)$$

with

$$B_2 := \int_{\partial\Omega} \frac{p'(\bar\rho)}{\rho} \frac{\sigma^2}{2} \mathbf{v}_b \cdot \mathbf{n} \, dS. \qquad (3.3.10)$$

We add (3.3.7) to (-1)(3.3.9), noting that the term

$$-\int_\Omega \mathbf{u} \cdot \nabla (p'(\bar\rho)\sigma) \, dx - \int_\Omega p'(\bar\rho)\sigma\nabla \cdot \mathbf{u} \, dx = -\int_{\partial\Omega} \mathbf{u} \cdot \mathbf{n} p'(\bar\rho)\sigma \, dS =: B_3, \ (3.3.11)$$

also vanishes due to the boundary conditions, we deduce that (3.3.7) reduces to the simpler form

$$\int_\Omega \left[(\lambda+\mu)(\nabla \cdot \mathbf{u})^2 + \mu |\nabla \ \mathbf{u}|^2 \right] dx = \mathcal{I}_1 - B_2 - B_3, \qquad (3.3.12)$$

where

$$\mathcal{I}_1 = \int_\Omega \mathbf{u} \cdot \mathbf{b}\, dx - \int_\Omega \frac{p'(\bar\rho)}{\rho}\sigma\mathbf{u} \cdot \nabla\rho\, dx - \int_\Omega \frac{p'(\bar\rho)}{\rho}\frac{\sigma^2}{2}\nabla \cdot \mathbf{v}_b\, dx + \int_\Omega \frac{\sigma^2}{2}\mathbf{v}_b \cdot \nabla\Big(\frac{p'(\bar\rho)}{\rho}\Big)\, dx.$$
(3.3.13)

since at the boundary $\partial\Omega$ it is $\mathbf{u} = \mathbf{v}_b = 0$, it follows that $B_2 = 0$, $B_3 = 0$, and (3.3.12) becomes

$$\int_\Omega \Big[(\lambda + \mu)(\nabla \cdot \mathbf{u})^2 + \mu|\nabla\,\mathbf{u}|^2\Big]dx = \mathcal{I}_1.$$
(3.3.14)

Now, from regularity assumptions taking into account (3.3.4), it is easy to check that

$$\mathcal{I}_1 \le c_1\|\sigma\|_{L^2}\|\mathbf{u}\|_{L^6} + c_2\|\sigma\|_{L^2}^2 + \|\rho_b\|_{L^\infty}\|\nabla\,\mathbf{v}_b\|_{L^{3/2}}\|\mathbf{u}\|_{L^6}^2$$
$$\le c_1\|\sigma\|_{L^2}\|\nabla\mathbf{u}\|_{L^2} + c_2\|\sigma\|_{L^2}^2 + c_0\|\nabla\mathbf{u}\|_{L^2}^2,$$
(3.3.15)

holds where

$$c_0 = c_e\|\rho_b\|_{L^\infty}\|\nabla\,\mathbf{v}_b\|_{L^{3/2}},$$
(3.3.16)

$$c_1 = c_e\Big(\|\mathbf{f}\|_{L^3} + \Big\|\frac{p'(\bar\rho)}{\rho}|\nabla\rho|\Big\|_{L^3} + \|\,\mathbf{v} \cdot \nabla\,\mathbf{v}_b\|_{L^3}\Big),$$

$$c_2 = \frac{1}{2}\Big\|\frac{p'(\bar\rho)}{\rho}|\nabla \cdot \mathbf{v}_b|\Big\|_{L^\infty} + \Big\||\mathbf{v}_b|\nabla\Big(\frac{p'(\bar\rho)}{\rho}\Big)\Big\|_{L^\infty}.$$

Hence we deduce the **energy inequality**

$$(\lambda + \mu)\|\nabla \cdot \mathbf{u}\|_{L^2}^2 + (\mu - c_0)\|\nabla\,\mathbf{u}\|_{L^2}^2 \le c_1\|\sigma\|_{L^2}\|\nabla\,\mathbf{u}\|_{L^2} + c_2\|\sigma\|_{L^2}^2.$$ (3.3.17)

To inequality (3.3.17), we add the hypothesis

$$c_0 < \mu,$$
(3.3.18)

with c_0 defined in (3.3.16)$_1$.

Free work equation

We look now for an estimate of $\|\sigma\|_{L^2}$ in terms of $\|\nabla\mathbf{u}\|_{L^2}$; to this end, we employ the free work equation. Notice that we can use Lemma 3.7.1, taking as σ the perturbation to density $\rho - \rho_b$, because the total mass and volume are fixed, and (3.7.3) holds. Let us multiply (3.3.3)$_2$ by $\mathbf{V}$ given in Lemma 3.7.1, and integrate by parts over Ω to receive the free work equation

$$\int_\Omega p'(\bar\rho)\sigma\nabla \cdot \mathbf{V}\, dx = -\int_\Omega \rho\,\mathbf{v} \cdot \nabla\,\mathbf{V} \cdot \mathbf{u}\, dx + \int_\Omega \Big[(\lambda + \mu)\nabla \cdot \mathbf{V}\nabla \cdot \mathbf{u} + \mu\nabla\,\mathbf{u} : \nabla\,\mathbf{V}\Big]dx$$

$$-\int_\Omega \mathbf{V} \cdot \mathbf{b}\, dx + B_4,$$

with

$$B_4 := \int_{\partial\Omega} k\sigma\mathbf{n}\cdot\mathbf{V}\,dS + \int_{\partial\Omega} \rho\,\mathbf{V}\cdot\mathbf{u}\,\mathbf{v}\cdot\mathbf{n}\,dS$$

$$- (\lambda+\mu)\int_{\partial\Omega} \nabla\cdot\mathbf{u}\mathbf{n}\cdot\,\mathbf{V}\,dS - \mu\int_{\partial\Omega} \mathbf{V}\cdot\nabla\,\mathbf{u}\cdot\mathbf{n}\,dS. \tag{3.3.19}$$

Employing the property $\mathbf{V}|_{\partial\Omega} = \mathbf{0}$ we deduce that $B_4 = 0$, and by $\nabla\cdot\mathbf{V} = \sigma$, we get

$$\int_{\Omega} p'(\bar{\rho})\sigma^2\,dx = \mathcal{I}_2, \tag{3.3.20}$$

where

$$\mathcal{I}_2 = -\int_{\Omega} \rho\,\mathbf{v}\cdot\nabla\mathbf{V}\cdot\mathbf{u}\,dx + \int_{\Omega} \Big[(\lambda+\mu)\nabla\cdot\mathbf{V}\nabla\cdot\mathbf{u} + \mu\nabla\,\mathbf{u} : \nabla\,\mathbf{V}\Big]dx - \int_{\Omega} \mathbf{V}\cdot\mathbf{b}\,dx. \tag{3.3.21}$$

We may estimate $\mathcal{I}_2$ as follows

$$\mathcal{I}_2 \leq \|\rho\mathbf{v}\|_{L^3}\|\nabla\,\mathbf{V}\|_{L^2}\|\,\mathbf{u}\|_{L^6} + (\lambda+\mu)\|\nabla\cdot\mathbf{u}\|_{L^2}\|\nabla\cdot\mathbf{V}\|_{L^2} + \mu\|\nabla\mathbf{V}\|_{L^2}\|\nabla\mathbf{u}\|_{L^2}$$

$$+ \|\rho_b\|_{L^\infty}\|\nabla\mathbf{v}_b\|_{L^{3/2}}\|\,\mathbf{V}\|_{L^6}\|\mathbf{u}\|_{L^6} + (\|\mathbf{f}\|_{L^3} + \|\mathbf{v}\cdot\nabla\mathbf{v}_b\|_{L^3})\|\sigma\|_{L^2}\|\,\mathbf{V}\|_{L^6}. \tag{3.3.22}$$

Using the limit Sobolev inequality and the properties of $\mathbf{V}$, we obtain

$$\mathcal{I}_2 \leq (\lambda+\mu)\|\nabla\cdot\mathbf{u}\|_{L^2}\|\sigma\|_{L^2} + l_2\mu\|\nabla\mathbf{u}\|_{L^2}\|\sigma\|_{L^2} + C_1\|\nabla\mathbf{u}\|_{L^2}\|\sigma\|_{L^2} + C_2\|\sigma\|_{L^2}^2, \tag{3.3.23}$$

with

$$C_1 = c_e(l_2\|\rho\mathbf{v}\|_{L^3} + l_1\|\rho_b\|_{L^\infty}\|\nabla\mathbf{v}_b\|_{L^{3/2}}), \tag{3.3.24}$$

$$C_2 = c_e l_1(\|\mathbf{f}\|_{L^3} + \|\mathbf{v}\cdot\nabla\mathbf{v}_b\|_{L^3}).$$

$$\square$$

Remark 3.3.1 *Notice that the constants C_1 and C_2* **may be arbitrarily small**, *provided that $\|\rho\mathbf{v}\|_{L^3}$ for C_1 and $\|\mathbf{v}\cdot\nabla\mathbf{v}_b\|_{L^3}$ for C_2 are sufficiently small. However, while the smallness assumption for $\nabla\mathbf{v}_b$ is reasonable, since $\mathbf{v}_b$ is given, it is not correct to assume $\|\rho\mathbf{v}\|_{L^3}$ to be small, because the perturbation is unknown, and we may use only regularity assumptions. Furthermore, we notice that $\lambda + \mu$ and μ* **are given phenomenological coefficients and in general are not small**.

From (3.3.20), we infer the following estimate for σ

$$m_1\|\sigma\|^2 \leq \int_{\Omega} p'(\bar{\rho})\sigma^2\,dx \leq (\lambda+\mu)\|\nabla\cdot\mathbf{u}\|_{L^2}\|\sigma\|_{L^2} + \mu\|\nabla\mathbf{V}\|_{L^2}\|\nabla\mathbf{u}\|_{L^2}$$

$$+ C_1\|\nabla\mathbf{u}\|_{L^2}\|\sigma\|_{L^2} + C_2\|\sigma\|_{L^2}^2, \tag{3.3.25}$$

where[2]

$$m_1 := \inf_{\Omega} p'(\rho(x)). \tag{3.3.26}$$

If C_2 is sufficiently small, then

$$C_2 < m_1. \tag{3.3.27}$$

We conclude that

$$\|\sigma\|_{L^2} \le \frac{(\lambda + \mu)}{(m_1 - C_2)}\|\nabla \cdot \mathbf{u}\|_{L^2} + \frac{\mu}{(m_1 - C_2)}\|\nabla\mathbf{u}\|_{L^2} + \frac{C_1}{(m_1 - C_2)}\|\nabla\mathbf{u}\|_{L^2}. \tag{3.3.28}$$

Inequality (3.3.28) furnishes the wanted estimate for σ. Substituting (3.3.28) into (3.3.17), we obtain

$$(\lambda + \mu)\|\nabla \cdot \mathbf{u}\|_{L^2}^2 + (\mu - c_0)\|\nabla\mathbf{u}\|_{L^2}^2 \tag{3.3.29}$$

$$\le \frac{c_1(\lambda + \mu)}{(m_1 - C_2)}\|\nabla\mathbf{u}\|_{L^2}\|\nabla \cdot \mathbf{u}\|_{L^2} + \frac{c_1\mu}{(m_1 - C_2)}\|\nabla\mathbf{u}\|_{L^2}^2 + \frac{c_1 C_1}{(m_1 - C_2)}\|\nabla\mathbf{u}\|_{L^2}^2$$

$$+ \left(\frac{2c_2(\lambda + \mu)^2}{(m_1 - C_2)^2}\|\nabla \cdot \mathbf{u}\|_{L^2}^2 + \frac{2c_2\mu^2}{(m_1 - C_2)^2}\|\nabla\mathbf{u}\|_{L^2}^2 + \frac{2c_2 C_1^2}{(m_1 - C_2)^2}\|\nabla\mathbf{u}\|_{L^2}^2.\right.$$

which implies that

$$\left(\mu - c_0 - \frac{c_1^2(\lambda + \mu)}{2(m_1 - C_2)^2} - \frac{c_1(\mu + C_1)}{(m_1 - C_2)} - \frac{2c_2((\lambda + \mu)^2 + \mu^2 + C_1^2)}{(m_1 - C_2)^2}\right)\|\nabla\mathbf{u}\|_{L^2}^2 \tag{3.3.30}$$

$$+ \frac{\lambda + \mu}{2}\left(1 - \frac{4c_2(\lambda + \mu)}{(m_1 - C_2)^2}\right)\|\nabla \cdot \mathbf{u}\|_{L^2}^2 \le 0.$$

Contradictions arise by assuming that

$$c_0 + \frac{c_1(\mu + C_1)}{(m_1 - C_2)} + \frac{c_1^2(\lambda + \mu) + 4c_2((\lambda + \mu)^2 + 2\mu^2 + 2C_1^2)}{2(m_1 - C_2)^2} < \mu, \tag{3.3.31}$$

$$\frac{4c_2(\lambda + \mu)}{(m_1 - C_2)^2} < 1.$$

We observe that c_1, and c_2 can be sufficiently small because they become small provided that the data $\mathbf{f}$, $\nabla\mathbf{v}_b$ and $\nabla\rho_b$ are small. The last assumption infers that the steady flow is slightly compressible. Thus from (3.3.30), we get a contradiction if (3.3.18) and (3.3.31) hold, and the theorem is completed.

---

[2]For isothermal fluids p' is a constant in space and time.

Remark 3.3.2 *Observe that the smallness of c_0, and C_2 is verified if the given L^2 norms of $\mathbf{f}$ and $\nabla\mathbf{v}_b$ are small. Furthermore, to prove uniqueness we require smallness hypotheses on c_1 and c_2. These constants contain also norms for ρ, $\mathbf{v}$ in the perturbed class of motions. On the other hand, the smallness of c_1, c_2 is ensured by the smallness of the L^3 norms of $\mathbf{f}$, $\nabla\mathbf{v}_b$, the L^2 norms of the gradients of ρ, ρ_b, and by the L^∞ norm of $\nabla\cdot\mathbf{v}_b$. This is physically meaningful because it requires the quality of uniqueness to hold in the class of slightly compressible homogeneous flows. A uniqueness theorem is given in [56].*

Remark 3.3.3 *Our uniqueness proof provides sufficient conditions for the solution be unique. Of course these conditions can be improved! For example, one may change the estimates of the nonlinear terms. Our estimates have been guided so as to obtain estimates in $L^6(\Omega)$ for $\mathbf{u}$ only, allowing us to avoid the Poincaré inequality. This gives us a uniform method equally applicable to unbounded domains.*

Remark 3.3.4 *Notice that the estimates (3.3.3) and (3.3.22) for $\mathcal{I}_1$ and $\mathcal{I}_2$ can be strongly weakened if the Poincare' inequality holds. The smallness of c_0 and C_2 implies smallness only to norms of $\mathbf{f}$ and $\nabla\mathbf{v}_b$, that are assumed given. In order to prove uniqueness, we need smallness hypotheses for c_1, c_2, and $\nabla\cdot(\rho\mathbf{v})$. These constants also contain norms for ρ, $\mathbf{v}$ for the perturbed class of motions. Moreover, smallness of c_1, c_2 is ensured by the assumption of smallness on different norms of gradients of ρ and ρ_b. This, together with the smallness hypothesis of $\nabla\cdot(\rho\mathbf{v})$, yields a clear physical interpretation, since under the above smallness hypothesis, uniqueness holds in a different class of slightly compressible homogeneous flows.*

Of course, the optimal basic condition allowing uniqueness to hold has to be studied using variational techniques, and this remains an open problem at this time.

3.3.2 Nonlinear Stability

Systems governing unsteady flows with initial data corresponding to perturbations of the basic steady flow ρ_b, $\mathbf{v}_b$ obey the following initial boundary value problem

$$\partial_t\rho + \nabla\cdot(\rho\mathbf{v}) = 0, \qquad (3.3.32)$$

$$\rho\partial_t\mathbf{v} + \rho\mathbf{v}\cdot\nabla\mathbf{v} = -\nabla p(\rho) + \mu\Delta\mathbf{v} + (\lambda+\mu)\nabla\nabla\cdot\mathbf{v} + \rho\mathbf{f},$$

$$\rho(x,0) = \rho_b(x) + \sigma_0(x), \qquad \mathbf{v}(x,0) = \mathbf{v}_b + \mathbf{u}_0(x),$$

$$\mathbf{v}|_{\partial\Omega} = 0, \qquad \int_\Omega \rho = M.$$

We assume that solutions to (3.3.32) exist globally in time, and that they are regular solutions in the class[3]

$$W = \{(\mathbf{v}, \rho) : \mathbf{v} \in L^\infty(0, \infty; L^3(\Omega)) \cap L^2(0, \infty; W^{1,2}(\Omega)); \quad (3.3.33)$$

$$\rho \in W^{1,\infty}(Q_T); \inf p'(\rho) > 0; \inf \rho > 0; \partial_t \rho \in L^\infty(0, \infty; L^\infty(\Omega))\}.$$

In this subsection we furnish an energy equation that, in spite of the analogous case for incompressible flows, does not provide a stability result for slow flows $\mathbf{v}_b$ close to incompressible ones, as it does not furnish an a priori estimate for the L^2 norms of perturbations. The difficulty arises from the fact that compressible fluids are governed by a mixed PDE system: partial parabolic for velocity, and hyperbolic for density. That is, in the energy equation a dissipative term for the density is missing.

Theorem 3.3.2 *Energy Equation. Let* $\mathbf{f} \in L^3$; *for* $\mathbf{v} = \mathbf{v}_b + \mathbf{u}$, $\rho = \rho_b + \sigma$, *solve* (3.2.9) *with* $(\mathbf{v}, \rho) \in W$. *Then, the energy of perturbations equation holds*

$$\frac{d}{dt}\Big[E_u + E_\sigma\Big] + \mu\, D_u(t) = \mathcal{I}_1', \qquad (3.3.34)$$

$$E_u = \frac{1}{2}\int_\Omega \rho \mathbf{u}^2\, dx, \qquad E_\sigma = \int_\Omega \frac{p'(\overline{\rho})}{\rho}\frac{\sigma^2}{2}\, dx,$$

$$\mu\, D_u = \int_\Omega \Big[(\lambda + \mu)(\nabla \cdot \mathbf{u})^2\, dx + \mu|\nabla\, \mathbf{u}|^2\Big]\, dx,$$

$$\mathcal{I}_1' = \mathcal{I}_1 - \int_\Omega \rho\frac{\sigma^2}{2}\frac{d}{dt}\Big(\frac{p'(\overline{\rho})}{\rho^2}\Big)\, dx,$$

$$\boldsymbol{b} = -\rho_b \boldsymbol{u} \cdot \nabla\, v_b + \sigma(\, \boldsymbol{f} - \, v \cdot \nabla\, v_b),$$

with $\mathcal{I}_1$ *defined in* (3.3.13).

Proof. Given $(\mathbf{v}_b, \rho_b)$ in $\mathcal{V}_b$, we assume that there exists a global solution $(\mathbf{v}, \rho)$ in W to (3.3.32), corresponding to the same force $\mathbf{f}$, the same mass M, and the same homogeneous boundary data. Let $\mathbf{u} = \mathbf{v} - \mathbf{v_b}$, $\sigma = \rho - \rho_b$ be the difference between the two solutions, which satisfies the Initial Boundary Value Problem for the difference system

$$\partial_t \sigma + \nabla \cdot (\rho\, \mathbf{u}) \quad + \nabla \cdot (\sigma\, \mathbf{v}_b) = 0, \qquad\qquad x \in \Omega,\, t \in (0, T)$$

$$\rho \partial_t \mathbf{u} + \rho\, \mathbf{v} \cdot \nabla\, \mathbf{u} \quad + (\rho_b \mathbf{u} + \sigma\, \mathbf{v}) \cdot \nabla\, \mathbf{v}_b =$$

$$-\nabla\Big(p(\rho) - p(\rho_b)\Big) + (\lambda + \mu)\nabla\nabla \cdot \mathbf{u} + \mu\Delta\, \mathbf{u} + \sigma \mathbf{f},$$

---

[3]Regularity class on ρ may be weakened; however, since there are no known existence theorems in this regularity class for large data, we prefer to give a simpler proof.

$$\sigma(x,0) = \sigma_0(x), \qquad \mathbf{u}(x,0) = \mathbf{u}_0(x),$$

$$\mathbf{u}|_{\partial\Omega} = 0, \qquad \int_\Omega \sigma = 0. \qquad (3.3.35)$$

In this calculation we have used the hypothesis $\partial_t \mathbf{v}_b = 0$.

Let us multiply $(3.3.35)_2$ by $\mathbf{u}$ and integrate over Ω to have

$$\frac{1}{2}\int_\Omega \left[\rho\partial_t|\mathbf{u}|^2 + \rho\mathbf{v}\cdot\nabla|\mathbf{u}|^2\right]dx - \int_\Omega\left[\mu\mathbf{u}\cdot\Delta\mathbf{u} + (\lambda+\mu)\mathbf{u}\cdot\nabla\nabla\cdot\mathbf{u}\right]dx$$

$$= -\int_\Omega \rho\mathbf{u}\cdot\nabla(p'(\bar\rho)\sigma)\,dx + \int_\Omega \mathbf{u}\cdot\mathbf{b}\,dx,$$

$$(3.3.36)$$

where $\bar\rho$ is a value of ρ between ρ, and ρ_b, and with $\mathbf{b}$ given in $(3.3.34)_{3,5}$.

Integrating by parts, and taking into account of boundary conditions, we get

$$\frac{1}{2}\frac{d}{dt}\int_\Omega \rho\mathbf{u}^2\,dx + \int_\Omega\left[(\lambda+\mu)(\nabla\cdot\mathbf{u})^2 + \mu|\nabla\,\mathbf{u}|^2\right]dx = -\int_\Omega \mathbf{u}\cdot\nabla(p'(\bar\rho)\sigma)\,dx$$

$$+ \int_\Omega \mathbf{u}\cdot\mathbf{b}\,dx. \quad (3.3.37)$$

Let us rewrite $(3.3.35)_1$ as follows

$$\partial_t\sigma + \mathbf{v}\cdot\nabla\sigma = -\rho\nabla\cdot\mathbf{u} - \mathbf{u}\cdot\nabla\rho_b - \sigma\nabla\cdot\mathbf{v_b}. \qquad (3.3.38)$$

Multiplying $(3.3.38)$ by $(p'(\bar\rho))\sigma/\rho$ we have

$$\int_\Omega \rho\frac{p'(\bar\rho)}{\rho^2}\frac{d}{dt}\left(\frac{\sigma^2}{2}\right)dx = -\int_\Omega p'(\bar\rho)\sigma\nabla\cdot\mathbf{u}\,dx \qquad (3.3.39)$$

$$-\int_\Omega \frac{p'(\bar\rho)}{\rho}\sigma\mathbf{u}\cdot\nabla\rho_b\,dx - \int_\Omega \frac{p'(\bar\rho)}{\rho}\sigma^2\nabla\cdot\mathbf{v}_b\,dx.$$

Integrating by parts, and using the Reynolds transport theorem, we get

$$\frac{d}{dt}\int_\Omega \frac{p'(\bar\rho)}{\rho}\frac{\sigma^2}{2}\,dx = \int_\Omega \frac{\sigma^2}{2}\rho\frac{d}{dt}\left(\frac{p'(\bar\rho)}{\rho^2}\right)dx - \int_\Omega p'(\bar\rho)\sigma\nabla\cdot\mathbf{u}\,dx$$

$$-\int_\Omega \frac{p'(\bar\rho)}{\rho}\sigma\mathbf{u}\cdot\nabla\rho_b\,dx \quad -\int_\Omega \frac{p'(\bar\rho)}{\rho}\sigma^2\nabla\cdot\mathbf{v}_b\,dx.$$

$$(3.3.40)$$

Thus, adding $(3.3.37)$–$(3.3.40)$, noticing that

$$-\int_\Omega \mathbf{u} \cdot \nabla(p'(\bar\rho)\sigma)\, dx - \int_\Omega p'(\bar\rho)\sigma \nabla \cdot \mathbf{u}\, dx = 0,$$

we deduce the simpler form of the perturbations energy equation

$$\frac{1}{2}\frac{d}{dt}\int_\Omega \rho \mathbf{u}^2\, dx + \frac{d}{dt}\int_\Omega \frac{p'(\bar\rho)}{\rho}\frac{\sigma^2}{2}\, dx + \int_\Omega \Big[(\lambda+\mu)(\nabla\cdot\mathbf{u})^2 + \mu|\nabla\ \mathbf{u}|^2\Big]dx = \mathcal{I}_1',$$
(3.3.41)

with $\mathcal{I}_1'$ given in $(3.3.34)_4$; the theorem is proved.   □

3.3.3 Nonlinear Exponential Stability

In previous subsections we have derived an energy equation for the perturbation $(\mathbf{u}, \sigma)$, and we have also remarked that such an equation cannot provide a stability result, since in such equations a dissipative term for σ is missing. In this subsection, we fill this gap by introducing the **free work equation**. More specifically, we prove the following theorem.

Theorem 3.3.3 Nonlinear Exponential Stability *Let the gas be perfect, i.e. let $p(\rho) = k\rho$, and Let $\boldsymbol{f}\in L^3(\Omega)$ and $(v_b, \rho_b) \in \mathcal{V}_b$ be a given solution to (3.3.1). Also let $\|\boldsymbol{f}\|_{L^3}$ and constants c_0, c_1, c_2, c_2', C_1, C_2, defined by (3.3.16), (3.3.42), (3.3.24), satisfy (3.3.18), (3.3.50) and (3.3.51), then (v_b, ρ_b) is nonlinearly exponentially stable in the class of perturbations solutions to (3.3.35) belonging to the regularity class $\mathcal{W}$.*

Proof. From regularity assumptions, the definition of $\mathcal{I}_1'$ given in $(3.3.34)_4$, and inequality (3.3.3), it is easy to check that

$$\mathcal{I}_1' \le c_1\|\sigma\|_{L^2}\|\mathbf{u}\|_{L^6} + \Big(c_2 + \Big\|\rho\frac{d}{dt}\Big(\frac{p'(\bar\rho)}{\rho^2}\Big)\Big\|_{L^\infty}\Big)\|\sigma\|_{L^2}^2 + c_0\|\nabla\mathbf{u}\|_{L^2}^2$$
$$\le c_1\|\sigma\|_{L^2}\|\nabla\mathbf{u}\|_{L^2} + c_2'\|\sigma\|_{L^2}^2 + c_0\|\nabla\mathbf{u}\|_{L^2}^2,$$

holds where c_1, c_2 are defined by (3.3.16) and

$$c_2' = c_2 + \Big\|\rho\frac{d}{dt}\Big(\frac{p'(\bar\rho)}{\rho^2}\Big)\Big\|_\infty.$$
(3.3.42)

Condition (3.3.18) allows us to write

$$\frac{d}{dt}\int_\Omega \Big[\frac{1}{2}\rho\mathbf{u}^2 + \frac{p'(\bar\rho)}{\rho}\frac{\sigma^2}{2}\Big]dx + (\lambda+\mu)\|\nabla\cdot\mathbf{u}\|_{L^2}^2 + (\mu - c_0)\|\nabla\ \mathbf{u}\|_{L^2}^2$$
(3.3.43)
$$\le c_1\|\sigma\|_{L^2}\|\nabla\ \mathbf{u}\|_{L^2} + c_2'\|\sigma\|_{L^2}^2.$$

Free work equation

We now derive an estimate for $\|\sigma\|_{L^2}$, to this end we employ the free work equation. In particular, we notice that we can use Lemma 3.7.4 because the total mass and the volume are fixed, and it holds

$$\int_\Omega \sigma\, dx = 0.$$

Let us multiply $(3.3.35)_2$ by $\mathbf{V}$ given in Lemma 3.7.4, and integrate by parts over Ω to get

$$\frac{d}{dt}\int_\Omega \rho\,\mathbf{u}\cdot\mathbf{V}\,dx - \int_\Omega p'(\bar\rho)\sigma\nabla\cdot\mathbf{V}\,dx = \int_\Omega \Big[(\lambda+\mu)\nabla\cdot\mathbf{V}\nabla\cdot\mathbf{u} + \mu\nabla\mathbf{u}:\nabla\mathbf{V}\Big]dx$$

$$- \int_\Omega \rho\Big(\mathbf{V}_t + \mathbf{v}\cdot\nabla\mathbf{V}\Big)\cdot\mathbf{u}\,dx$$

$$- \int_\Omega \mathbf{V}\cdot\mathbf{b}\,dx.$$

$$(3.3.44)$$

Employing the properties of $\mathbf{V}$ we have

$$-\frac{d}{dt}\int_\Omega \rho\,\mathbf{u}\cdot\mathbf{V}\,dx + \int_\Omega p'(\bar\rho)\sigma^2\,dx = \mathcal{I}_2', \qquad (3.3.45)$$

where

$$\mathcal{I}_2' = \mathcal{I}_2 - \int_\Omega \rho\mathbf{V}_t\cdot\mathbf{u}\,dx,$$

with $\mathcal{I}_2$ defined in (3.3.21). By (3.3.23) and estimate (3.3.23), we deduce

$$\mathcal{I}_2' \le (\lambda+\mu)\|\nabla\cdot\mathbf{u}\|_{L^2}\|\sigma\|_{L^2} + l_2\mu\|\nabla\mathbf{u}\|_{L^2}\|\sigma\|_{L^2} + C_1\|\nabla\mathbf{u}\|_{L^2}\|\sigma\|_{L^2}$$

$$+ C_2\|\sigma\|_{L^2}^2 + \|\rho\|_{L^3}\|\mathbf{V}_t\|_{L^2}\|\nabla\mathbf{u}\|_{L^2}. \qquad (3.3.46)$$

Using the properties of $\mathbf{V}$ we obtain

$$\mathcal{I}_2' \le (\lambda+\mu)\|\nabla\cdot\mathbf{u}\|_{L^2}\|\sigma\|_{L^2} + l_2\mu\|\nabla\mathbf{u}\|_{L^2}\|\sigma\|_{L^2} + C_1\|\nabla\mathbf{u}\|_{L^2}\|\sigma\|_{L^2}$$

$$+ C_2\|\sigma\|_{L^2}^2 + c_e\, l_3\|\rho\|_{L^3}\|\nabla\mathbf{u}\|_{L^2}^2,$$

where C_1 is given in (3.3.24). Hence, we can deduce the following series of inequalities providing a dissipative term for σ

$$-\frac{d}{dt}\int_\Omega \rho\,\mathbf{u}\cdot\mathbf{V}\,dx + m_1\|\sigma\|^2 \le -\frac{d}{dt}\int_\Omega \rho\,\mathbf{u}\cdot\mathbf{V}\,dx + \int_\Omega p'(\bar\rho)\sigma^2\,dx$$

$$\le (\lambda+\mu)\|\nabla\cdot\mathbf{u}\|_{L^2}\|\sigma\|_{L^2} + l_2\mu\|\nabla\mathbf{u}\|_{L^2}\|\sigma\|_{L^2} + C_1\|\nabla\mathbf{u}\|_{L^2}\|\sigma\|_{L^2}$$

$$+ C_2\|\sigma\|_{L^2}^2 + c_e\, l_3\|\rho\|_{L^3}\|\nabla\mathbf{u}\|_{L^2}^2,$$

$$(3.3.47)$$

where m_1 is defined in (3.3.26), and C_2 is given in (3.3.24)$_2$. *Notice that the constant C_1 **cannot be assumed arbitrarily small** because the term $\|\rho \mathbf{v}\|_{L^3}$ depends strictly upon the perturbed flow on which we only make regularity assumptions.* On the other hand, the constant C_2 can be assumed to be arbitrarily small because it is composed of two terms, each of them containing either the force or basic motion $\nabla \mathbf{v}_b$, which is given data in the stability problem. If C_2 is sufficiently small,

$$C_2 < m_1,$$

we conclude that

$$-\frac{d}{dt}\int_\Omega \rho\, \mathbf{u} \cdot \mathbf{V}\, dx + (m_1 - C_2)\|\sigma\|_{L^2}^2 \le (\lambda + \mu)\|\nabla \cdot \mathbf{u}\|_{L^2}\|\sigma\|_{L^2}$$

$$\tag{3.3.48}$$

$$+ \mu\|\nabla \mathbf{u}\|_{L^2}\|\sigma\|_{L^2} + C_1\|\nabla \mathbf{u}\|_{L^2}\|\sigma\|_{L^2} + c_e\|\rho\|_{L^3}\|\nabla \mathbf{u}\|_{L^2}^2.$$

The equation (3.3.48) furnishes a dissipative term for σ. Adding (3.3.48) multiplied by the constant ν to (3.3.43), we obtain the **modified energy inequality**

$$\frac{d}{dt}\mathbb{E}(t) + (\lambda + \mu)\|\nabla \cdot \mathbf{u}\|_{L^2}^2 + (\mu - c_0)\|\nabla \mathbf{u}\|_{L^2}^2 + \nu(m_1 - C_2)\|\sigma\|_{L^2}^2$$

$$- \nu(\lambda + \mu)\|\nabla \cdot \mathbf{u}\|_{L^2}\|\sigma\|_{L^2} - \nu\mu\|\nabla \mathbf{u}\|_{L^2}\|\sigma\|_{L^2}$$

$$- \left(\nu C_1 + c_1\right)\|\nabla \mathbf{u}\|_{L^2}\|\sigma\|_{L^2} - c_2'\|\sigma\|_{L^2}^2 - \nu\, c_e\|\rho\|_{L^3}\|\nabla \mathbf{u}\|_{L^2}^2 \le 0,$$

$$\tag{3.3.49}$$

where we have introduced the **modified energy** $\mathbb{E}$

$$\mathbb{E}(t) = E_u(t) + E_\sigma - \nu\int_\Omega \rho\, \mathbf{u} \cdot \mathbf{V}\, dx.$$

We choose

$$C_2 < \frac{m_1}{4}, \qquad \nu < \frac{\mu - c_0}{c_e\|\rho\|_{L^3}}, \qquad c_2' < \nu\frac{m_1}{4}, \tag{3.3.50}$$

requiring the compatibility condition

$$\frac{4c_2'}{m_1} < \frac{\mu - c_0}{c_e\|\rho\|_{L^3}}. \tag{3.3.51}$$

Condition (3.3.51) can be read as a smallness assumption on c_2', and as largeness assumption on μ. Thus the quadratic form appearing in (3.3.49) is decreased by

$$\left(\lambda + \mu\right)\|\nabla \cdot \mathbf{u}\|_{L^2}^2 + \nu \frac{m_1}{4}\|\sigma\|_{L^2}^2 - \nu(\lambda + \mu)\|\nabla \cdot \mathbf{u}\|_{L^2}\|\sigma\|_{L^2}$$

$$+ \left(\mu - c_0 - \nu c_e\|\rho\|_{L^3}\right)\|\nabla \mathbf{u}\|_{L^2}^2 + \nu \frac{m_1}{4}\|\sigma\|_{L^2}^2 - \left(\nu\mu + \nu C_1 + c_1\right)\|\nabla \mathbf{u}\|_{L^2}\|\sigma\|_{L^2},$$

$$(3.3.52)$$

which is positive definite under suitable assumptions on the coefficients.

In order to give an idea of how to find stability, we present only a sufficient condition for stability to hold. Assume the following inequalities are satisfied

$$\nu(\lambda + \mu) < m_1,$$

$$\nu c_e\|\rho\|_{L^3} + c_0 < \frac{\mu}{2}, \tag{3.3.53}$$

$$\nu C_1 + c_1 < \nu\mu.$$

The first condition is a condition on ν to be small. The last two conditions are conditions on μ to be large. So we decrease (3.3.52) with the quadratic form

$$\frac{\mu}{2}\|\nabla \mathbf{u}\|_{L^2}^2 + \nu \frac{m_1}{4}\|\sigma\|_{L^2}^2 - 2\nu\mu\|\sigma\|_{L^2}\|\nabla \mathbf{u}\|_{L^2},$$

which is positive definite if

$$8\nu\mu < m_1. \tag{3.3.54}$$

Considering all previous conditions as smallness assumptions on the coefficients, and largeness assumptions on μ, and observing that $\mu < 3(\lambda + \mu)$ we may summarize the conditions on ν in the unique condition

$$\nu \leq \frac{m_1}{24(\lambda + \mu)}. \tag{3.3.55}$$

Substituting this relation in (3.3.49), we get

$$\frac{d}{dt}\mathbb{E}(t) + b\mathbb{E}(t) \leq 0, \tag{3.3.56}$$

and by the use of Gronwall's Lemma we obtain

$$\mathbb{E}(t) \leq \mathbb{E}(0)\, e^{-bt}. \tag{3.3.57}$$

Since $\mathbb{E}(t)$ is positive definite for small values of ν, we have the decay for the L^2 norms of perturbations $\mathbf{u}$, and σ, and this completes the theorem.   □

3.4 Ω Bounded, Non Zero Boundary Data

Consider the system

$$\nabla \cdot (\rho \mathbf{v}) = 0,$$

$$\nabla(\rho \, \mathbf{v} \otimes \, \mathbf{v}) = -\nabla p + \lambda \nabla \nabla \cdot \mathbf{v} + 2\mu \nabla \cdot D(\mathbf{v}) + \rho \mathbf{f}, \qquad x \in \Omega, \; t \in (0, T),$$

$$\mathbf{v}|_{\partial\Omega} = \mathbf{w} \qquad \int_{\Omega} \rho \, dx = M, \qquad \rho \geq 0,$$

$$\int_{\partial\Omega} \rho \mathbf{w} \cdot \mathbf{n} dS = 0,$$

$$(3.4.1)$$

with

$$p = p(\rho).$$

Let $\mathbf{f} \in L^3(\Omega)$, and let us define the regularity class where stability is proved

$$\mathcal{V} = \{\mathbf{v}, \rho \in L^3(\Omega) \times W^{1,\infty}(\Omega);$$

$$\inf p'(\rho) > m_1 > 0; \quad \inf \rho > m > 0\}.$$

Remark 3.4.1 *Recall that at times the momentum field $\rho \mathbf{v}$ is prescribed at boundary instead of velocity field $\mathbf{v}$ this is for us an open problem.*

3.4.1 Uniqueness

Theorem 3.4.1 *Uniqueness of Steady Flow Let $\mathbf{f} \in L^3(\Omega)$, and let $(\mathbf{v}_b, \rho_b) \in \mathcal{V}_b$ be a given solution to boundary value problem (3.4.1). Then, if $\|\mathbf{f}\|_{L^3}$ and constants c_0, c_1, c_2, C_1, C_2 defined by (3.3.16), and (3.3.24) satisfy (3.3.18), (3.3.27) and (3.3.31), thus $(\mathbf{v}_b, \rho_b)$ is the unique solution to (3.4.1) in the regularity class of solutions to the boundary value problem (3.3.1) with $(\mathbf{v}, \rho) \in \mathcal{V}$.*

Proof. All reasoning outlined in the previous theorems continue to hold. We may still apply the free work equation because of Lemma 3.7.1, where the auxiliary function $\mathbf{V}$ has a zero value at boundary. Specifically, we are again looking for solutions in the class of fluid flows with the same total mass and volume, assuming the compatibility condition $\int_{\Omega} \sigma \, dx = 0$ to be satisfied.

In this way, the boundary terms, deriving from integration by parts in the second step of the proof, vanish (as in the previous theorem).

The only difficulty lies in the boundary terms appearing in the first step of the proof, where we derive a generalized energy equation. To deal with these terms, we follow Serrin, [128], and notice that $\mathbf{u}$ vanishes on $\partial\Omega$ and has the same regularity as in the previous theorem; as such, all terms containing $\mathbf{u}$ at the boundary will vanish. Only the term B_2 doesn't contain $\mathbf{u}$, that is

$$B_2 = \int_\Sigma \frac{p'(\bar{\rho})}{\rho} \frac{\sigma^2}{2} \mathbf{v}_b \cdot \mathbf{n}\, dS. \qquad (3.4.2)$$

$$B_2 = \int_\Sigma \frac{p'(\bar{\rho})}{\rho} \mathbf{v}_b \cdot \mathbf{n}\left(\frac{\sigma^2}{2}\right) dS = \int_{\Sigma_{out}} \frac{p'(\bar{\rho})}{\rho} \mathbf{v}_b \cdot \mathbf{n}\left(\frac{\sigma^2}{2}\right) dS \qquad (3.4.3)$$

$$+ \int_{\Sigma_{in}} \frac{p'(\bar{\rho})}{\rho} \mathbf{v}_b \cdot \mathbf{n}\left(\frac{\sigma^2}{2}\right) dS = I_{out} + I_{in},$$

where Σ_e, Σ_{in} denote the parts of the boundary where the fluid exits and enters, respectively. We must deal with the part of the term B_2 in Σ_{in}, and Σ_e:

$$\mathbf{v}_i \cdot \mathbf{n}|_{\Sigma_{in}} \le 0, \quad \rho|_{\Sigma_{in}} = \rho_b|_{\Sigma_{in}}, \quad i = 1, 2$$

$$\mathbf{v}_i \cdot \mathbf{n}|_{\Sigma_e} \ge 0, \qquad i = 1, 2$$

On Σ_{out}, $\mathbf{v}_b \cdot \mathbf{n} > 0$, thus $I_{out} > 0$, whereas on Σ_{in} it is σ zero because the density is prescribed, hence it is always $B_2 > 0$.

In (3.4.2) we have on the side Σ_{in} $\sigma = 0$, while on the side Σ_e it is

$$-\int_{\Sigma_e} \frac{p'(\bar{\rho})}{\rho} \frac{\sigma^2}{2} \mathbf{v}_b \cdot \mathbf{n}\, dS \le 0. \qquad (3.4.4)$$

Substituting these informations in (3.3.12) we receive

$$\int_\Omega \left[(\lambda+\mu)(\nabla \cdot \mathbf{u})^2 + \mu|\nabla\, \mathbf{u}|^2\right] dx = I_1 - B_2 - B_3 \le I_1. \qquad (3.4.5)$$

Thus the conclusion of the previous theorem continues to hold.   □

Remark 3.4.2 Body forces *The uniqueness result holds for any barotropic fluid. It does not contradict the first counterexample, cf. Sect. 1.3.3, because it states uniqueness once solutions exist. Furthermore, the counterexample concerns the existence of real positive densities for small forces, and our uniqueness requires the smallness of forces.*

3.4.2   *Nonlinear Exponential Stability*

Consider system[4]

$$\partial_t \rho + \nabla \cdot (\rho \mathbf{v}) = 0, \qquad (3.4.6)$$

$$\rho \partial_t \mathbf{v} + \rho \mathbf{v} \cdot \nabla \mathbf{v} = -\nabla p(\rho) + \mu \Delta \mathbf{v} + (\lambda + \mu) \nabla \nabla \cdot \mathbf{v} + \rho \mathbf{f},$$

$$\rho(x,0) = \rho_b(x) + \sigma_0(x), \qquad \mathbf{v}(x,0) = \mathbf{v}_b + \mathbf{u}_0(x),$$

$$\mathbf{v}|_{\partial\Omega} = \mathbf{w}, \qquad \int_\Omega \rho = M.$$

Theorem 3.4.2 *Rigid Porous and Moving Domain* $\Omega(t) = \Omega$ *Let* $\mathbf{f} \in L^3(\Omega)$, *and let* $(\mathbf{v}_b, \rho_b) \in \mathcal{V}_b$ *be a given solution to* (3.3.1). *Also let* $\|\mathbf{f}\|_3$ *and constants* c_0, c_1, c_2, c_2', C_1, C_2, *defined by* (3.3.16), (3.3.42), (3.3.24), *satisfy* (3.3.18), (3.3.50) *and* (3.3.51), *then* $(\mathbf{v}_b, \rho_b)$ *is exponentially stable in the class of perturbation solutions to* (3.4.6) *belonging to the regularity class* $\mathcal{W}$ *defined in Subsect. 3.2.2.*

Proof. In the wake of Lemma 3.7.4, the reasonings outlined in Theorems 3.3.2, 3.3.3 continue to hold. We recall that we are looking for solutions in the class of fluid flows with the same total mass and volume, hence the compatibility condition $\int_\Omega \sigma \, dx = 0$ is satisfied. We may still apply free work equation because Lemma 3.7.4 holds. Thus, we may construct the auxiliary function $\mathbf{V}$, with zero value at the boundary. As in the previous theorem, the boundary terms, derived from the integration by parts in the second step of the proof, vanish.

Finally, asymptotic stability can also be proven, and we leave it as an open problem for the interested reader.

Remark 3.4.3 -**Isothermal Gas** *Notice that for isothermal gases*

$$p'(\rho) = k\theta = constant.$$

Therefore, in the case of bounded domains no hypothesis relating to the existence of a minimum density that is nonnegative $\rho \geq 0$. *The terms*

$$\left\| \frac{p'(\bar{\rho}}{\rho} \nabla \rho \right\|_{L^3} < b_1, \quad \left\| \frac{p'(\bar{\rho}}{\rho} \nabla \cdot \mathbf{v}_b \right\|_{L^\infty} < b_1, \quad \left\| \mathbf{v}_b \cdot \nabla \frac{p'(\bar{\rho}}{\rho} \right\|_{L^\infty} < b_1,$$

may be bounded for $\rho \geq 0$ *provided* $\nabla \rho$, $\nabla \cdot \mathbf{v}_b$, *and* $\mathbf{v}_b$ *vanish suitably at points where* $\rho = 0$. *Regularity assumptions can also be weakened.*

---

[4]Sometime it is prescribed the momentum field $\rho\mathbf{v}$ at boundary instead of velocity field $\mathbf{v}$ this is for us an open problem.

Remark 3.4.4 -**Meaning of hypotheses in Theorem 3.5.2** *If we set* $\mathbf{f} = \nabla U + \mathbf{f}_1$, *then for* $\mathbf{f}_1 = 0$ *we can prove uniqueness for any large potential* U. *However if* $\mathbf{f}_1 \neq 0$ *we must assume smallness of the total force* $\mathbf{f}$. *In cf. [8], and [72], existence and uniqueness have been proven in the class of steady solutions corresponding to large potential forces and small non-potential ones. Following the same procedure employed in [8], and reconsidering the proof of Theorem 3.5.1, it appears possible to prove a uniqueness theorem for a larger class of perturbations; we leave it a challenging open problem.*

Remark 3.4.5 - *Notice that as it stands, the hypotheses of Theorem 3.5.1 have no physical meaning, as they represent smallness assumptions on solutions*

$$c_1 = \left\| \frac{p'(\bar\rho)}{\rho} |\nabla \rho| \right\|_{L^3} + \| \boldsymbol{v} \cdot \nabla \boldsymbol{v}_b \|_{L^3} \leq \epsilon,$$

$$c_2 = \left\| \frac{p'(\bar\rho)}{\rho} |\nabla \cdot \boldsymbol{v}_b| \right\|_{L^\infty} + \left\| |\boldsymbol{v}_b| \nabla \left(\frac{p'(\bar\rho)}{\rho} \right) \right\|_{L^\infty} \leq \epsilon.$$

In order for the hypotheses to have physical meaning, there is a need for a priori estimates *of an upper bound in certain norms of the solution, depending upon data. Once a priori estimates are proven, the smallness requirement on* c_1, c_2 *is translated into a smallness requirement for the forces only. Currently, this is an open problem.*

Remark 3.4.6 *If we set* $\rho \mathbf{f} = \mathbf{F}$, *with* $\mathbf{F}$ *force per unit of volume in the difference system, it disappears. Also in the case of non zero boundary data, there appear to be no smallness conditions on the boundary data* $\mathbf{w}$. *Again this absurdum disappears once "a priori estimates" are known, actually "a priori estimates" provide control on the norms of solutions in terms of the norms of* $\mathbf{F}$ *and* $\mathbf{w}$. *The problem of finding a priori estimates is one of the most challenging open question in the theory of compressible fluids.*

Remark 3.4.7 *Notice that smallness of* c_0 *and* C_2 *requires smallness only of the norms of* $\mathbf{f}$ *and* $\nabla \mathbf{v}_b$, *and these are assumed given. However, to prove uniqueness we need a hypothesis of smallness on* c_1, c_2, *and* $\nabla \cdot (\rho \mathbf{v})$. *These constants contain norms of* ρ, $\mathbf{v}$ *in the perturbed class of motions. Smallness of* c_1, c_2 *is only ensured by the assumption of smallness on the gradient of* ρ, *and* ρ_b. *This is physically meaningful due to the fact that uniqueness holds in the class of slightly compressible homogeneous flows.*

3.5 Ω Exterior, Fixed Compact Region

Let Ω be a domain exterior to a compact region $\mathcal{C} \subseteq R^3$. Note that in this text we deal only with three-dimensional domains; the two-dimensional case has been omitted. To deal with two dimensional exterior domains, one should

follow the lines already sketched for other uniqueness proofs, paying attention
to estimates of functionals $\mathcal{I}_1$, $\mathcal{I}_2$; [38]. We leave the proof as a challenging
problem to the willing reader.

Under the assumption of smallness of external forces, we shall prove
uniqueness of steady solutions, as well as asymptotic stability of these
solutions.

3.5.1   *Uniqueness*

Consider the **boundary value problem**

$$\nabla \cdot (\rho \mathbf{v}) = 0,$$

$$\nabla(\rho\, \mathbf{v} \otimes \mathbf{v}) = -\nabla p + \lambda \nabla \nabla \cdot \mathbf{v} + 2\mu \nabla \cdot D(\mathbf{v}) + \rho \mathbf{f}, \qquad x \in \Omega,\ t \in (0, T),$$

$$p = p(\rho)$$

$$\mathbf{v}|_{\partial\Omega} = 0, \qquad \lim_{x \to \infty} \mathbf{v} = 0 \qquad \lim_{x \to \infty} \rho = \rho_\infty.$$

$$\text{(3.5.1)}$$

We assume that regular steady solutions to (3.5.1) exist globally in time, and
that they are regular solutions in the class

$$\mathcal{V} = \{(\mathbf{v}, \rho) : \mathbf{v} \in L^3(\Omega) \cap W^{1,2}(\Omega); \qquad\qquad \text{(3.5.2)}$$

$$\rho \in W^{1,\infty}(\Omega) \cap L^2(\Omega)\ \inf p'(\rho) > 0;\ \inf \rho > 0\}.$$

Extending the method of proof of previous uniqueness theorems, we shall
prove the following

Theorem 3.5.1 Uniqueness of the Steady Flow *Let* $\mathbf{f} \in L^3(\Omega)$, *and*

Then, if $\|\mathbf{f}\|_{L^3}$ *and constants* c_1, c_2 *defined by* (3.3.16) *are suitably small,*
the solution $(\mathbf{v}_b, \rho_b) \in \mathcal{V}_b$ *is unique in the regularity class* $\mathcal{V}$.

Proof. Let us remark that in exterior domains, there are two main difficulties.
The first lies in the estimates of nonlinear terms in Sobolev spaces, because
the Poincare' inequality is no longer valid. The second difficulty concerns the
boundary terms in the integration by parts, because new boundary terms
appear at infinity.

It is easy to overcome the first difficulty, since we notice that in the
proofs of Theorems 3.3.1–3.4.1 we have used the limit Sobolev inequality
that continue to hold in exterior domains, cf. Remarks 3.4.2, 3.4.4, 3.4.8. In
this way in the volume integrals we use the same inequalities as in the case
of a bounded domain because only the limit embedding inequality

$$\|u\|_{L^6} \le c_e \|\nabla u\|_{L^2}$$

has been used.

Since this inequality continues to hold in unbounded domains, the first difficulty has been avoided.

The second difficulty must be studied. Let $\mathcal{C}$ be the compact region to which Ω is exterior. To deal with unbounded regions, it is customary to integrate the equations over the bounded domain $\Omega_R = \Omega \cap B_R$, where B_R is the ball centered at the origin $O \in \mathcal{C}$, of radius R, with $\mathcal{C} \subseteq B_R$. Such procedure produces boundary terms over $\partial B_R = S_R$ which must go to zero as R goes to infinity. This holds true if the integrands over ∂B_R decay to zero sufficiently fast at infinity. Therefore in the wake of the proof of Theorem 3.3.2 we study the behavior as $R \to \infty$ of the boundary terms and, once we prove that boundary terms on S_R tend to zero, by use of summability assumptions as in Lemma 3.7.6, using the same smallness assumptions as in Theorem 3.3.2 we get again contradiction.

Boundary terms on S_R. We have to deal with several boundary terms. The first series of terms comes from multiplication of momentum equation times $\mathbf{u}$, and integration over Ω_R

$$
\begin{aligned}
B_{1R} = & \int_{S_R} \frac{1}{2} |\mathbf{u}|^2 \rho \mathbf{v} \cdot \mathbf{e}_R \, dS_R + \int_{S_R} p'(\bar\rho)\sigma \mathbf{u} \cdot \mathbf{e}_R \, dS_R \\
& + \int_{S_R} \lambda \mathbf{u} \cdot \mathbf{e}_R \nabla \cdot \mathbf{u} dS_R + 2\mu \int_{S_R} \mathbf{u} \cdot D(\mathbf{u}) \cdot \mathbf{e}_R \, dS_R.
\end{aligned}
\tag{3.5.3}
$$

The second series of terms derives from the multiplication $(3.5.1)_1$ by $p'(\bar\rho)/\rho$, and with integration over Ω_R, we have

$$
B_{2R} - \int_{S_R} \frac{p'(\bar\rho)}{\rho} \frac{\sigma^2}{2} \mathbf{v}_b \cdot \mathbf{e}_R \, dS_R.
\tag{3.5.4}
$$

The third series of boundary terms is derived by the multiplication of $(3.5.1)_2$ by $\mathbf{V}$ given in Lemma 3.7.5, and integration over Ω_R to produce the following chain of boundary terms

$$
\begin{aligned}
& - \int_{S_R} p'(\bar\rho)\sigma \mathbf{e}_R \cdot \mathbf{V} \, dS_R - \int_{S_R} \rho \, \mathbf{V} \cdot \mathbf{u} \, \mathbf{v} \cdot \mathbf{e}_R dS_R \\
& + (\lambda + \mu) \int_{S_R} \nabla \cdot \mathbf{u} \mathbf{e}_R \cdot \mathbf{V} \, dS_R + \mu \int_{S_R} \mathbf{V} \cdot \nabla \, \mathbf{u} \, \mathbf{e}_R \, dS_R.
\end{aligned}
\tag{3.5.5}
$$

We may additionally employ the Lemma 3.7.6, which ensures the decay to zero of the surface integral of a product of functions f g belonging to L^p, L^q respectively, with $1/p + 1/q > 1/2$. Actually, by hypothesis we have that $\mathbf{u} \in L^3$, $\sigma \in L^2$, thus the boundary terms approach zero as $R \to \infty$. We analyze two terms containing many of the difficulties.

$$\int_{S_R} p'(\bar\rho)\sigma\mathbf{u}\cdot\mathbf{e}_R\,dS_R, \qquad -\int_{S_R} p'(\bar\rho)\sigma\mathbf{e}_R\cdot\mathbf{V}\,dS_R \qquad (3.5.6)$$

Lemma 3.7.6 in addition with summability properties $\mathbf{u}\in L^3$, $\sigma\in L^2$ ensure that (3.5.6), (3.5.3), (3.5.4) and (3.5.5) go to zero at infinity. $\square$

Remark 3.5.1 *Notice that in this proof we are using the hypothesis on the density*

$$\min p'(\rho) =: m_1 > 0.$$

This is certainly true for isothermal gases, and for isentropic gases with infinite mass, in particular for densities having positive infimum. The proof of uniqueness for a finite amount of gas remains open. This last problem has relevance in concrete problems, and there are known explicit solutions; cf. Lane Emden equation in Nishida [74]. We leave it as open challenging problem.

3.5.2   *Nonlinear Exponential Stability*

Despite the incompressible case, for exterior domains it is still possible to prove exponential stability results provided a summability hypothesis on the density $\rho\in L^{3/2}(\Omega)$ is introduced. Summability of density may be physically verified if the gas is not too dense. We give a stability proof for isothermal fluids, and leave the problem of proving exponential stability for fluids with different state equation on the pressure as an open problem.

Consider the **initial boundary value problem**

$$\partial_t\rho\nabla\cdot(\rho\mathbf{v}) = 0,$$

$$\partial_t(\rho\mathbf{v}) + \nabla(\rho\,\mathbf{v}\otimes\mathbf{v}) = -\nabla p + \lambda\nabla\nabla\cdot\mathbf{v} + 2\mu\nabla\cdot D(\mathbf{v}) + \rho\mathbf{f}, \qquad x\in\Omega,\ t\in(0,T),$$

$$p = p(\rho)$$

$$\rho(x,0) = \rho_0(x), \qquad\qquad \mathbf{v}(x,0) = \mathbf{v}_0(x),$$

$$\mathbf{v}|_{\partial\Omega} = 0, \qquad \lim_{x\to\infty}\mathbf{v} = 0 \qquad \lim_{x\to\infty}\rho = \rho_\infty.$$

$$(3.5.7)$$

We assume that regular steady solutions to (3.5.7) exist globally in time, and that they are regular solutions in the class

$$\mathcal{W} = \{(\mathbf{v},\rho) : \mathbf{v}\in L^\infty(0,\infty;L^3(\Omega))\cap L^2(0,\infty;W^{1,2}(\Omega)); \qquad (3.5.8)$$

$$\rho\in W^{1,\infty}\cap L^2(\Omega)\ \inf p'(\rho) > 0;\ \inf\rho > 0;\ \partial_t\rho\in L^\infty(0,\infty;L^\infty(\Omega))\}.$$

Theorem 3.5.2 Rigid Domains Exterior to a Fixed Obstacle *Let* $f \in L^3(\Omega)$, *and consider the system* (3.2.1) *to which we append the following boundary conditions*

$$\mathbf{v}|_{\partial\Omega} = 0, \qquad \lim_{x \to \infty} \boldsymbol{v} = 0 \qquad \rho \in L^{3/2}(\Omega). \qquad (3.5.9)$$

Then, if $\|\boldsymbol{f}\|_{L^3}$ *and constants* c_0, c_1, c_2, c_2', C_1, C_2, *defined by* (3.3.16), (3.3.42), (3.3.24), *satisfy* (3.3.18), (3.3.50) *and* (3.3.51), *then the solution* $(\boldsymbol{v}_b, \rho_b) \in X_b$ *is unique in the regularity class* $\mathcal{V}$.

Proof. The first difficulty is represented by the loss of Poincare' inequality. To solve this problem we notice that in the proof of Theorem 3.3.3, Poincare' inequality has not been used. We have employed only the limit embedding inequality

$$\|u\|_{L^6} \le c_e \|\nabla u\|_{L^2}.$$

Therefore, since this inequality continues to hold in unbounded domains, one difficulty has been avoided.

   The second difficulty must be studied: to deal with unbounded regions it is a common strategy to integrate the equations over the bounded domain $\Omega_R = \Omega \cap B_R$, where B_R is the ball centered at the origin, of radius R. Next, after integration by parts, one studies the behavior as $R \to \infty$ of the boundary terms. This produces boundary terms over $\partial B_R = S_R$ that must be proven to go to zero. Therefore, once we prove that the boundary terms are zero, using the same smallness assumptions as in Theorem 3.3.3, we get again exponential decay. $\square$

Boundary terms on S_R. We have to deal with several boundary terms. The first series of boundary terms comes by the multiplication of the momentum equation by $\mathbf{u}$, and integration over Ω_R

$$\int_{S_R} \frac{1}{2}|\mathbf{u}|^2 \rho \mathbf{v} \cdot \mathbf{e}_R \, dS_R + \int_{S_R} k\sigma \mathbf{u} \cdot \mathbf{e}_R \, dS_R + \int_{S_R} \lambda \mathbf{u} \cdot \mathbf{e}_R \nabla \cdot \mathbf{u} dS_R$$

$$+ 2\mu \int_{S_R} \mathbf{u} \cdot D(\mathbf{u}) \cdot \mathbf{e}_R \, dS_R. \qquad (3.5.10)$$

   The second series of boundary terms is derived from the multiplication of (3.5.7)$_1$ by k/ρ, and integration over Ω_R, so that we have

$$-\int_{S_R} \frac{k}{\rho} \frac{\sigma^2}{2} \mathbf{v}_b \cdot \mathbf{e}_R \, dS_R. \qquad (3.5.11)$$

The third and last series of boundary terms derives by multiplication of $(3.5.7)_2$ by $\mathbf{V}$, given in Lemma 3.7.5, and integration over Ω_R to receive the following chain of boundary terms

$$
-\int_{S_R} k\sigma\mathbf{e}_R \cdot \mathbf{V}\, dS_R - \int_{S_R} \rho\, \mathbf{V} \cdot \mathbf{u}\, \mathbf{v} \cdot \mathbf{e}_R dS_R
$$

$$
+ (\lambda + \mu)\int_{S_R} \nabla \cdot \mathbf{u}\mathbf{e}_R \cdot \mathbf{V}\, dS_R + \mu\int_{S_R} \mathbf{V} \cdot \nabla\, \mathbf{u} \cdot \mathbf{e}_R\, dS_R.
$$

(3.5.12)

We may employ the decay properties of functions on $L^3(\Omega)$, see Lemma 3.7.6. By hypothesis we have that $\mathbf{v} \in L^3$, $\sigma \in L^2$, thus the boundary terms go to zero as $R \to \infty$. We analyze two terms that contain most of the difficulties.

$$
\int_{S_R} k\sigma\mathbf{u} \cdot \mathbf{e}_R\, dS_R, \qquad -\int_{S_R} k\sigma\mathbf{e}_R \cdot \mathbf{V}\, dS_R
$$

(3.5.13)

Employing Lemma 3.7.6 and summability for $\mathbf{v} \in L^3$, $\sigma \in L^2$ we find that (3.5.10), (3.5.11), (3.5.12) and (3.5.13) go to zero at infinity.

Finally, we have obtained an equation analogous to (3.3.49) that we rewrite for reader's convenience

$$
\frac{d}{dt}\mathbb{E}(t) + (\lambda + \mu)\|\nabla \cdot \mathbf{u}\|_{L^2}^2 + (\mu - c_0)\|\nabla\mathbf{u}\|_{L^2}^2 + \nu(m_1 - C_2)\|\sigma\|_{L^2}^2
$$

$$
- \nu(\lambda + \mu)\|\nabla \cdot \mathbf{u}\|_{L^2}\|\sigma\|_{L^2} - \nu\mu\|\nabla\mathbf{u}\|_{L^2}\|\sigma\|_{L^2}
$$

(3.5.14)

$$
- \left(\nu C_1 + c_1\right)\|\nabla\, \mathbf{u}\|_{L^2}\|\sigma\|_{L^2} - c_2'\|\sigma\|_{L^2}^2 - \nu\, c_e\|\rho\|_{L^3}\|\nabla\mathbf{u}\|_{L^2}^2 \leq 0,
$$

where al symbols keep previous definitions. Using the same reasoning as for bounded domains, under smallness hypotheses (3.3.18), (3.3.27) and (3.3.31), on the constants we again deduce the first part of the inequality (3.3.55) inequality, i.e.

$$
(\mu - c_0)\|\nabla\mathbf{u}\|_{L^2}^2 + \left(\nu(m_1 - C_2) - c_2'\right)\|\sigma\|_{L^2}^2 - \nu(\lambda + \mu)\|\nabla \cdot \mathbf{u}\|_{L^2}\|\sigma\|_{L^2}
$$

$$
- \left(\nu C_1 + c_e c_1\right)\|\sigma\|_{L^2}\|\nabla\, \mathbf{u}\|_{L^2} > \left(b_1\|\nabla\mathbf{u}\|_{L^2}^2 + b_2\|\sigma\|_{L^2}^2\right)
$$

(3.5.15)

and as a consequence we have

$$\frac{d}{dt}\mathbb{E}(t) + b_1\|\nabla\mathbf{u}\|_{L^2}^2 + b_2\|\sigma\|_{L^2}^2 \leq 0, \tag{3.5.16}$$

with b_1, b_2 positive constants. The assumption on density implies

$$\int_\Omega \rho\mathbf{u}^2\,dx \leq \|\rho\|_{L^{3/2}}\|\mathbf{u}\|_{L^6}^2 \leq c_e\|\rho\|_{L^{3/2}}\|\nabla\mathbf{u}\|_{L^2}^2,$$

which infers

$$b_1\|\nabla\mathbf{u}\|_{L^2}^2 + b_2\|\sigma\|_{L^2}^2 \geq b\mathbb{E}(t).$$

Therefore, again we deduce that

$$\frac{d}{dt}\mathbb{E}(t) + b\mathbb{E}(t) \leq 0, \tag{3.5.17}$$

and Gronwall's Lemma furnishes the exponential decay.

Remark 3.5.2 *Notice that in the above proof we have used the following state equation for the pressure*

$$p'(\rho) = k.$$

This is certainly true for isothermal gases, and for isentropic gas with positive density, and thus with infinite mass. The proof of uniqueness for a finite amount of gas remains open for isentropic gases. This last problem has relevance in concrete problems, and there are known explicit solutions; cf. Lane Emden equation in Nishida [74]. We leave it as open problem.

3.6   Instability

In this section we prove a nonlinear instability theorem. To write the instability theorem we again use, as thermodynamic potential, the Helmholtz free energy per unit of mass

$$\Psi = \int^\rho \frac{p(s)}{s^2}ds.$$

In this case, the instability assumption may be written as follows:

Hypothesis of instability (HI)

(1) In the absence of external forces, there exists more than one equilibrium configuration, that is the condition $p(\rho) = c$ where c is fixed, that is satisfied by several values of ρ, for example ρ_i, $i = 1, \ldots, N$, corresponding to the same given mass[5]

---

[5]We give an example to clarify the assumption. Let $c > 0$, $\rho_* > 0$ be given positive constants. For a fixed given mass M, we choose as pressure the following one

$$p(\rho) = \left((\rho - \rho_*)^2 - c^2 \right)^2,$$

where c is a positive constant. Thus there are three values ρ_i, $i = 1, 2, 3$ of ρ, satisfying the two conditions

$$4(\rho - \rho_*) \left((\rho - \rho_*)^2 - c^2 \right) \nabla \rho = 0, \qquad \int_\Omega \rho = M. \qquad (3.6.1)$$

We find three factors in $(4.2.7)_1$ whose product must be zero. If one also considers discontinuous functions, equation $(4.2.7)$ is more delicate to deal with, thus we look for piecewise functions such that

$$4(\rho - \rho_*) \left((\rho - \rho_*)^2 - c^2 \right) = 0, \qquad \int_\Omega \rho = M. \qquad (3.6.2)$$

The solutions are constant. For $\rho_* = M/|\Omega|$, the solution is

$$\rho_1 = \rho_*.$$

Furthermore, for $\rho_* \neq M/|\Omega|$, functions

$$\rho_\pm = \rho_* \pm |c|,$$

are solutions if

$$\rho_* \pm |c| = \frac{M}{|\Omega|}.$$

Both solutions $\rho_\pm$ satisfy

$$\int_\Omega \rho \, dx = M.$$

The above constants $\rho_\pm$ are taken by solving the PDE

$$\nabla \left((\rho - \rho_*)^2 - c^2 \right)^2 = 0, \qquad \int_\Omega \rho = M. \qquad (3.6.3)$$

If Ω is an interval $\Omega = (0, d)$, system $(3.6.3)$ is satisfied also by piecewise solutions defined in pieces of $\Omega = (0, d/2) \cup (d/2, d)$ as follows

$$\rho_1 = \begin{cases} \rho_* + |c|, & x \in (0, d/2) \\ \rho_* - |c|, & x \in (d/2, d), \end{cases}$$

with $\rho_* = M/d$.

(2) In at least one equilibrium configuration, say ρ_1, it holds that

$$\frac{d^2 \rho \Psi}{d\rho^2}\bigg|_{\rho_1} < 0.$$

In this case, we shall prove the following theorem:

Theorem 3.6.1 Instability If the second order derivative of the Helmholtz free energy $\Psi(\rho)$ satisfies property (HI), then the equilibrium position S_1, at ρ_1 is unstable.

As will be apparent from the proof, the instability hypothesis can be weakened; however, we consider it for the sake of simplicity.

We give a proof of nonlinear instability, Theorem 3.6.1, by contradiction, cf. [105]. Assume by absurdum that the motion is stable in $W^{3,2}$ norm. This allows us to choose initial data $\mathbf{v}_0$, σ_0 so small that $\mathbf{u}$, σ are less than arbitrarily small constant ϵ for all times t, in the norms $W^{3,2}$, $W^{2,2}$, respectively. First, we recall that a balance theorem for the total energy holds for the motion $\mathbf{u}$, σ, (3.2.14)

$$\frac{d}{dt}\int_\Omega \left(\frac{1}{2}\rho \mathbf{v}^2 + \rho\Psi(\rho)\right)dx = -\mu \mathcal{D}_u(t), \tag{3.6.4}$$

where Ψ is the Helmholtz free energy per unit of mass. Following our work in Sect. 3.3.2, we obtain

$$\frac{d}{dt}\int_\Omega \left\{\frac{1}{2}\rho \mathbf{v}^2 + \frac{1}{2}\frac{d^2(\rho\Psi)}{d\rho^2}\bigg|_{\rho_*}\sigma^2 + o(\sigma^2))\right\}dx = -\mu \mathcal{D}_u(t), \tag{3.6.5}$$

but now the total energy has no definite sign, because $\frac{\partial^2 \Psi}{\partial \rho^2}\big|_{\rho_*}$ is negative for $\rho_* \in (\rho_1 - \epsilon, \rho_1 + \epsilon)$. To deduce a priori estimates of perturbations, we use the free work equation. In this case, the procedure will be slightly different. We denote by ρ_1 the constant equilibrium density such that $E(\rho_1)$ is an isolated local maximum.

Another solution is given in $\Omega = (0, d/4) \cup (d/4, d/2) \cup (d/2, 3d/4) \cup (3d/4, d)$ as follows

$$\rho_2 = \begin{cases} \rho_* + |c|, & x \in (0, d/4) \\ \rho_* - |c|, & x \in (d/4, d/2), \\ \rho_* + |c|, & x \in (d/2, 3d/4) \\ \rho_* - |c|, & x \in (3d/4, d), \end{cases}$$

with $\rho_* = M/d$.

A numerable number of solutions of this type can be constructed by allowing the division of $(0, d)$ to vary. Notice that these solutions have the same prescribed total mass (M) and volume.

Let us choose the test function $\mathbf{V}$ as the one constructed in Lemma 3.7.4. Multiplying $(3.2.9)_2$ by $\mathbf{V}$, we obtain the **free work equation** for compressible fluids

$$\frac{d}{dt} \int_\Omega \rho \mathbf{v} \cdot \mathbf{V} dx = \int_\Omega (p(\rho) - p(\rho_1)) \nabla \cdot \mathbf{V} dx$$
$$- \int_\Omega \Big[\mu \nabla \mathbf{v} \cdot \nabla \mathbf{V} + (\lambda + \mu) \nabla \cdot \mathbf{v} \nabla \cdot \mathbf{V} \Big] dx \qquad (3.6.6)$$
$$+ \int_\Omega \rho \mathbf{v} \cdot \Big[\frac{\partial \mathbf{V}}{\partial t} + \mathbf{v} \cdot \nabla \mathbf{V} \Big] dx.$$

The first term on the right hand side of (3.6.6) is the most important one because it has a definite sign. Recalling the definition of Hemholtz free energy,

$$\int_\Omega (p(\rho) - p(\rho_1)) \nabla \cdot \mathbf{V} dx = \int_\Omega \Big[\frac{d^2(\rho \Psi)}{d\rho^2} \Big|_{\rho_1} \sigma^2 + o(\sigma^2) \Big] dx,$$

and by instability hypothesis,

$$- \int_\Omega (p(\rho) - p(\rho_1)) \nabla \cdot \mathbf{V} dx = \int_\Omega \Big[a^2 \sigma^2 + o(\sigma^2) \Big] dx,$$

with

$$a^2 = \min_{x,t} \Big(- \frac{d^2(\rho \Psi)}{d\rho^2} \Big|_{\rho_1} \Big).$$

By subtracting (3.6.5) from (3.6.6) multiplied by $-\nu$, with ν a positive constant, we deduce

$$\frac{1}{2} \frac{d}{dt} \int_\Omega \Big[a^2 \sigma^2 - \rho \mathbf{v}^2 - 2\nu \rho \mathbf{v} \cdot \mathbf{V} + o(\sigma^2) \Big] dx = \mu \mathcal{D}_u + \nu \int_\Omega a^2 \sigma^2 dx$$
$$+ \nu \int_\Omega \Big[\mu \nabla \mathbf{v} \cdot \nabla \mathbf{V} + (\lambda + \mu) \nabla \cdot \mathbf{v} \nabla \cdot \varphi \Big] dx - \nu \int_\Omega \rho \mathbf{v} \cdot \Big[\mathbf{V}_t + \mathbf{v} \cdot \nabla \mathbf{V} \Big] dx.$$
$$(3.6.7)$$

We set

$$z(t) = \int_\Omega \Big[a^2 \sigma^2 - \rho \mathbf{v}^2 - 2\nu \rho \mathbf{v} \cdot \mathbf{V} + o(\sigma^2) \Big] dx \le 2 \, a^2 \, \|\sigma\|_{L^2}^2,$$

and notice that for ν small enough, employing the properties of $\mathbf{V}$ proved in Lemma 3.7.4, we get

$$\mathcal{D}_u + \nu \int_\Omega a^2 \sigma^2 dx + \nu \int_\Omega \Big[\mu \nabla \mathbf{v} \cdot \nabla \mathbf{V} + (\lambda + \mu) \nabla \cdot \mathbf{v} \nabla \cdot \mathbf{V} \Big] dx$$
$$- \nu \int_\Omega \rho \mathbf{v} \cdot \Big[\mathbf{V}_t + \mathbf{v} \cdot \nabla \mathbf{V} \Big] dx \ge \nu \frac{a^2}{2} \|\sigma\|_{L^2}^2 \ge \frac{\nu \, a^2}{4} z(t). \qquad (3.6.8)$$

Therefore, (3.6.7) yields

$$\frac{d}{dt} z(t) \geq \frac{\nu a^2}{4} z(t),$$

which implies

$$z(t) \geq z(0)\, e^{\frac{\nu a^2}{4} t}.$$

This relation infers a contradiction if $z(0) > 0$, for example if $\mathbf{v}(x,0) = 0$ and $o(\sigma^2) < a^2 \sigma^2$.

Remark 3.6.1 *Our instability proof doesn't provide any information on the growth of perturbations, we have proven only that there exists a continuum set of initial data, whose amplitude δ goes to zero, such that the perturbation cannot be controlled for all times, however δ is. Equivalently, we claim that there exists a continuum set of initial data, whose amplitude δ goes to zero, and whose corresponding motion doesn't rest.*

3.7 Auxiliary Lemmas

In this section we give some Lemmas that have been employed in the proofs of main Theorems. The first three Lemmas have been used in the proof of uniqueness of basic steady flows in the class of steady flows. The second block of Lemmas has been used in the proof of asymptotic stability of basic steady flows. The last Lemma concerns decay properties of summable functions in exterior domains. We will address explicitly the Theorem where these functions are used. The section ends with some bibliographic comments.

3.7.1 Function V for Uniqueness

Let us give three Lemmas for the auxiliary functions needed to prove uniqueness; we shall specify where these lemmas are used.

The proofs of Lemmas 3.7.1 and 3.7.2 are well known; cf. Chap. 3 in [36], therefore only some remarks are given here.

Lemma 3.7.1 V for Uniqueness, Ω bounded. *Let Ω be a bounded domain with boundary class C^1. Given a regular positive function $\rho(x)$ and function $\sigma \in L^2(\Omega)$, with*

$$\int_{\Omega} \sigma \, dx = 0. \tag{3.7.1}$$

Then there exists a vector field $\mathbf{V} \in W^{1,2}(\Omega))$ which satisfies the following boundary value problem

$$\nabla \cdot \mathbf{V} = \sigma, \quad x \in \Omega,$$

$$\mathbf{V}(x)|_{\partial\Omega} = 0.$$

Furthermore, there exists positive constants l_0, l_1, l_2, such that the following estimates hold true:

$$\|\mathbf{V}\|_{L^2} \le l_0 \|\sigma\|_{L^2},$$

$$\|\mathbf{V}\|_{L^6} \le l_1 \|\sigma\|_{L^2}, \tag{3.7.2}$$

$$\|\nabla\mathbf{V}\|_{L^2} \le l_2 \|\sigma\|_{L^2},$$

Proof. Notice that the compatibility condition

$$\int_\Omega \nabla \cdot \mathbf{V}\, dx = \int_\Omega \sigma\, dx$$

must be satisfied because Ω is bounded, and we must explicitly add the condition (3.7.10). There are many references to this result; cf. [16] and Lemma 3.1 of Sect. 3 of the monograph by [36], suggested for further reference.

The last two inequalities follow from classical embedding inequalities and the Poincaré inequality,

$$\|\mathbf{V}\|_{L^6} \le c_e \|\nabla\mathbf{V}\|_{L^2}, \ \|\mathbf{V}\|_{L^2} \le c \|\nabla\mathbf{V}\|_{L^2}.$$

$\square$

Lemma 3.7.2 V for Uniqueness of the Steady Flow Ω Exterior. *Let $\Omega \in C^1$ be exterior to a compact region $\mathcal{C}$, and let it be given that $\sigma \in L^2(\Omega)$. There exists a vector field $\mathbf{V} \in W^{1,1}(\Omega) \cap W^{1,2}(\Omega)$ which satisfies the following boundary value problem*

$$\nabla \cdot \mathbf{V} = \sigma, \quad x \in \Omega,$$

$$\mathbf{V}(x)|_{\partial\Omega} = 0,$$

$$\lim_{x \to \infty} \mathbf{V} = 0.$$

$(3.7.2)_{1,2}$ *continues to hold. Finally, there exists a positive constant such that, for any compact set Ω_c in Ω containing $\partial\Omega$, the following estimate holds true:*

$$\|\mathbf{V}\|_{L^2(\Omega_c)} \le l_c \|\sigma\|_{L^2}.$$

Proof. First of all we remark that since the domain is exterior, direct integration of $(3.7.2)_1$ over the sphere B_R implies Actually, since the domain is exterior, by applying the Gauss theorem to $\Omega_R = \Omega \cap B_R$, with B_R a ball of radius R centered at the origin, we have

$$\int_{\Omega_R} \nabla \cdot \mathbf{V} \, dx = \int_{S_1} \mathbf{e}_r \cdot \mathbf{V} R^2 \, d\omega,$$

where S_1 is the unitary ball, and $\mathbf{e}_r$ is the unit vector in the radial direction; namely, we have $\mathbf{e}_r = \frac{\mathbf{x}}{|x|}$. Computing the limits of both sides of this identity as $R \to \infty$ we get

$$\lim_{R \to \infty} \int_{\partial \Omega_R} \mathbf{V} \cdot \mathbf{e}_R dS_R) = \lim_{R \to \infty} \int_{S_1} \sigma R^2 \, d\omega.$$

Such identity doesn't imply anymore the condition

$$\int_{\Omega} \sigma \, dx = 0.$$

Namely, the hypothesis of zero mean of σ is not needed in exterior domains.

There are many references to this result, see cf. [16, 36]. The last two inequalities follow by classical embeddings inequality

$$\|\mathbf{V}\|_{L^6} \le c_e \|\nabla \mathbf{V}\|_{L^2},$$

$$\|\mathbf{V}\|_{L^2(\Omega_c)} \le l_c \|\nabla \mathbf{V}\|_{L^2}.$$

$\square$

3.7.2 Function V for Stability

In this subsection we construct the auxiliary functions used in the proof of asymptotic stability in Sect. 3.5.2. These Lemmas concern functions defined in the space time domain $Q_T = \Omega \times (0, T)$. Let

$$\partial_t \sigma = -\nabla \cdot (\rho \mathbf{u}) \in L^2(0, \infty; L^2(\Omega)). \tag{3.7.3}$$

Lemma 3.7.3 V for the Decay to Rest State, Ω bounded. *It is given that the fields $(\mathbf{u}, \sigma) \in \mathcal{V}$, with $\partial_t \sigma$ satisfying (3.7.3), and ρ_1 a constant. Then there exists a vector field $\mathbf{V} \in L^\infty(0, \infty; W_0^{1,2}(\Omega))$ with $\partial_t \mathbf{V} \in L^\infty(0, \infty; L^2(\Omega))$, which satisfies the following problem*

$$\nabla \cdot (\rho_b \mathbf{V}) = \frac{\sigma}{\rho_1}, \quad x \in \Omega,$$

$$\mathbf{V}(x)|_{\partial \Omega} = 0. \tag{3.7.4}$$

Moreover, there exist positive constants l_1, l_2 l_3 depending on ρ_b, ρ_1, and Ω, such that the following estimates hold true:

$$\|\mathbf{V}\|_{L^6} \leq l_1 \|\sigma\|_{L^2}$$

$$\|\nabla\mathbf{V}\|_{L^2} \leq l_2 \|\sigma\|_{L^2},$$

$$\|\partial_t\mathbf{V}\|_{L^2} \leq l_3 \|\nabla\mathbf{u}\|_{L^2}.$$

We consider Ω to be a rigid bounded domain.

Lemma 3.7.4 V for the Decay to the Steady Flow, Ω bounded. *Given the fields $(\mathbf{u}, \sigma) \in \mathcal{V}$ with $\partial_t\sigma$ satisfying (3.7.3), then there exists a vector field $\mathbf{V} \in L^\infty(0, \infty; W_0^{1,2}(\Omega))$ with $\partial_t\mathbf{V} \in L^\infty(0, \infty; L^2(\Omega))$ which satisfies the following problem*

$$\nabla \cdot \mathbf{V} = \sigma, \quad x \in \Omega,$$
$$\mathbf{V}(x)|_{\partial\Omega} = 0. \tag{3.7.5}$$

There also exists positive constants l_1, l_2 l_3 depending on ρ_b, ρ_1, and Ω, such that the following estimates hold true:

$$\|\mathbf{V}\|_{L^2} \leq l_1 \|\sigma\|_{L^2},$$

$$\|\nabla\mathbf{V}\|_{L^2} \leq l_2 \|\sigma\|_{L^2},$$

$$\|\partial_t\mathbf{V}\|_{L^2} \leq l_3 \|\nabla\mathbf{u}\|_{L^2}.$$

First of all we remark that the compatibility condition

$$\int_\Omega \nabla \cdot \mathbf{V} \, dx = \int_\Omega \sigma \, dx = 0, \tag{3.7.6}$$

is satisfied for bounded domains, because the conservation of mass and the initial conditions on the density require

$$\int_\Omega \sigma(x, 0) \, dx = \int_\Omega \sigma(x, t) \, dx.$$

We solve such a problem by constructing an explicit solution. Assume that the domain Ω is the union of domains Ω_i, $i = 1, \ldots, N$, star shaped with respect to certain balls $B_i \subseteq \Omega_i$. Let us take $N = 1$, (though our arguments hold for arbitrary N). We define $\rho_b\mathbf{V}$ according to the formula of Bogovski, cf. [16],

$$\rho_b(x)\mathbf{V}(x, t) = \int_{\Omega_\eta} (\sigma(y, t) - \bar{\sigma}(t))\left[\frac{\mathbf{x} - \mathbf{y}}{|x - y|^n} \int_{|x-y|}^\infty w\left(y + \xi\frac{\mathbf{x} - \mathbf{y}}{|x - y|}\right)\xi^2 d\xi\right] dy$$

$$= \int_{\Omega_\eta} (\sigma(y, t) - \bar{\sigma}(t))\mathbf{N}(\mathbf{x}, \mathbf{y})dy,$$

$$\tag{3.7.7}$$

and, by periodicity extend $\mathbf{V}$ onto the whole domain $x_* \in R^2$, $0 < z < \eta$.

A simple derivation in time yields

$$\partial_t \mathbf{V}(x,t) = -\frac{1}{\rho_b(x)} \int_{\Omega_\eta} \nabla \cdot (\rho \mathbf{u}) \mathbf{N}(\mathbf{x}, \mathbf{y}) dy, \qquad (3.7.8)$$

thus by classical theorems on singular kernels, it is not difficult verify that $\partial_t \mathbf{V}$ satisfies

$$\nabla \cdot (\rho_0 \partial_t \mathbf{V}) = -\nabla \cdot (\rho \mathbf{u}), \quad x \in \Omega,$$

$$\mathbf{V}(x)|_{\partial\Omega} = 0 \qquad\qquad\qquad\qquad (3.7.9)$$

$$\mathbf{V}(x,0) = \int_\Omega \sigma_0(y) \mathbf{N}(x,y) dy.$$

Inequalities (3.7.3) follow the Calderon–Zygmund theorem; cf. [104].

The next Lemma holds for exterior domains with rigidly bounded boundaries.

Lemma 3.7.5 V for the Decay to the Steady Flow, Ω Exterior. *Let there be given the fields $(\mathbf{u}, \sigma) \in \mathcal{V}$, with $\partial_t \sigma$ satisfying (3.7.3). Then there exists a vector field $\mathbf{V} \in L^\infty(0, \infty; W_0^{1,2}(\Omega))$ with $\partial_t \mathbf{V} \in L^\infty(0, \infty; L^2(\Omega))$, which satisfies the following problem*

$$\nabla \cdot \mathbf{V} = \sigma, \quad x \in \Omega,$$

$$\mathbf{V}(x)|_{\partial\Omega} = 0, \qquad\qquad (3.7.10)$$

$$\lim_{x \to \infty} \mathbf{V} = \mathbf{0}.$$

Furthermore, there exists a positive constant l_1, l_2 l_3 depending on ρ_b, ρ_1, and Ω, such that the following estimates hold true:

$$\|\mathbf{V}\|_{L^6} \le l_1 \|\sigma\|_{L^2},$$

$$\|\nabla \mathbf{V}\|_{L^2} \le l_2 \|\sigma\|_{L^2},$$

$$\|\partial_t \mathbf{V}\|_{L^2} \le l_3 c_* \left(\|\mathbf{u}\|_{W^{1,2}} + \|\sigma\|_{L^2} \right),$$

where Ω_c is a compact set in Ω, containing $\partial\Omega$.

Lemma 3.7.6 *Let $f \in L^p(\Omega)$, $g \in L^q(\Omega)$, with $1/p + 1/q \ge 2/3$, then*

$$\int_{S_R} f(x)g(x) dS_R \to 0, \qquad |x| \to \infty. \qquad (3.7.11)$$

The proof is based on the property stating that if $\int_0^\infty \frac{h(R)}{R}dR < \infty$, the function $h(R)$ must go to zero at infinity. We fix $h(R) = \int_{S_R} f(x)g(x)dS_R$ and compute the previous integral, applying the Holder inequality with three exponents p, q, r:

$$\frac{1}{p} + \frac{1}{q} + \frac{1}{r} = 1,$$

$$\int_0^\infty R^{-1} \int_{S_1} fgR^2 dS_1 dR \tag{3.7.12}$$

$$\leq \left(\int_0^\infty \int_{S_1} f^p R^2 dS_1 dR \right)^{1/p} \left(\int_0^\infty \int_{S_1} g^q R^2 dS_1 dR \right)^{1/q} \left(\int_0^\infty \int_{S_1} R^{-r} R^2 dS_1 dR \right)^{1/r}.$$

The first two integrals at right are finite by hypothesis, hence the integral at left will be finite if the last integral is finite. This means that $r > 3$, which implies

$$\frac{1}{p} + \frac{1}{q} \geq \frac{2}{3}. \tag{3.7.13}$$

The Lemma is completely proved.

Some inequalities

The following inequalities hold true.

Lemma 3.7.7 *Let* $\mathbf{u}$ *be a solenoidal vector field in* $W^{1,2}(\Omega)$, *with* Ω *a bounded domain. If* $\mathbf{u}$ *is orthogonal to rigid motions, then the following inequalities hold true*

$$\|\mathbf{u}\|_{L^2(\Omega)} \leq c_P \|\nabla\mathbf{u}\|_{L^2(\Omega)},$$

$$\|\mathbf{u}\|_{L^6(\Omega)} \leq c_S \|\nabla\mathbf{u}\|_{L^2(\Omega)}, \tag{3.7.14}$$

where c_P, *and* c_S *are the Poincare' and Sobolev constants.*

Proof. Inequalities (3.7.14) are true in a domain, bounded at least in one direction, when $\nabla\mathbf{u} = 0$ implies $\mathbf{u} = 0$, cf. [5]. This statement is true due to the hypothesis that $\mathbf{u}$ vanishes on $\Sigma \times \{0\}$. □

3.7.3 Bibliographical Notes

The well-posedness question for barotropic models of fluid motions is a challenging one; indeed, in some cases is not even known yet how to formulate the boundary value problem correctly. In Chap. 1 it has been observed that under the action of potential forces, such as isentropic gas in a bounded domains under the action of large forces, or isothermal gas in an exterior domains under the action of small forces, the rest state may not exist; cf. [95, 99, 102]. Obviously, in order to solve the problem of existence and uniqueness of a steady solution, the boundary value problem should be correctly formulated.

With regards to this, we quote some results on the existence of weak unsteady solutions for large data proven in cf. [26, 27, 66] and [28], in the isentropic case, and in cf. [120] for the isothermal case. In these theorems there is no proof of uniqueness. Concerning regular solutions close to the rest state, there have been existence theorems since 1981 under assumptions of viscosity coefficients, cf. [90, 91, 95], and without restrictions on the coefficients, cf. [13, 54, 56, 70–72, 80, 130, 146].

Chapter 4
Isothermal Fluids with Free Boundaries

ed ecco, qual sul presso del mattino
per li grossi vapor Marte rosseggia
giu' nel ponente sopra il suol marino,
13, II, Purgatorio, A. Dante

The reason is a very small island in the ocean of irrational.

4.1 Introduction

In this chapter we prove uniqueness, stability and instability theorems for the
rest state of heavy isothermal viscous fluids filling a portion of horizontal
layer. A new definition, *initial data control*, of a solution will be introduced.
We provide a priori estimates for a given spatial norm of the difference
between a given flow that may be either steady, or unsteady and the rest
state, provided the given flow belongs to a suitable regularity class. The lines
of proof of nonlinear stability and instability theorems follow those given by
the author in cf. [107], with Solonnikov in [113, 115, 116], and with Massari
and Shimizu in [68].

To clarify the concept of control and loss of control of a solution from
initial data, we begin with a concrete example.

Let a capillary liquid drop F be pending below a rigid surface S with air at
rest all around. Let us name F_b its equilibrium position, and assume F_b to be
linearly stable. Usually, a blast perturbs the drop F that changes its position
from F_b into F_0, that we name *initial data*. If the blast is sufficiently light,
then F_0 will be sufficiently close to F_b and the possibility exists that F will
oscillate around F_b. In such a case $F - F_b$ is controlled by initial data $F_0 - F_b$,
at least when $F_0 - F_b$ is sufficiently small. However, if the blast is sufficiently

M. Padula, *Asymptotic Stability of Steady Compressible Fluids*,
Lecture Notes in Mathematics 2024, DOI 10.1007/978-3-642-21137-9_4,
© Springer-Verlag Berlin Heidelberg 2011

intense then it will move the fluid in the initial position F_0 sufficiently far from F_b, and possibly in next times $t > 0$, and F will follow down. In such a case the perturbation $F - F_b$ loses the control in terms of its initial data $F_0 - F_b$ for large enough times.

We address the following equivalent questions:

Question Q_1 *Is it possible to compute the lower bound on a blast's intensity, such that the liquid motion F, corresponding to the blast's perturbation with initial data F_0, represents the detachment and the fall down of a liquid drop?*

Question Q_2 *Is it possible to compute a lower bound on the geometrical measure of the free boundary of F_0 such that the liquid drop F, corresponding to initial data F_0, will fall down at a certain time t?*

These questions arise when the rest state is linearly stable, and therefore the rest state is also nonlinearly stable for sufficiently small initial data. However, in reality the initial data are large, thus often information on linear stability is not sufficient to ensure nonlinear stability of the rest state, and the rest state is not observed.

Remark 4.1.1 *Notice that sometimes a mathematically small physical quantity cannot be controlled in the laboratory. That is, in 'the real world' mathematically small initial data cannot be controlled, and the basic state cannot be observed, even though it is nonlinearly stable.*

In Sects. 4.6 and 4.7 of this chapter we address the study of motions corresponding to initial data far from the rest state.

Even though the question appears elementary, we have not yet found a correct mathematical position for the question, though the mathematical formulation of such a problem is not trivial. Therefore, in this chapter we limit ourselves to the mathematical formulation of problem Q_2, where instead of a liquid pending drop we have a section of a liquid layer, so that there are no contact angles, and the geometry is Cartesian.

Main goal *Our aim is to* **reduce the study of stability, linear and not, to the sign of a suitable functional $\mathbb{E}$ called the modified energy functional.** To this aim, we provide a qualitative procedure to compute a non-dimensional number, say $G_{NL} = G_L$, below which the motion is nonlinearly stable for sufficiently small initial data. In this method it is not necessary to know the solution. Here we give a direct method to compute: (a) the nonlinear stability of the equilibrium configuration S_b for initial data sufficiently small; (b) the loss of control of the equilibrium configuration S_b for initial data sufficiently large.

Denote by $S(S_0, t)$ the motion of the fluid section of layer corresponding to initial data S_0. Assume that the function $S(S_b, t) = S_b$ is an exact steady solution of the fluid layer, corresponding to the initial data S_b. In this chapter

we reduce the problem of stability, linear and not, to the sign of the given functional called 'energy of perturbation'

$$\mathcal{E}(t) := E(S(S_0, t)) - E(S_b), \qquad (4.1.1)$$

where $E(S(S_0, t))$ denotes the total energy of the physical system, at time t, corresponding to initial data S_0. We furnish the above proofs through Lyapunov direct methods. Our main tool consists in the introduction of a new function using the idea of free work cf. [104, 106, 108]. It is worth emphasizing that our method is simple, straightforward, and does not use complicated analysis.

Chapter 4 contains a modified version of the results proved by the author and Solonnikov in cf. [113]. It is straightforward to extend the above theorems to a general domain, bounded either entirely by a deformable boundary (b), or by a rigid boundary (c), cf. [113–115]. In domains partially filled with the fluid, the problem of contact angles appears, this problem remains open.

In general, *a stability threshold computed by linear theory **does not** ensure nonlinear stability for large initial data*: in fact, *stability thresholds computed using a nonlinear approach (large initial data) generally **do not** coincide with stability thresholds computed by linear theory*. A basic flow judged linearly stable could be found to be non-observable for large initial perturbations. Therefore, the direct study of nonlinear stability or instability of a given flow without linearized procedure makes sense; cf. Padula [107], Padula and Solonnikov [113, 116], Frolova and Padula [34], and Massari et al. [68].

The plan of the chapter is as follows:

Section 4.2 Classical definitions of equilibrium figures and rest state are given. Equations of motion are introduced in a very general form. The known relation between stability of equilibrium configurations and the sign of the total energy is explicitly analyzed.

Section 4.3 Some definitions related to stability are introduced. In particular the definition of control of a solution via initial data is given.

Section 4.4 Uniqueness of the rest state of a piece of horizontal layer of heavy isothermal gas with an upper free boundary is studied.

Section 4.5 Nonlinear stability of the rest state of a piece of horizontal layer of heavy isothermal gas with a free boundary is studied. Take a frame with the origin on the rigid plane π and the z axis normal to π, directed toward the fluid. The stability result depends on the z-component of the gravity. If gravity is opposite to the z-axis direction, then nonlinear asymptotic stability is proved; if on the contrary gravity is directed along the z-axis, then loss of initial data control is proved.

Section 4.6 A instability theorem is proved when there exists a negative total energy, and the loss of control from initial data is proved when initial data are far from the equilibrium configuration.

Section 4.7 It is calculated an initial energy $\mathcal{E}(0)$ of perturbations, positive
         definite for all small initial data $S_0 - S_b$, and negative for some
         suitable large initial data $S_0 - S_b$.
Section 4.8 Auxiliary Lemmas are given where suitable test functions are
         constructed.

4.2   Position of the Problem

In this section we introduce the basic equations governing steady and non-
steady motions for a fixed portion of viscous isothermal gas filling a section
Ω of a horizontal layer of viscous fluid bounded *above or below*,(respectively),
by a rigid plane Γ_B, and *below or above*,(respectively), by a free surface Γ_t[1].

4.2.1   Geometrical Tools

Introduce an ortho-normal reference frame $\mathcal{R} = \{O, \mathbf{i}, \mathbf{j}, \mathbf{k}\}$ with $O \in \Gamma_B$,
$\mathbf{k}$ orthogonal to Σ, and directed inside the fluid toward the free surface.
Let us call (x, y, z), $z > 0$, the coordinates of a point of Ω_t in $\mathcal{R}$. We set
$x' \equiv (x, y)$ to denote a point $x \equiv (x', 0)$ belonging to Σ; in general, given the
three-dimensional vector $\mathbf{u}$, the two-dimensional vector $\mathbf{u}'$ will be written as
$\mathbf{u}' \equiv (v_x, v_y)$, and $\mathbf{u} \equiv (\mathbf{u}', v_z)$. We denote $\nabla' = (\partial_x, \partial_y)$, $\nabla' \cdot \mathbf{u}' = \partial_x u_x + \partial_y u_y$.
For the sake of simplicity we assume periodicity conditions on lateral walls.
     The use of Cartesian representation of surfaces Γ_t, Γ_ζ means implicitly
that we are not considering the formation of reversal flows.
     The periodicity cell is given by $\Sigma = (0, a) \times (0, b) \subseteq R^2$. Let us call
by $W_c^{1,2}(\Omega)$ the subspace of $W^{1,2}(\Omega)$ of functions η satisfying the following
property

$$\eta = c = constant \qquad \longrightarrow \qquad \eta = 0.$$

We recall that in the subspace of $W_c^{1,2}(\Omega)$ there exists a constant C_P,
independent of η, such that the Poincaré inequality holds true, cf. [5],

$$\|\eta\|_{L^2(\Sigma)}^2 \leq C_P \|\nabla' \eta\|_{L^2(\Sigma)}^2. \tag{4.2.1}$$

We adopt the Cartesian representation both for rigid plane Γ_B and free
surface Γ_t:

---

[1]We assume the liquid below the rigid plane Γ_B because we need negative potential energy.
We notice that a negative potential energy can be easily realized by rapidly accelerating a
beaker of water downwards (and standing clear!).

$$\Gamma_B = \{x = (x', z) \in R^3 : x' \in \Sigma, \ z = 0\},$$

$$\Gamma_t = \{x = (x', z) \in R^3 : x' \in \Sigma, \ z = h + \eta(x', t)\}.$$

In order to have a simply connected domain for all time, we must assume that

$$\zeta(x', t) := h + \eta(x', t) > 0 \quad \text{for all } t \geq 0. \tag{4.2.2}$$

The domain Ω_t occupied by the fluid is represented by

$$\Omega_t = \{x = (x', z) : \ x' \in \Sigma, \ 0 < z < \zeta(x', t)\}.$$

In the steady case, the domain and its boundary will not depend on t, hence we shall adopt the conventions

$$\Gamma_\zeta = \{x = (x', z) : \ x' \in \Sigma, \ z = \zeta(x')\}$$

$$\Omega_\zeta = \{x = (x', z) : \ x' \in \Sigma, \ 0 < z < \zeta(x')\}.$$

Here $\mathbf{n}$ denotes the exterior unit normal vector at the point of free surface Γ_ζ, and it holds

$$\mathbf{n} = \frac{1}{\mathcal{G}}(-\nabla'\zeta, 1) \qquad \mathcal{G}(\zeta) = \sqrt{1 + |\nabla'\zeta|^2}.$$

Moreover, to describe a infinitesimal element dS of the surface Γ_ζ we shall use the notation

$$dS = \sqrt{1 + |\nabla'\zeta|^2}\, dx' = \mathcal{G}dx'.$$

As integral elements, we use the notations

$$\int_{\Gamma_t} dS = \int_\Sigma \mathcal{G}dx',$$

$$\int_{\Omega_t} dx = \int_\Sigma \int_0^{\zeta(x', t)} dz dx', \qquad dx = dx' dz = dz dx'.$$

Let be given a massic potential U, and let $\mathcal{H}$ be the double mean curvature of a surface Γ_t.

As known, the double mean **curvature** $\mathcal{H}$ of a surface Γ_ζ is expressed by

$$\mathcal{H}(\zeta) = \nabla' \cdot \left(\frac{\nabla'\zeta}{\sqrt{1 + |\nabla'\zeta|^2}} \right). \tag{4.2.3}$$

The double mean curvature is known as the nonlinear positive **Laplace (1749-1827)-Beltrami (1836-1900) operator.**

Concerning equilibrium figures of viscous capillary fluids, we will mention two capillary prototypes: a self-gravitating liquid drop, and a section of a horizontal layer of heavy liquid. In capillarity theory the unknown domain Ω_b is occupied by the liquid, referred to as the equilibrium figure, defined by the equation of its boundary $\partial\Omega_b$.

Let Ω, Γ denote either Ω_ζ, Γ_ζ, or Ω_t, Γ_t, Ω_b, Γ_b, respectively. Here it is $\partial\Omega = \Gamma \cup \Gamma_B \cup \Gamma_l$, where Γ_B the rigid planar part of the boundary $\partial\Omega$, Γ the deformable part of the boundary $\partial\Omega$, free surface, Γ_l and the fixed lateral flat, vertical walls of $\partial\Omega$. In both of our examples, $\overline{\Gamma} \cap \overline{\Gamma_B} = \emptyset$; namely, we do not consider contact angles. As this implies either $\Gamma_B = \emptyset$ liquid drop, thus we have $\Gamma_l = \emptyset$, or $\Gamma_B \neq \emptyset$, layer, thus $\Gamma_l \neq \emptyset$. For $\Gamma_l \neq \emptyset$ in two dimensional domains $\partial\Omega_*$ is the perimeter of the area Ω_*, it will be the union of two segments $\Gamma_l = \Gamma_1 \cup \Gamma_2$, where we prescribe periodicity conditions, on Γ_ζ, and on Γ_B.

We conjecture that all proofs and considerations in this section continue to hold in the case of an infinite layer. We leave it as an open problem.

4.2.2 Equations of Motion

We write now the well posedness problem for the equations governing steady and non-steady motion of viscous isothermal fluids.

Find a domain $\Omega_\zeta = \{\mathbf{x} = (x', z) : x' \in \Sigma, 0 < z < \zeta(x')\}$, and a triple $(\mathbf{u}, \rho, \zeta)$ defined in Ω_ζ, periodic with periodicity cell Σ, and satisfying the following Free Boundary Value Problem

$$\nabla \cdot (\rho\mathbf{u}) = 0, \qquad\qquad \Omega_\zeta,$$

$$\rho(\mathbf{u} \cdot \nabla)\mathbf{u} - \mu\nabla \cdot S(\mathbf{u}) = -\nabla p(\rho) + \rho\nabla U, \quad \Omega_\zeta,$$

$$\mathbf{u} \cdot \mathbf{n} = 0, \qquad\qquad \Gamma_\zeta, \qquad (4.2.4)$$

$$\mu S(\mathbf{u})\mathbf{n} - p(\rho)\mathbf{n} = (-p_e + \kappa\mathcal{H}(\zeta))\mathbf{n}, \qquad \Gamma_\zeta$$

$$\mathbf{u}(x', 0) = 0, \qquad\qquad \Sigma$$

$$\textstyle\int_{\Omega_\zeta} \rho(x)dx = M,$$

with

$$S(\mathbf{u}) := 2D(\mathbf{u}) + \frac{(\lambda + \mu)}{\mu}\nabla \cdot \mathbf{u}\mathbf{I},$$

and $D(\mathbf{u})$ is the velocity deformation tensor. In (4.2.4), $\mathcal{H}(\zeta)$ is the double mean curvature of Γ_t, $\mathbf{n}$ denotes the exterior unit normal vector at the point of free surface Γ_t, $\tilde{\mathbf{n}} = \mathcal{G}\mathbf{n}$.

Let there be given the cell $\Sigma \subseteq R^2$, the total mass M, and the constants $\mu > 0$, λ, $3\lambda + 2\mu \geq 0$, (shear and bulk viscosities), $\kappa > 0$ (surface tension), *g (gravity), p_e (external pressure). Find a domain $\Omega_t = \{\mathbf{x} = (x', z) : x' \in \Sigma, \ 0 < z < \zeta(x', t)\}$, and a triple $(\mathbf{u}, \rho, \eta)$ defined in Ω_t, periodic with periodicity cell Σ, satisfying the Initial Free Boundary Value Problem*

$$\rho_t + \nabla \cdot (\rho \mathbf{u}) = 0, \qquad\qquad \Omega_t$$

$$\rho \left(\mathbf{u}_t + \mathbf{u} \cdot \nabla \mathbf{u}\right) = -k\nabla \rho + \mu \nabla \cdot S(\mathbf{u}) + \rho g \mathbf{k}, \quad \Omega_t$$

$$\zeta_t \left(x', t\right) = \mathbf{u} \cdot \tilde{\mathbf{n}}\left(x', \zeta(x', t), t\right), \qquad \Sigma,$$

$$-k\rho \mathbf{n} + \mu S(\mathbf{u}) = \kappa \mathcal{H}(\zeta)\mathbf{n} - p_e \mathbf{n}, \qquad \Gamma_t,$$

$$\mathbf{u}\left(x', 0, t\right) = 0, \qquad\qquad \Sigma, \qquad (4.2.5)$$

$$\mathbf{u}\left(x', z, 0\right) = \mathbf{u}_0\left(x', z\right), \qquad\qquad \Omega_0,$$

$$\rho\left(x', z, 0\right) = \rho_0\left(x', z\right), \qquad\qquad \Omega_0,$$

$$\eta\left(x', 0\right) = \eta_0\left(x'\right), \qquad\qquad \Sigma,$$

$$M = \int_{\Omega_b} \rho_b dx' dz = \int_{\Omega_0} \rho_0 dx' dz,$$

$$\mu S(\mathbf{u}) = 2\mu D(\mathbf{u}) + \lambda \nabla \cdot \mathbf{u} \, \mathbf{I}.$$

In (4.2.5), $\mathcal{H}(\zeta)$ is the double mean curvature of Γ_t, $\mathbf{n}$ denotes the exterior unit normal vector at the point of free surface Γ_t, $\tilde{\mathbf{n}} = \mathcal{G}\mathbf{n}$. We assume here that the initial density ρ_0, and height ζ_0 are always positive.

4.2.3   Rest State and Equilibrium Configurations

We compute exact solutions for barotropic fluids in a layer, and see that they are described by rest states, or equivalently, by equilibrium configurations. External force $\mathbf{f}$ per unit of mass is derived from a potential U, $\mathbf{f} = \nabla U$. If U is the gravitational potential, the orientation of the vertical axis z is done in the direction of the free surface Γ_b **upward or downward** according to whether Σ is situated *below or above* the free surface Γ_b, respectively.

Remark 4.2.1 *For a horizontal layer of heavy fluid we have: $U = -gz$, if Γ_t is above Σ, $\mathbf{k}$ upward oriented. $U = gz$, if Γ_t is below Σ, $\mathbf{k}$ downward oriented. The potential energy has the opposite sign to U.*

It is worth to distinguishing **rest state from equilibrium configuration.** In subsequent sections we will adopt the following definitions.

Let $\mathbf{u}(x, t)$, $\rho(x, t)$ denote the velocity and the density of a fluid particle, $x \in \Omega_t$, $t \in (0, T)$ the rest state for a fluid is defined by the motions having zero velocity.

Definition 4.2.1 Rest State $\mathbf{u}_b = 0$, $\rho = \rho_b$, $\zeta = \zeta_b$. *A fluid motion* $(\mathbf{u}_b, \rho_b, \zeta_b)$ *will be said to be in a* **rest state** *if the following system of indefinite equations is satisfied:*

$$\mathbf{u}_b(x) = 0, \qquad\qquad x \in \Omega_{\zeta_b}$$
$$-\nabla p(\rho_b) + \rho_b \nabla U(x) = 0 \qquad\qquad x \in \Omega_{\zeta_b},$$
$$-p(\rho_b(\zeta_b)) = \kappa \mathcal{H}(\zeta_b) - p_e(x'), \qquad\qquad x' \in \Sigma,$$
$$\int_\Sigma \int_0^{\zeta_b} \rho_b(x) dx = M. \qquad\qquad (4.2.6)$$

To (4.2.6), boundary conditions must be added. At the boundary $\partial\Sigma$ we use periodicity conditions to simplify the problem.

Definition 4.2.2 Equilibrium Configuration (ρ_b, ζ_b) *A fluid configuration* (ρ_b, ζ_b) *will be said to be an* **equilibrium configuration** *if the following system of indefinite equations is satisfied:*

$$-\nabla p(\rho_b) + \rho_b \nabla U(x) = 0 \qquad\qquad x \in \Omega_{\zeta_b}$$
$$-p(\rho_b(\zeta_b)) = \kappa \mathcal{H}(\zeta_b) - p_e(x'), \qquad\qquad x' \in \Sigma$$
$$\int_\Sigma \int_0^{\zeta_b} \rho_b(x) dx = M. \qquad\qquad (4.2.7)$$

To (4.2.7), boundary conditions must be added. At the boundary $\partial\Sigma$ we use periodicity conditions to simplify the problem.

We quote as alternative definition of equilibrium configuration in capillarity theory the one given by Finn in [31] where density is supposed a given function of ζ, the surface density is assumed to be non homogeneous, and the equilibrium configuration is described by the function $\zeta_b = \zeta_b(x')$ [32]. In the case of homogeneous surface tension, the following definition is adopted to study the equilibrium configurations:

Definition 4.2.3 Capillary Equilibrium Configurations $\zeta_b = \zeta_b(x')$ The equation of the free surface Γ_b is given as solution $\zeta_b = \zeta_b(x')$ of the elliptic equation

$$\kappa \mathcal{H}(\zeta_b) + U(\zeta_b) + \Lambda = 0, \qquad x' \in \Sigma$$
$$p(\rho_b(\zeta_b)) = p_e(\zeta_b), \qquad\qquad (4.2.8)$$

with κ the surface tension, Λ a Lagrange multiplier, and p_e a given function. Solutions $\zeta_b = \zeta_b(x')$ to (4.2.8) denote capillary equilibrium figures.

Definition 4.2.8 is deduced by considering the stationary points of the total energy functional $E_U(\mathbf{u}, \rho, \zeta)$ computed when $\mathbf{u} = 0$, and the density is a given function of the space $\rho_b = \rho_b(x)$, under the constraint on the total mass to be prescribed.

Remark 4.2.2 *From now on we extend to R^3 the functions velocity $\mathbf{u}_b = 0$ and density $\rho_b = \rho_b(x)$. The extension is done simply taking the extension of $\mathbf{u}$ to be zero outside Ω_b, taking the same function $\rho_b(x)$ for points not in Ω_b. In this way $\mathbf{u}(x)$ is zero everywhere, while the extended density is continuous. With this position we may compute the difference between the two solutions in Ω_t.*

Remark 4.2.3 *In all Definitions 4.2.6, 4.2.7 and 4.2.8, the pressure is known as function of ρ.*

In Definitions 4.2.7 and 4.2.8, velocity fields are implicitly assumed to be zero, the main difference lies in the density variable. Specifically, in Definition 4.2.7 density is supposed an unknown function of x, while in Definition 4.2.8 density is a known function of x, the only unknown being the geometry of the domain $\zeta_b = \zeta_b(x')$.

The main difference between the two definitions of equilibrium configuration lies in the definition of energy as functional of one or two variables. However, as it will appear in the sequel, final computations do not differ, namely, equilibrium configurations will coincide.

4.2.4 First and Second Variations of Energy

Choose a barotropic gas:

$$p = p(\rho).$$

We recall that solutions to (4.2.4) with $\mathbf{u} = 0$ provide stationary points of the nonlinear functional, representing the total energy at rest

$$E(0, \rho, \zeta) = \kappa|\Gamma_\zeta| + \Pi_\Psi + \Pi_U + \Lambda M,$$

where $|.|$ denotes the measure of the set Γ_ζ, Λ is a Lagrange multiplier, and

$$M = \int_{\Omega_\zeta} \rho\, dx$$

is the total mass. The term $\kappa|\Gamma_\zeta|$ represents the surface energy. Moreover, given the Helmholtz free energy per unit of mass $\Psi := \int^\rho \frac{p(s)}{s^2} ds$ we have defined the global Helmholtz free energy

$$\Pi_\Psi(\Omega_\zeta) = \int_{\Omega_\zeta} \rho \int^\rho \frac{p(s)}{s^2} ds\, dx,$$

and for the density of potential $U(x)$ per unit of mass, we have defined the global potential energy

$$\Pi_U(\Omega_\zeta) = - \int_{\Omega_\zeta} \rho U(x)dx.$$

The nonlinear functional Π_U is considered in the set of domains Ω_ζ close to Ω_b, for gases having the same total mass and position of the barycenter as that of Ω_b, with $\partial\Omega_\zeta = \Gamma_\zeta \cup \Gamma_B \cup \Gamma_l$, $\partial\Omega_b = \Gamma_b \cup \Gamma_B \cup \Gamma_l$, and Γ_ζ close to Γ_b.

Notice that the sign of $E(0,\rho,\zeta)$ is a function of the external potential that may be gravitational, centrifugal etc., indeed the external forces rule the stability of an equilibrium figure.

To be more clear, we compute explicitly the first and second variations of $E(0,\rho,\zeta)$ in the simplified case of a layer, $\Gamma_l \neq 0$. Set $\zeta_{,i}$ to denote the partial derivative of ζ with respect to x_i, $i = 1,2$, $x_1 = x$, $x_2 = y$, $\zeta_{,i} = \frac{\partial\zeta}{\partial x_i}$. In the case of a layer we may write[2]

$$E(0,\rho,\zeta) = \mathcal{F}(\rho,\zeta,\nabla'\zeta) := \int_\Sigma \Big(F(\nabla'\zeta) + G(\rho,\zeta)\Big)dx, \qquad (4.2.9)$$

where

$$F(\nabla'\zeta) := \kappa\sqrt{1 + |\nabla'\zeta|^2},$$

$$G(\rho,\zeta) := \int_0^\zeta \rho\Big[\int^{\rho(x)} \frac{p(s)}{s^2}ds - U(x) + \Lambda\Big]dz.$$

Notice that for $\rho = \rho_b + \sigma$ satisfying the principle of conservation of mass it holds

$$\int_{\Omega_\zeta} \rho\,dx - \int_{\Omega_b} \rho_b dx = 0. \qquad (4.2.10)$$

To compute the partial derivative of E with respect to ρ, we take ζ fixed and perturb ρ as $\widetilde{\rho} = \rho + t\sigma$, with $\sigma \in L^\infty$ satisfying (4.2.10). We compute its **Frechet (1873-1973)** derivative with respect to t, to find

$$\partial_\rho E(0,\widetilde{\rho},\zeta)[\sigma,\eta] = \int_{\Omega_\zeta}\Big[\int^\rho \frac{p(s)}{s^2}ds - U + \Lambda + \frac{p(\rho)}{\rho}\Big]\sigma\,dx \qquad (4.2.11)$$

$$= \int_{\Omega_\zeta}\Big[\int^\rho \frac{p'(s)}{s}ds - U + \Lambda\Big]\sigma\,dx,$$

---

[2]We remark that in the case of a drop it holds

$$E(0,\rho,\zeta) = \mathcal{F}(\rho,\zeta,\nabla'\zeta) := \int_\Sigma \Big(F(\zeta,\zeta_{,i}) + G(\rho,\zeta)\Big)dx,$$

and computations are more involved. In particular $b(x)$ involves also surface's terms.

with η perturbation to the height ζ_b. Equation (4.2.11) holds for all σ with

$$\int_{\Omega_b} \sigma\, dx = -\int_{\Omega_\zeta/\Omega_b} \rho\, dx.$$

At the stationary point ρ_b, ζ_b the first variation with respect to ρ of E must vanish for all σ satisfying (4.2.10), and $\eta \in W_c^{1,2}(\Omega_\zeta)$.

Set $\rho = \rho_b$, $\zeta = \zeta_b$ in (4.2.11) we get

$$\partial_\rho E(0, \rho_b, \zeta_b)[\sigma, \eta] = \int_{\Omega_b} \Big[\int^{\rho_b} \frac{p'(s)}{s} ds - U + \Lambda \Big] \sigma\, dx = 0, \qquad (4.2.12)$$

for all σ with

$$\int_{\Omega_b} \sigma\, dx = 0.$$

From (4.2.12) we deduce that σ is varying in L_c^2 the quotient Lebesgue space L^2 with respect the constants. Finally, recalling that Λ is constant, assuming (4.2.12) true for all σ in L_c^2 we obtain

$$\int^{\rho} \frac{p'(s)}{s} ds - U = C,$$

where C, is an arbitrary constant. Equivalently, one may state that the motion equation $(4.2.7)_1$ implies that the partial derivative of E in the equilibrium configuration (ρ_b, ζ_b) vanishes

$$\partial_\rho E(0, \rho_b, \zeta_b)[\sigma, \eta] = 0. \qquad (4.2.13)$$

Considering the momentum equation written for the rest state, one could make the equivalent assumption that the Frechét derivative of E with respect to ρ calculated in $(0, \rho_b, \zeta_b)$ is zero.

Analogously, the Frechét derivative with respect to ζ of $\mathcal{F}$ in $(0, \rho_b, \zeta_b)$ is given by

$$\delta_\zeta E(0, \rho, \zeta)[\sigma, \eta] = \sum_{i=1,}^{3} \partial_{\zeta,i} \mathcal{F}(\rho, \zeta, \nabla'\zeta)[\sigma, \eta] + \delta_\zeta \mathcal{F}(\rho, \zeta, \nabla'\zeta)[\sigma, \eta] \quad (4.2.14)$$

$$= \int_\Sigma \Big(\sum_{i=1, i\neq j}^{3} \frac{d}{dt} F(\zeta_{,i} + t\eta_{,i}, \zeta_{,j})) + \frac{d}{dt} G(\rho, \zeta + t\eta) \Big) dx' \Big|_{t=0}$$

$$= \int_\Sigma \Big(\sum_i \frac{\partial F}{\partial \zeta_{,i}} (\nabla'\zeta)\eta_{,i} + \partial_\zeta G(\rho, \zeta)\eta \Big) dx'.$$

We observe that

$$\partial_\zeta G(\rho, \zeta) := \rho(x', \zeta) \left[\int^{\rho(x',\zeta)} \frac{p'(s)}{s} ds - U(x', \zeta) + \Lambda \right].$$

(4.2.15)

This relation computed in the equilibrium configuration (ρ_b, ζ_b) must vanish, for all η, which furnishes

$$\delta_\zeta E(0, \rho_b, \zeta_b)[\sigma, \eta] = \int_\Sigma \left(\sum_i \frac{\partial F}{\partial \zeta_{,i}} (\nabla' \zeta_b) \eta_{,i} + \partial_\zeta G(\rho_b, \zeta_b) \eta \right) dx = 0.$$

(4.2.16)

Equations (4.2.12) and (4.2.16) are the Euler–Lagrange equations associated to E, and within our notations it holds

$$\nabla \left(\int^{\rho_b} \frac{p'(s)}{s} ds - U(x) \right) = 0, \qquad x \in \Omega_\zeta,$$

(4.2.17)

$$-\sum_i \frac{\partial}{\partial x_i} \frac{\partial F}{\partial \zeta_{,i}} (\nabla' \zeta_b) + \partial_\zeta G(\rho_b, \zeta_b) = 0, \qquad x' \equiv (x_i) \in \Sigma, \quad i = 1, 2.$$

In the simplified case of a layer $(4.2.17)_2$ can be represented as the **nonlinear operator B_n**

$$B_n(\rho, \zeta, \zeta_{,i}) := -\sum_i \frac{\partial}{\partial x_i} \frac{\partial F}{\partial \zeta_{,i}} (\nabla' \zeta) + \partial_\zeta G(\rho, \zeta) = -\kappa \Delta_\Gamma \zeta(x', t) + b(x),$$

(4.2.18)

where

$$-\Delta_\Gamma \zeta = -\sum_i \frac{\partial}{\partial x_i} \frac{\partial F(\nabla' \zeta)}{\partial \zeta_{,i}} = -\sum_i \frac{\partial}{\partial x_i} \left(\frac{\zeta_{,i}}{\sqrt{1 + |\nabla' \zeta|^2}} \right),$$

(4.2.19)

is the nonlinear positive Laplace–Beltrami operator on Γ, and

$$b(\rho) = \partial_\zeta G(\rho, \zeta) = -\rho \partial_\zeta U(\zeta) = -(\pm 1) \rho g.$$

is related to the potential U; [114, 115, 132–136].

In conclusion, the equations governing the rest state of a compressible fluid with a free boundary are given by

$$\partial_\rho F(\rho_b, \zeta_b, \nabla' \zeta_b)[\sigma, \eta] = \partial_\rho E(0, \rho_b, \zeta_b)[\sigma, \eta] = 0$$

(4.2.20)

$$\partial_\zeta F(\rho_b, \zeta_b, \nabla' \zeta_b)[\sigma, \eta] = \partial_\zeta E(0, \rho_b, \zeta_b)[\sigma, \eta] = 0.$$

A variation of E with respect to ρ, computed in ρ_b and η_b, is a quadratic form in the difference $\sigma = \rho - \rho_b$

$$\partial^2_{\rho^2} E(0, \rho, \zeta)[\sigma, \eta] = \int_{\Omega_\zeta} \frac{p'(\rho)}{\rho} \sigma^2 \, dx. \tag{4.2.21}$$

Another variation of E with respect ρ and ζ, computed in ρ_b and η_b, is a quadratic form in the difference $\sigma = \rho - \rho_b$, $\eta = \zeta - \zeta_b$

$$\partial^2_{(\rho,\zeta)} E(0, \rho, \zeta)[\sigma, \eta] = \int_\Sigma \Big(\int^{\rho(x',\zeta)} \frac{p'(s)}{s} ds - U(x', \zeta) + \Lambda \Big) \sigma \eta \, dx'. \tag{4.2.22}$$

Yet another variation of E with respect to ζ, computed in ζ_b, is a quadratic form of the difference $\eta = \zeta - \zeta_b$.

$$\partial^2_{\zeta^2} E(0, \rho_b, \zeta_b)[\sigma, \eta] = \tag{4.2.23}$$

$$\sum_j \delta_{\zeta_j} \int_\Sigma \Big(\sum_i \frac{\partial F(\nabla'\zeta)}{\partial \zeta_{,i}} \Big|_{((\zeta_b),1,(\zeta_b),j + t\zeta_{,j},(\zeta_b),n)} \eta_{,i} + \partial_\zeta G(\rho_b, \zeta) \Big|_{\zeta_b + t\zeta} \eta \Big) \, dx' \Big|_{t=0}$$

$$= \int_\Sigma \Big(\sum_{i,j} \frac{\partial^2 F(\nabla'\zeta)}{\partial \zeta_{,i} \partial \zeta_{,j}} \Big|_{\zeta_b} \eta_{,i} \eta_{,j} + \partial^2_{\zeta^2} G(\rho_b, \zeta) \Big|_{\zeta_b} \eta^2 \Big) \, dx'.$$

Conversely we have

$$\sum_{i,j} \partial^2_{\zeta_{,j},\zeta_{,i}} \mathcal{F}(\zeta_b + t\zeta, \nabla'(\zeta_b + t\zeta))[\sigma, \eta] =$$

$$\sum_{i,j} \Big[\partial^2_{\zeta_{,j},\zeta_{,i}} \mathcal{F}(\zeta_b, \nabla'\zeta_b)[\sigma, \eta] + o(\eta, \eta_{,j} \eta_{,i}) \Big] \tag{4.2.24}$$

$$=: \partial^2_{\zeta^2} E(0, \rho_b, \zeta_b)[\sigma, \eta] + o(\eta^2, \nabla'\eta^2).$$

Subtract $B_n(\rho_b, \zeta_b, (\zeta_b)_{,i})$ to $B_n(\rho, \zeta, \zeta_{,i})$, defined in (4.2.18), and expand the difference by Taylor polynomial with initial point ρ_b, ζ_b. For $\sigma = \rho - \rho_b$, $\eta = \zeta - \zeta_b$ we obtain

$$B_n(\rho, \zeta, \nabla'\zeta) - B_n(\rho_b, \zeta_b, \nabla'\zeta_b) =: B_1(\sigma, \eta, \nabla'\eta) = \mathcal{L}(\sigma, \eta) + R(\sigma, \eta, \nabla'\eta), \tag{4.2.25}$$

where we have distinguished the linear part $\mathcal{L}(\sigma, \eta)$ from the differential nonlinear operator of an order higher that one, $R(\sigma, \eta, \nabla'\eta) = o(\sigma, \eta, \nabla'\eta)$. We have thus set

$$\mathcal{L}(\sigma, \eta) = -\kappa \Delta_{\Gamma_b} \eta + b_b(x') \eta \tag{4.2.26}$$

where $b_b(x') = b(\rho_b, \zeta_b, x')$, and Δ_{Γ_b} is the Laplace–Beltrami operator computed along the given geometry ζ_b. It yields

$$\partial_\zeta G(\rho_b, \zeta_b) := \rho_b \Big[\int^{\rho_b(x)} \frac{p'(s)}{s} ds - U(\zeta_b) + \Lambda \Big]. \qquad (4.2.27)$$

$$\Delta_{\Gamma_b}\zeta := \sum_i \frac{\partial}{\partial x_i} \Big(\frac{\zeta_{,i}}{\sqrt{1 + |\nabla'\zeta_b|^2}} \Big). \qquad (4.2.28)$$

Let $\big(. , . \big)$ be the scalar product on Γ, where $\bar\rho$ denotes a value between ρ and ρ_b, as does $\bar\zeta$. The scalar products

$$\Big(B_1(\sigma, \eta, \nabla'\eta), \eta \Big) = \Big(\mathcal{L}(\sigma, \eta), \eta \Big) + \Big(R(\sigma, \eta, \nabla'\eta), \eta \Big) = \partial^2_{\zeta^2} E(0, \bar\rho, \bar\zeta)[\sigma, \eta],$$
$$(4.2.29)$$

$$\Big(\mathcal{L}(\sigma, \eta), \eta \Big) = \partial^2_{\zeta^2} E_l(\rho_b, \zeta_b)[\sigma, \eta],$$

represent the second variations of the total energy E, and of the quadratic part E_l, with

$$\delta^2 E(0, \rho_b, \zeta_b)[\sigma, \eta] = \partial^2_{\rho^2} E(0, \rho_b, \zeta_b)[\sigma, \eta] + \partial^2_{\zeta^2} E(0, \rho_b, \zeta_b)[\sigma, \eta]$$

$$+2\partial^2_{\rho\zeta} E(0, \rho_b, \zeta_b)[\sigma, \eta]$$

$$= \int_{\Omega_b} \frac{p'(\bar\rho)}{\bar\rho} \sigma^2(x)\, dx - \kappa \int_\Sigma \Big(\Delta_{\Gamma_b}\eta + b(\rho_b, x)\eta \Big)\eta(x')\, dx'$$

$$+ \int_\Sigma \Big(\int^{\rho_b} \frac{p'(s)}{s} ds - U + \Lambda \Big) \sigma(x', \zeta(x'))\eta(x')\, dx'.$$

Again employing the momentum equation for the rest we obtain

$$\int^{\rho_b} \frac{p'(s)}{s} ds - U + \Lambda = 0,$$

and we may reduce (4.2.30) to the simpler relation

$$\delta^2 E(0, \rho_b, \zeta_b)[\sigma, \eta] = \qquad\qquad\qquad\qquad\qquad (4.2.30)$$

$$\int_{\Omega_b} \frac{p'(\bar\rho)}{\bar\rho} \sigma^2(x)\, dx + \int_\Sigma \Big(\kappa \frac{|\nabla_{\Gamma_b}\eta|^2}{\sqrt{1 + |\nabla'\zeta_b|^2}} - b(\rho_b)\eta^2(x') \Big) dx'.$$

furnishing a quadratic form in σ and η.

Stable Equilibrium Figure

In Capillarity theory an equilibrium figure $\mathcal{F}_b$ is regarded as

(1) *Nonlinearly stable* if the second variation $\delta^2 E$ of the total energy E, computed in $\mathcal{F}_b$, furnishes a positive quadratic form

(2) unstable if $\delta^2 E$, computed in $\mathcal{F}_b$, can take negative values.

Of course, since the energy is defined up to a constant, one can state the previous assertions in a slightly different way: an equilibrium figure in a neighborhood of S_b may be regarded as nonlinearly stable if $E(S_0) - E(S_b)$ is positive and nonlinearly unstable if $E(S_0) - E(S_b)$ can take negative values.

If we write the total energy E in the variables $(0, \rho - \rho_b, \zeta - \zeta_b) \equiv (0, \sigma, \eta)$, there are perturbations to the equilibrium figure S_b. We denote by E_l the part of the total energy E quadratic in the perturbation $(0, \sigma, \eta)$.

An equilibrium figure is regarded as

(3) Linearly stable if the second variation $\delta^2 E_l$ of E_l is positive
(4) Unstable if $\delta^2 E_l$ can take negative values

4.3 Definitions Related to Stability

As known the rest state $\mathbf{u}_b = 0$, $\rho_b = \rho_b(x', z)$, $\zeta_b = h$, where h is unknown, is solution to

$$k\nabla \rho_b = -\rho_b g \nabla z, \qquad\qquad \Omega_h = \Sigma \times (0, h),$$

$$- p(\rho_b)(\zeta_b) = -p_e + \kappa \mathcal{H}(\zeta_b)), \qquad\qquad \Sigma$$

$$\mathcal{H}(\zeta_b) = \nabla'\left(\frac{\nabla'\zeta_b}{\sqrt{1 + |\nabla'\zeta_b|^2}}\right), \qquad\qquad \Sigma. \qquad (4.3.1)$$

Actually, in the parallelepiped $\Omega_h = \Sigma \times (0, h)$ there exists only the rest state S_b, with

$$S_b = \left\{ \mathbf{u}_b = 0, \quad \rho_b = \rho_* \exp\left(-\frac{gz}{k}\right), \quad \zeta_b = h = \frac{k}{g}\ln\left(1 + \frac{Mg}{p_e|\Sigma|}\right) \right\}, \quad (4.3.2)$$

and ρ_* given by

$$\rho_* = \frac{Mg}{k|\Sigma|\left(1 - \exp\left(-\frac{gh}{k}\right)\right)}. \qquad (4.3.3)$$

We shall continue to denote by ρ_b the extension of the density of the rest state to R^3, and we set

$$\Gamma_t = \{(x', z) : x' \in \Sigma, z = \zeta(x', t)\}.$$

Let S_b denote the rest state (4.3.2) of a portion of heavy liquid filling a horizontal layer. Traditionally the triple $S(t) = (\mathbf{u}, \sigma, \eta)$ defined by $\mathbf{u}(x, t) = \mathbf{u}(x, t) - \mathbf{u}_b(x)$, $\sigma(x, t) = \rho(x, t) - \rho_b(x)$, $\eta = \zeta - \zeta_b$, is called the perturbation to S_b. The perturbation $(\mathbf{u}, \sigma, \eta)$ solves the following initial boundary value problem

$$\sigma_t + \nabla \cdot (\rho \mathbf{u}) = 0, \qquad \Omega_t,$$

$$\rho \left(\mathbf{u}_t + \mathbf{u} \cdot \nabla \mathbf{u} \right) = -k\nabla(p(\rho) - p(\rho_b)) + \mu\nabla \cdot S(\mathbf{u}) + \sigma\nabla U, \quad \Omega_t$$

$$\eta_t \left(x', t \right) = \mathbf{u} \cdot \tilde{\mathbf{n}} \left(x', \eta(x', t), t \right), \qquad \Sigma,$$

$$-kp(\rho_b + \sigma)\mathbf{n} + \mu S(\mathbf{u}) = \kappa\mathcal{H}(\eta)\mathbf{n} - p_e\mathbf{n}, \qquad \Gamma_t, \quad (4.3.4)$$

$$\mathbf{u} \left(x', 0, t) \right) = 0, \qquad \Sigma,$$

$$\mathbf{u} \left(x', z, 0 \right) = \mathbf{u}_0 \left(x', z \right), \qquad \Omega_0,$$

$$\sigma \left(x', z, 0 \right) = \sigma_0 \left(x', z \right), \qquad \Omega_0,$$

$$\eta \left(x', 0 \right) = \eta_0 \left(x' \right) \qquad \Sigma,$$

$$M = \int_{\Omega_0} \rho_0 dx' dz,$$

$$\mu S(\mathbf{u}) = 2\mu D(\mathbf{u}) + \lambda\nabla \cdot \mathbf{u} \, \mathbf{I}.$$

Notice that the following chain of identity holds

$$M = \int_{\Omega_b} \rho_b dx' dz = \int_{\Omega_0} \rho_0 dx' dz = \int_{\Omega_t} \rho(x, t) dx' dz.$$

In (4.3.4) $\mathcal{H}(\zeta)$ is the double mean curvature of Γ_t, and $\mathbf{n}$ denotes the exterior unit normal vector at the point of free surface Γ_t, $\tilde{\mathbf{n}} = \sqrt{1 + |\nabla'\eta|^2}\mathbf{n}$. We assume the following conditions on initial density perturbation $|\sigma_0| < \rho_0/2$, height perturbation $|\eta_0| < \zeta_0/2$.

For regular solutions, linear and nonlinear stabilities **do depend** on external forces.

We may reduce the study of stability of S_b to the study of stability of the zero solution $S_0 = \{\mathbf{u} = 0, \quad \sigma = 0, \quad \eta = 0\}$; stability studies the control for all time $t > 0$ of a suitable norm of perturbation $S(t)$.

Let $(\mathbf{u}(x, t), \sigma(x, t), \eta(x', t))$ be a triple of regular functions, and set $\mathbf{u}_0(x) = \mathbf{u}(x, 0)$, $\sigma_0(x) = \sigma, (x, 0)$, $\eta_0(x') = \eta(x', 0)$. In the Cartesian product $X := W^{1,2}(\Omega_t) \times L^2(\Omega_t) \times W^{1,\infty}(\Sigma)$ we introduce the norm

$$\|(\mathbf{u}, \sigma, \eta)\|_X(t) := \|\mathbf{u}(\cdot, t)\|_{W^{1,2}(\Omega_t)} + \|\sigma(\cdot, t)\|_{L^\infty(\Omega_t)} + \|\eta(\cdot, t)\|_{W^{1,\infty}(\Sigma)}. \tag{4.3.5}$$

and we set $Y := W^{1,2}(\Omega_0) \times L^2(\Omega_0) \times W^{1,\infty}(\Sigma)$, and set

$$\|(\mathbf{u}_0, \sigma_0, \eta_0)\|_Y := \|(\mathbf{u}, \sigma, \eta)\|_X(0). \tag{4.3.6}$$

With this norm we may define the set of solutions $(\mathbf{u}, \sigma, \zeta)$ to (4.3.4) in the neighborhood of the zero $\mathcal{I}_r \subseteq X$ as the set of functions such that $\|(\mathbf{u}, \sigma, \eta)\|_X(t) < r$.

4.3.1 Initial Data Control

Here we give some basic definitions concerning the control in time of $S(t)$ through initial data, or stability with respect to initial data.

Definition 4.3.1 *The rest state S_b is said to be nonlinearly stable in the norm $\|(\mathbf{u}, \sigma, \eta)\|_X(t)$ if for all $\epsilon > 0$ there exists a $\delta > 0$ such that for all initial data $(\mathbf{u}_0, \sigma_0, \eta_0)$ verifying $\|(\mathbf{u}_0, \sigma_0, \eta_0)\|_Y < \delta$ it happens that*

$$\|(\mathbf{u}, \sigma, \eta)\|_X(t) < \epsilon, \qquad\qquad \forall t > 0.$$

Definition 4.3.1 depends on the norm $\|.\|_X(t)$, and on the radius δ of initial data, it is a local statement.

The difference between continuous dependence and stability lies on the time intervals $(0, T)$, and $(0, \infty)$, respectively.

Definition 4.3.2 *The rest state is said to be unstable if it is not nonlinearly stable.*

For the linearized problem associated to (4.3.4), all definitions can be repeated.

Definition 4.3.3 *The rest state S_b is said to be linearly stable in the norm $\|(\mathbf{u}, \sigma, \eta)\|_X(t)$ if there exists a constant $b \geq 0$ such that*

$$\|(\mathbf{u}, \sigma, \eta)\|_X(t) < \|(\mathbf{u}_0, \sigma_0, \eta_0)\|_Y \exp^{-bt}, \qquad\qquad \forall t > 0.$$

(1) Definition 4.3.3 is independent of the size of initial data, it is a global statement.
(2) We name the stability with $b = 0$ **marginal stability**.
(3) We name the stability with $b > 0$ **exponential stability**.

Definition 4.3.4 *The rest state S_b is said to be linearly unstable in the norm $\|(\mathbf{u}, \sigma, \eta)\|_X(t)$ if there exist a constant $b > 0$, and certain initial data $(\mathbf{u}_0, \sigma_0, \eta_0)$ such that*

$$\|(\mathbf{u}, \sigma, \eta)\|_X(t) > \|(\mathbf{u}_0, \sigma_0, \eta_0)\|_Y \exp^{bt}, \qquad\qquad \forall t > 0.$$

Let us observe that *the difference between linear and nonlinear stability with respect to initial data in classical fluid-dynamics is due to the size of the distance between initial data $S(0)$ and S_b.* In our case we are studying directly the difference between two motions $S(t) - S_b$.

Remark 4.3.1 *Denote by R a physical non-dimensional number ruling the linear stability of a given motion.*

(i) Let R_N the critical number such that for $R < R_N$ there is nonlinear stability, for initial data less than a finite computable constant δ_N.

(ii) Let R_L the critical number such that for $R < R_L$ there is linear stability, and for $R > R_L$ there holds linear instability. Applying the linearization principle, defined in Sect. 2.1 of Chap. 2, we know that for $\|(\mathbf{u}_0, \sigma_0, \eta_0)\|_X < \delta_L$ the null solution is also nonlinearly stable.

(iii) It results

$$R_N < R_L, \qquad\qquad \delta_L < \delta_N.$$

Usually the radius δ_L of initial data appropriate to ensure nonlinear stability is function of R, in general δ_L approaches zero as R approaches R_L.

(iv) In general while for the critical value R_N, δ_N is physically controllable, in the interval (R_N, R_L) the radius $\delta(R)$ may be very small, hence no more physically controllable.

(v) In the reality initial data may be large, see our example in the introduction, and we want to know what happens in this circumstance, cf. (4.6.5).

Previous definitions on nonlinear stability say nothing about the control for large periods of time of solutions, in a given norm, with initial data far from S_b. Indeed in nonlinear phenomena, for large initial data, a solution $S(t)$ may lose its control from initial data, even though S_b is nonlinearly stable (for small initial perturbations). This fact occurs frequently, and it constitutes the real discrepancy between linear and nonlinear stability. Currently it appears that there are no rigorous definitions of this phenomena, thus we introduce here two new definitions:

Definition 4.3.5 A perturbation $(\mathbf{u}, \sigma, \eta)$ to the rest state S_b is said to have **initial data control** in the range $\mathcal{I}_{2a}/\mathcal{I}_a$ if, and only if for all initial data $(\mathbf{u}_0, \sigma_0, \eta_0)$ satisfying

$$a < \|(\mathbf{u}_0, \sigma_0, \eta_0)\|_Y < 2a, \tag{4.3.7}$$

the solution $(\mathbf{u}, \sigma, \eta)$ to (4.3.4) corresponding to $(\mathbf{u}_0, \sigma_0, \eta_0)$ is bounded for all time, namely if there exists a suitable constant $\alpha = \alpha(a) > 0$ such that

$$\|(\mathbf{u}, \sigma, \eta)\|_X(t) \le \alpha, \qquad \forall t > 0. \tag{4.3.8}$$

Definition 4.3.6 The rest state is said to **lose the initial data control** if there exists a positive large number a and there exist initial data $(\mathbf{u}_0, \sigma_0, \eta_0)$ satisfying (4.3.7), such that the corresponding solution $(\mathbf{u}(x,t), \sigma(x,t), \eta(x,t))$ has no initial data control. Namely, given $\alpha > 0$, there exists $T > 0$ such that the solution $(\mathbf{u}, \sigma, \eta)$ of the problem (4.3.4) with initial data satisfying (4.3.7) satisfies the inequality

$$\|(\mathbf{u}, \sigma, \eta)\|_X(T) \ge \alpha. \tag{4.3.9}$$

Remark 4.3.2 *Definitions 4.3.2, 4.3.6 are meaningful only for nonlinear systems, because linear stability is global.*

The aim of the present chapter is to relate the stability of the rest to the sign of initial total energy when the rigid surface is either upward or downward situated.

Let $\mathcal{W}$ be the following regularity class

$$\mathcal{W} = \{\mathbf{u}(x,t),\ \sigma(x,t), \eta(x',t) :$$

$$\mathbf{u} \in L^2(0,\infty; W^{1,2}_\sharp(\Omega_t)) \cap L^\infty(0,\infty; L^3(\Omega_t)),$$

$$\sigma \in C^0(0,\infty; L^\infty(\Omega_t)), \quad \eta \in L^\infty(0,T; W^{1,\infty}_\sharp(\Sigma)),\ |\eta| < h/4,\ \},$$

where symbol $\sharp$ denotes periodicity in the horizontal direction. For a layer with a rigid surface below the liquid, we prove the following theorems:

Theorem 4.3.1 *[Theorem 2.3 in [113]] Assume the fluid is heavy and situated above a horizontal rigid plane, suppose there exist global solutions $(\mathbf{u}, \sigma, \eta) \in \mathcal{W}$ to (4.3.4) corresponding to large initial data $(\mathbf{u}_0, \sigma_0, \eta_0)$. Then the rest state $S_b = \{\mathbf{u}_b = 0, \sigma_b = \sigma(z), \eta_b = 0\}$ is exponentially nonlinearly stable in the energy norm in the class $\mathcal{W}$ of perturbations, for all initial data.*

*Notice that the hypothesis of Theorem 4.3.1 on the fluid above the rigid plane implies that the **total energy E is always positive**. This strong result is true because the weight is the only acting force, and has a stabilizing effect.*

Drawing upon results proved in [114], for a spherical geometry one can easily deduce in the simpler geometry of the layer the following instability result.

Theorem 4.3.2 *[From Theorem 1 in [116]] Let there exist a set of functions $(\mathbf{u}_0, \sigma_0, \eta_0)$ such that the **linearized part of the total energy E_l becomes negative**, then problem (4.3.4) has solutions growing exponentially as $t \to \infty$. This means that the rest state S_b is unstable.*

*Notice that the hypotheses of Theorem 4.3.2 imply that all the **total energy E may be negative for small initial data**. Thus our Theorem allows the claim that S_b is unstable. In this case, there is no need of a linearization principle.*

Before giving our result we introduce the non-dimensional characteristic Grashof number

$$G := \frac{g\,|\Sigma|}{\kappa}.$$

Remark 4.3.3 *The assumptions of Theorem 4.3.2 are realized assuming the fluid below the horizontal layer and under suitable assumptions on $G = (g|\Sigma|/\kappa)$ and $|\Omega_b|$.*

4.3.2 Present Results

In Sect. 4.4 we prove uniqueness of the rest state in the class of steady solutions.

In Sect. 4.5 we prove an energy identity and notice that, if the initial energy is negative, the energy equation doesn't provide an a priori estimate for the L^2 norms of solutions. In the remaining part of Sect. 4.5 we prove nonlinear exponential stability of the rest state if the energy of perturbation at initial time has only positive value in the neighborhood of zero.

In Sect. 4.6, we prove the instability of the rest state and the loss of initial data control. Precisely, assuming there exists a set of initial data $\mathbf{u}_0$, σ_0, η_0 such that the total energy at $t = 0$ $E(\mathbf{u}_0, \sigma_0, \eta_0)$ is negative, we prove a instability Theorem 4.3.3, *without using a linearization principle*; this is done in the wake of what done in [116]. Thus we introduce the definition of *control of a solution via its initial data for all times $t > 0$*, and we prove Theorem 4.3.4 claiming that if there exists (may be large) initial data $\mathbf{u}_0$, σ_0, η_0 such that $E(\mathbf{u}_0, \sigma_0, \eta_0)$ becomes negative, then the solution is not controlled by initial data for all times t.

In Sect. 4.7 we exhibit data κ, g h, such that: (a) for all infinitesimal initial data $(\mathbf{u}_0, \sigma_0, \eta_0)$ the total energy $E_0 = E(\mathbf{u}_0, \sigma_0, \eta_0)$ is positive, thus by Theorem 4.3.3 solutions corresponding to such E_0 are nonlinearly stable; (b) it is possible to construct initial data $\mathbf{u}_0, \sigma_0, \eta_0$ larger than a computable constant such that the total energy $E_0 = E(\mathbf{u}_0, \sigma_0, \eta_0)$ is negative. As such, recalling Theorem 4.3.4 of Sect. 4.6, we claim that solutions $(\mathbf{u}, \sigma, \eta)$ corresponding to such initial data lose the initial data control.

Result I

Theorem 4.3.3 *Assume that for all $\delta > 0$, there exists at least one initial data $(\mathbf{u}_0, \sigma_0, \eta_0)$ with $\|(\mathbf{u}_0, \sigma_0, \eta_0)\|_Y < \delta$, such that the initial energy E_0 is negative,*

$$- E_0 = g\|\eta_0\|^2_{L^2(\Sigma)} - 2\kappa > 0, \qquad (4.3.10)$$

then the rest state S_b is nonlinearly unstable. Precisely, there exists $\epsilon > 0$ satisfying that for all $\delta > 0$ there is an initial value $(\mathbf{u}_0, \sigma_0, \eta_0) \in W^{1,2}(\Omega_0) \times W^{1,\infty}(\Sigma)$ less than δ, and there exists a $T > 0$ such that the solution $(\mathbf{u}, \sigma, \eta)$ to the problem (4.3.4) satisfies the inequality

$$\|(\mathbf{u}, \sigma, \eta)\|_X(T) \geq \epsilon. \qquad (4.3.11)$$

Remark 4.3.4 *Notice that the existence of $(\mathbf{u}_0, \sigma_0, \eta_0)$ satisfying (4.3.10) is equivalent to the condition $G > 1$; we leave this proof to interested reader.*

We observe that the solution $(\mathbf{u}, \sigma, \eta)$ to problem (4.3.4) is a *perturbation to the rest state.*

Theorem 4.3.4 *Given a constant a, let there exist $(\mathbf{u}_0, \sigma_0, \eta_0)$ with $a < |(\mathbf{u}_0, \sigma_0, \eta_0)|_Y < 2a$, which renders negative the initial energy E_0,*

$$- E_0 = g\|\eta_0\|_{L^2(\Sigma)}^2 - 2\kappa \int_\Sigma \left(\sqrt{1 + |\nabla'\eta_0|^2} - 1\right) dx' - \|\sqrt{\rho_0}\mathbf{u}_0\|_{L^2(\Omega_0)}^2$$

$$- \int_{\Omega_0} \frac{p'(\overline{\rho})}{\overline{\rho}} \sigma^2 \, dx > 0, \tag{4.3.12}$$

where $\overline{\rho}$ is a value between ρ_0 and ρ_b, then the solution $(\mathbf{u}, \sigma, \eta)$ to problem (4.3.4) with initial data (4.3.7) has no initial data control.

Remark 4.3.5 *Notice that, since the total energy $E(\mathbf{u}, \sigma, \eta)$ is a non quadratic functional of $(\mathbf{u}, \sigma, \eta)$, $E(\mathbf{u}_0, \sigma_0, \eta_0)$ may be negative for fixed functions $(\mathbf{u}_0 = \mathbf{u}_b, \sigma_0 = \sigma_b, \eta_0 = \eta_b)$, while its quadratic part may still be positive computed at the same functions $(\mathbf{u}_b, \sigma_b, \eta_b)$.*

Result II

We describe a practical situation where the rest state is linearly stable, and for the same situation we give a computable number A such that the motions $(\mathbf{u}(x,t), \sigma(x,t), \eta(x',t))$ corresponding to initial data larger than A, lose the initial data control.

Theorem 4.3.5 *We assume that the linear stability hypothesis holds,*

$$G < G_d, \tag{4.3.13}$$

where $G_d := 4\pi^2 \min\{1, 1/d^2\}$, cf. Sect. 6.1. We construct the initial values $(\mathbf{u}_0, \sigma_0, \eta_0) \in Y$ that furnish negative initial energy E_0:

$$- E_0 = g\|\eta_0\|_{L^2(\Sigma)}^2 - 2\sigma \int_\Sigma \left(\sqrt{1 + |\nabla'\eta_0|^2} - 1\right) dx' - \|\mathbf{u}_0\|_{L^2(\Omega_0)}^2 - \int_{\Omega_0} \frac{p'(\overline{\rho_0})}{\overline{\rho_0}} \sigma_0^2 \, dx > 0. \tag{4.3.14}$$

Following upon Theorem 4.3.4 we may claim:

There exists a positive large number a, such that the solution $(\mathbf{u}(x,t), \sigma(x,t), \eta(x,t))$, corresponding to initial data $(\mathbf{u}_0, \sigma_0, \eta_0)$ satisfying (4.3.7), is not controlled by the initial data. Namely, however fixed $\alpha > 0$ there exists a initial data $(\mathbf{u}_0^\alpha, \sigma^\alpha, \eta_0^\alpha)$ satisfying (4.3.7), and an instant $T^\alpha > 0$ such that the corresponding solution $(\mathbf{u}^\alpha, \sigma^\alpha, \eta^\alpha)$ to problem (4.3.4) satisfies the inequality

$$\|(\mathbf{u}^\alpha, \sigma^\alpha, \eta^\alpha)\|_X(T^\alpha) \geq \alpha. \tag{4.3.15}$$

4.4   Uniqueness

We begin by recalling the correct boundary problem introduced in Sect. 4.2.2
Boundary Value Problem when the Domain has a Free Boundary.
   Find a domain Ω_ζ and a triple $(\mathbf{v}, \rho, \zeta)$ defined in $\Omega_\zeta \times \Omega_\zeta \times \Sigma$, periodic
in x' with period $\mathbf{a}' \equiv (a, b)$, which solves the following free boundary value
problem

$$
\begin{aligned}
\nabla \cdot (\rho \mathbf{u}) &= 0, & \Omega_\zeta, \\
\rho(\mathbf{u} \cdot \nabla)\mathbf{u} - \mu \nabla \cdot S(\mathbf{u}) &= -\nabla p - \rho g \nabla z, & \Omega_\zeta, \\
\mathbf{u} \cdot \mathbf{n} &= 0, & \Gamma_\zeta, & \qquad (4.4.1) \\
\mu S(\mathbf{u})\mathbf{n} - p\mathbf{n} &= (-p_e + \kappa \mathcal{H}(\zeta))\mathbf{n}, & \Gamma_\zeta, \\
\mathbf{u}(x', 0) &= 0, & \Sigma, \\
\textstyle\int_{\Omega_\zeta} \rho(x)dx &= M,
\end{aligned}
$$

where $\mathbf{u} = (\mathbf{u}', v_z)$ is the velocity, p is the pressure of isothermal fluid, $p = k\rho$,
and $\mathcal{H}(\zeta)$ is the nonlinear double mean curvature introduced in (4.2.3) of
Sect. 4.2.1. The positive constants κ surface tension, g gravity acceleration,
p_e the uniform pressure of atmosphere, are considered to be given.
   We now construct the rest state, $\mathbf{u}_b = 0$, $\rho_b = \rho_b(x', z)$, $\zeta_b = h$ with h
unknown, solution to

$$
\begin{aligned}
\nabla p(\rho_b) &= -\rho_b g \mathbf{k} \text{in} & \Omega_{\zeta_b}, & \qquad (4.4.2) \\
-p(\rho_b) &= -p_e + \kappa \mathcal{H}(\zeta_b)), & \text{in } \Sigma \\
\mathcal{H}(\zeta_b) &= \nabla' \cdot \left(\frac{\nabla' \zeta_b}{\sqrt{1 + |\nabla' \zeta_b|^2}} \right), & \text{in } \Sigma.
\end{aligned}
$$

We deduce that the density must have the form

$$
\rho_b = \rho_* \exp\left(-\frac{gz}{k} \right), \qquad \rho_* = const. \qquad (4.4.3)
$$

$$
\rho_*(\zeta_b) = \frac{Mg}{k|\Sigma|\left(1 - \exp\left(-\frac{g\zeta_b}{k}\right)\right)} = \frac{M}{\int_\Sigma \int_0^{\zeta_b} \exp\left(-\frac{gz}{k}\right)dzdx'},
$$

where ζ_b is unknown. On the free surface of $\Sigma \times (0, h)$, the condition $(4.4.1)_4$
reduces to the continuity condition on the pressure

$$
p_e = k\rho_b(\zeta_b) = k\rho_*(\zeta_b) \exp\left(-\frac{g\zeta_b}{k} \right). \qquad (4.4.4)
$$

Substituting the value for ρ_*, given in (4.4), into (4.4.4), and reminding that $\mathcal{H}(\zeta_b) = 0$, we get

$$p_e = \frac{M \exp\left(-\frac{gh}{k}\right)}{|\Sigma|\left(1 - \exp\left(-\frac{gh}{k}\right)\right)}. \tag{4.4.5}$$

From this we compute the height of the layer, given by

$$\zeta_b = \frac{k}{g} \ln\left(1 + \frac{M}{p_e|\Sigma|}\right) =: h, \tag{4.4.6}$$

so we find ζ_b as function of $|\Sigma|$, M, g and p_e as expected. Hence we claim that the set of stationary solutions of a compressible fluid in a layer with periodicity cell Σ under a given external pressure p_e is not empty. Actually, in the parallelepiped $\Omega_h = \Sigma \times (0, h)$ there exists at least the rest state S_b, with

$$S_b = \left\{ \mathbf{u}_b = 0, \rho_b = \rho_* \exp\left(-\frac{gh}{k}\right), h = \frac{k}{g} \ln\left(1 + \frac{M}{p_e|\Sigma|}\right) \right\},$$

and ρ_* given by (4.4).

Remark 4.4.1 *Explicit calculations of the rest state for isentropic gases are not known in general, and this is why we deal the isothermal case. It would be interesting to solve the uniqueness and stability problems for some particular isentropic fluid, provided the rest state is given in explicit form. Of course, we leave it as an open problem.*

We now prove uniqueness of S_b in the class of three-dimensional regular solutions, corresponding to the same force g, the same total mass M, the same periodicity Σ, and to the same external pressure p_e. Precisely, our uniqueness theorem concerns uniqueness of the rest state of an isothermal fluid under gravity action; for the steady solutions $(\mathbf{u}(x', z), \rho(x', z), \zeta(x'))$ we introduce the following regularity class[3]

$$\mathcal{V} = W_\sharp^{1,2}(\Omega_\zeta) \times L^\infty(\Omega_\zeta) \times W_\sharp^{1,\infty}(\Sigma),$$

where $\sharp$ means periodicity in the horizontal direction.

Below we prove the following uniqueness theorem, see also [113].

Theorem 4.4.1 *The rest state S_b is the unique solution to system (4.4.1) in the class of steady motions $\mathcal{V}$, corresponding to the same external data.*

Proof. Assume by absurdum that there exists another solution ρ, $\mathbf{u}$, ζ to problem (4.4.1), we prove that the perturbation

---

[3]The regularity class $\mathcal{V}$ can be weakened.

$$\mathbf{u}, \qquad \sigma = \rho - \rho_b, \qquad \eta = \zeta - \zeta_b,$$

is identically zero.

Let us multiply $(4.4.1)_2$ by $\mathbf{u}$ and integrate over Ω_ζ. Taking into account the solenoidality of $\rho\mathbf{u}$, we obtain

$$\frac{\mu}{2}\int_{\Omega_\zeta} S(\mathbf{u}) : S(\mathbf{u})dx'dz - k\int_{\Omega_\zeta} \rho\nabla\cdot\mathbf{u}dx'dz = -g\int_{\Gamma_\zeta}\rho\,\zeta\mathbf{u}\cdot\mathbf{n}dS \quad (4.4.7)$$

$$-\int_{\Gamma_\zeta}\frac{|\mathbf{u}|^2}{2}\rho\mathbf{u}\cdot\mathbf{n}dS + \int_{\Gamma_\zeta}\mathbf{u}\cdot(\mu S(\mathbf{u})\mathbf{n} - p\mathbf{n})dS.$$

By virtue of $(4.4.1)_3$ and $(4.4.1)_4$ all boundary terms vanish, so $(4.4.7)$ reduces to

$$\frac{\mu}{2}\int_{\Omega_\zeta} S(\mathbf{u}) : S(\mathbf{u})dx - k\int_{\Omega_\zeta}\rho\nabla\cdot\mathbf{u}dx = 0. \qquad (4.4.8)$$

From $(4.4.1)$ we know that

$$\rho\nabla\cdot\mathbf{u} = -\rho\mathbf{u}\cdot\nabla\ln\rho + \nabla\cdot(\rho\mathbf{u}).$$

Substituting this identity in $(4.4.8)$, and employing the solenoidality of $\rho\mathbf{u}$, and $(4.4.1)_3$, we find that the second integral in $(4.4.8)$ is zero. This yields

$$\frac{\mu}{2}\int_{\Omega_\zeta} S(\mathbf{u}) : S(\mathbf{u})dx = 0. \qquad (4.4.9)$$

From Korn's (1870-1945) inequality proved in [9] Lemma 4.7, we deduce that also $\nabla\mathbf{u} = 0$, and boundary condition $(4.4.1)_6$ implies that $\mathbf{u} = 0$. Next, rewriting $(4.4.1)_2$ with $\mathbf{u} = 0$ we get

$$\rho = \rho_b + \sigma = \widehat{\rho}\exp\left(-\frac{gz}{k}\right), \qquad (4.4.10)$$

where due to condition $(4.4.1)_6$ $\widehat{\rho}$ is given by

$$\widehat{\rho}(x',\zeta) = \frac{M}{\int_\Sigma\int_0^\zeta\exp\left(-\frac{gz}{k}\right)dzdx'}. \qquad (4.4.11)$$

The proof $\sigma = 0$, or $\rho = \rho_b = \rho_*\exp\left(-\frac{gz}{k}\right)$, will be achieved once one proves that $\zeta(x') = h$, and that as a consequence, $\rho_* = \hat{\rho}$. To this end we rewrite $(4.4.1)_4$ with $\mathbf{u} = 0$; inserting, for the pressure at boundary, the notation $\mathcal{P}(\zeta)$

$$\mathcal{P}(\zeta) = \frac{Mk\exp\left(-\frac{g\zeta}{k}\right)}{\int_\Sigma \int_0^\zeta \exp\left(-\frac{gz}{k}\right)dzdx'} = -\kappa\nabla'\cdot\left(\frac{\nabla'\zeta}{\sqrt{1+|\nabla'\zeta|^2}}\right) + p_e. \quad (4.4.12)$$

Moreover, we recall that

$$\mathcal{P}(h) = \frac{Mk\exp\left(-\frac{gh}{k}\right)}{\int_\Sigma \int_0^h \exp\left(-\frac{gz}{k}\right)dzdx'} = p_e.$$

Subtracting $\mathcal{P}(h)$ to $\mathcal{P}(\zeta)$, since $\nabla'\zeta = \nabla'\eta$, we get

$$\mathcal{P}(\zeta) - \mathcal{P}(h) = -\kappa\nabla'\cdot\left(\frac{\nabla'\eta}{\sqrt{1+|\nabla'\eta|^2}}\right). \quad (4.4.13)$$

Multiply (4.4.13) by η and integrate over Σ to obtain

$$\kappa\int_\Sigma \frac{|\nabla'\eta|^2}{\sqrt{1+|\nabla'\eta|^2}}dx' - \int_\Sigma \eta\left(\mathcal{P}(\zeta) - \mathcal{P}(h)\right)dx' = 0. \quad (4.4.14)$$

By the Lagrange mean value theorem we deduce

$$\kappa\int_\Sigma \frac{|\nabla'\eta|^2}{\sqrt{1+|\nabla'\eta|^2}}dx' - \int_\Sigma \eta^2|\mathcal{P}'(\bar\zeta)|dx' = 0, \quad (4.4.15)$$

where we have used the identity $\mathcal{P}(\zeta) - \mathcal{P}(h) = \mathcal{P}'(\bar\zeta)\eta$, and from (4.4.12) $\mathcal{P}'$ is negative. From (4.4.14), and convexity of the function $\frac{|\nabla'\zeta|^2}{\sqrt{1+|\nabla'\zeta|^2}}$ in $\nabla'\zeta$ it follows $|\nabla'\zeta| = 0$, so $\zeta = const.$, and the side condition that the total mass is prescribed (4.4.10), and $(4.4.1)_6$ allow us to claim that $\zeta = h$. Finally, again (4.4.11) tells us that $\rho = \rho_b$ hence $\sigma = 0$, and the theorem is completely proved. $\qquad\square$

Remark 4.4.2 *We leave as open the extension of the proof of uniqueness of rest state under general large potential forces, and in general three-dimensional domains, in the case when the explicit rest state is known. Also it is interesting and challenging the problem to give a proof when ρ_b is not explicitly given, see e.g. the case of isentropic flows. In this contest it would be interesting to consider also self gravitating forces in domains either bounded, or exterior to a compact set. In case of self gravitating forces there is the possibility to fix zero boundary condition on the density, this represents the case of a star atmosphere, [32, 74].*

4.5 Nonlinear Stability

In previous section, in the class of steady, regular perturbations, we have
studied uniqueness of the rest state for a free boundary problem of a
horizontal layer of compressible fluids corresponding to gravity as external
force. In this section, we study the evolution in time of non-steady flows by
perturbing initially the rest state S_b assuming the same constant external
pressure p_e. We follow the lines of stability proof given in [113].

In order to state our theorem, we introduce one class of weak solutions

$$(\mathbf{u}(x', z, t), \rho(x', z, t), \zeta(x', t)) \in \mathcal{W},$$

$$\mathcal{W} = \left[L^\infty \left(0, \infty; L^3_\sharp(\Omega_t) \right) \cap L^2 \left(0, \infty; W^{1,2}_\sharp(\Omega_t) \right) \right) \times$$

$$L^\infty \left(0, \infty; C^0_\sharp(\bar{\Omega}_t) \right) \times L^\infty \left(0, \infty; W^{1,\infty}_\sharp(\Sigma) \right) \right],$$

where $\sharp$ means periodicity in the horizontal directions. For the norm of a
triple $(\mathbf{u}, \sigma, \eta)$ in this space we use the symbol

$$\|(\mathbf{u}, \sigma, \eta)\|_{\mathcal{W}} := \sup_t \|\mathbf{u}\|_{L^3(\Omega_t)} + \left(\int_0^\infty \|\mathbf{u}\|^2_{W^{1,2}(\Omega_t)} dt \right)^{1/2} + \tag{4.5.1}$$

$$\sup_{x,t} |\sigma| + \sup_{x,t} |\nabla' \eta|.$$

These weak solutions satisfy the energy identity.

4.5.1 Energy Equation

Let us assume there exist globally in time regular solutions to the initial
boundary value problem (4.3.4), with $U = -gz$. Under the hypothesis that
the rest state S_b is a minimum for the energy functional, namely $E(0, \rho_b, h) <
E(\mathbf{u}, \rho, \zeta)$, for all $(\mathbf{u}, \rho, \zeta)$ in a suitable ball of $\mathcal{W}$, having radius R, we prove
that in the bounded domain Ω_t the L^2-norm of any *regular* perturbation to
the unique basic state S_b, decays to zero as $t \to \infty$, with exponential rate.
In particular, the domain tends to Ω_b. First we use, in non-standard way,
the energy method to find the behavior in time of the difference between the
energy $E(t)$ of the non steady motion and the energy E_b of the rest state.
We observe that $E(t) - E_b > 0$ dominates a suitable spatial norm of the
difference between the solutions $(\mathbf{u}, \rho, \zeta)$, and the rest state $(0, \rho_b, h)$. Next,
to get the decay to zero in time of a suitable norm of perturbation, we use
the free work identity that furnishes dissipative terms for the perturbations
in density and height. This will be achieved by using the FWE with the
auxiliary function constructed in Lemma 4.8.2 in the Appendix. Finally we
construct an equation governing the evolution of the modified energy $\mathbb{E}$.

We recall that when possible, in integral terms of volume we are using the simplified notation $dx = dx'dz$. Let us derive the energy identity that represents the difference of energy equations between the non steady motion and the rest state (Dirichlet method).

Theorem 4.5.1 *Energy Equation I. Let* $\mathbf{u}$, $\rho = \rho_b + \sigma$, $\zeta = h + \eta$ *solve* (4.3.4) *with* $(\mathbf{u}, \sigma, \zeta) \in \mathcal{W}$. *Then, the energy equation holds*

$$\frac{d}{dt}\Big[E_u + E_\sigma + E_\eta\Big] + D_u(t) = 0, \tag{4.5.2}$$

$$E_u = \frac{1}{2}\int_{\Omega_t} \rho \mathbf{u}^2\, dx,$$

$$E_\sigma = k\int_{\Omega_t} \rho(\ln\rho - \ln\rho_b)dxdz,$$

$$E_\eta = \kappa\int_\Sigma \frac{|\nabla'\eta|^2}{2(1+|\nabla'\bar\eta|^2)^{3/2}}dx' + k\rho_b(h)\int_\Sigma \eta dx',$$

$$D_u = \frac{\mu}{2}\int_{\Omega_t} |S(\mathbf{u})|^2)\, dx,$$

where $\bar\rho$ *is a value of* ρ *between* ρ_b *and* $\rho = \rho_b + \sigma$, *and* $\bar\zeta$ *is a value between* h *and* $\zeta = h + \eta$.

In subsequent sections the identity $\nabla'\eta = \nabla'\zeta$, will be important. For capillary tubes ζ_b is no longer constant, and the problem remains open, because

$$\nabla'\zeta \neq \nabla'\eta.$$

From (4.3.4) we derive the perturbations equations

$$\begin{array}{ll} \sigma_t + \nabla\cdot(\rho\mathbf{u}) = 0, & \Omega_t, \\[2mm] \rho\Big(\mathbf{u}_t + \mathbf{u}\cdot\nabla\mathbf{u}\Big) = \mu S(\mathbf{u}) - \nabla p(\rho) - \rho g\nabla z, & \Omega_t, \\[2mm] \eta_t = \mathbf{u}\cdot\tilde{\mathbf{n}}, & \Gamma_t \\[2mm] \mu\nabla\cdot S(\mathbf{u})\mathbf{n} - p\mathbf{n} = (-p_e + \kappa\mathcal{H}(\eta))\mathbf{n}, & \Gamma_t, \quad (4.5.3) \\[2mm] \mathbf{u}(x',0,t) = 0, & \Sigma, \\[2mm] \mathbf{u}(x',z,0) = \mathbf{u}_0(x',z), & \Omega_0, \\[2mm] \sigma(x',z,0) = \sigma_0(x',z), & \Omega_0, \\[2mm] \eta(x',0) = \eta_0(x'), & \Sigma, \\[2mm] \displaystyle\int_{\Omega_t} \rho(x',z,t)dx = M, & \end{array}$$

where $\tilde{\mathbf{n}} = \mathbf{n}\sqrt{1 + |\nabla'\eta|^2}$. We assume here that the initial density ρ_0, ζ_0 is positive.

Notice that $(4.5.3)_9$ expresses the condition that the total mass of the fluid, at initial time, coincides with that of the basic rest state S_0.

Multiplying $(4.5.3)_2$ by $\mathbf{u}$ and integrate over Ω_t, we obtain

$$\frac{1}{2}\int_{\Omega_t} \rho\frac{d\mathbf{u}^2}{dt}\,dx = \int_{\Gamma_t} \mathbf{u}\cdot(\mu S(\mathbf{u}) - k\rho\mathbf{I})\mathbf{n}\,dS$$

$$- D_u(t) + k\int_{\Omega_t}\rho\nabla\cdot\mathbf{u}\,dx - \int_{\Omega_t}\rho g\mathbf{u}\cdot\nabla z\,dx. \qquad (4.5.4)$$

By the Gauss (1777-1855) lemma, the Reynolds transport theorem 1.3.2, using (1.3.2), the continuity equation, and the condition $\mathbf{V}\cdot\mathbf{n} = \mathbf{u}\cdot\mathbf{n}$, where $\mathbf{V}$ is the velocity of points at boundary, it holds

$$\int_{\Omega_t}\rho\frac{df}{dt}\,dx = \int_{\Omega_t}\rho(\partial_t f + \mathbf{u}\cdot\nabla f)\,dx = \int_{\Omega_t}\Big(\partial_t(\rho f) + \nabla\cdot(\rho f\mathbf{u})\Big)\,dx =$$

$$\frac{d}{dt}\int_{\Omega_t}\rho f\,dx - \int_{\Gamma_t}\rho f\mathbf{V}\cdot\mathbf{n}\,dS + \int_{\Gamma_t}\rho f\mathbf{u}\cdot\mathbf{n}\,dS = \frac{d}{dt}\int_{\Omega_t}\rho f\,dx. \qquad (4.5.5)$$

Hence we deduce

$$\frac{1}{2}\int_{\Omega_t}\rho\frac{d\mathbf{u}^2}{dt}\,dx = \frac{d}{dt}\int_{\Omega_t}\frac{1}{2}\rho\mathbf{u}^2\,dx. \qquad (4.5.6)$$

Next, since the rest state satisfies $g\nabla z = -k\nabla\ln\rho_b$, we have

$$-\int_{\Omega_t}\rho g\mathbf{u}\cdot\nabla z\,dx = k\int_{\Omega_t}\rho\mathbf{u}\cdot\nabla\ln\rho_b\,dx = \qquad (4.5.7)$$

$$k\int_{\Omega_t}\rho\frac{d\ln\rho_b}{dt}\,dx = \frac{d}{dt}k\int_{\Omega_t}\rho\ln\rho_b\,dx.$$

By continuity equation and by transport theorem, we deduce the following identity

$$k\int_{\Omega_t}\rho\nabla\cdot\mathbf{u}\,dx = -k\int_{\Omega_t}\frac{d\rho}{dt}\,dx = -k\int_{\Omega_t}\rho\frac{d\ln\rho}{dt}\,dx = -\frac{d}{dt}\int_{\Omega_t}k\rho\ln\rho\,dx. \qquad (4.5.8)$$

Moreover, by using the continuity equation and Reynolds transport theorem, it is possible to verify that the following identity holds

$$\int_{\Omega_t} \Big[\rho \nabla \cdot \mathbf{u} - \rho g \mathbf{u} \cdot \nabla z \Big] dx = -\frac{d}{dt} \int_{\Omega_t} k\, \rho \Big(\ln \rho - \ln \rho_b \Big) dx.$$

By virtue of $(4.5.3)_3$, the periodicity in x', and the identity $p_e = k\rho_b(h)$, boundary terms furnish

$$\int_{\Gamma_t} \mathbf{u} \cdot (\mu\, S(\mathbf{u}) - k\rho \mathbf{I}) n\, dS_x = \int_{\Sigma} \mathbf{u} \cdot \tilde{\mathbf{n}} (\kappa \mathcal{H}(\eta) - p_e) dx' \qquad (4.5.9)$$

$$= \kappa \int_{\Sigma} \nabla' \cdot \Big(\frac{\nabla' \eta}{\sqrt{1 + |\nabla' \eta|^2}} \Big) \partial_t \eta\, dx' - k\rho_b(h) \int_{\Sigma} \partial_t \eta\, dx'$$

$$= -\frac{d}{dt} \Big\{ \kappa \int_{\Sigma} (\sqrt{1 + |\nabla' \eta|^2} - 1) dx' - k\rho_b(h) \int_{\Sigma} \eta\, dx' \Big\}.$$

Notice that continuity equation, together with $(4.5.3)_7$ yields

$$\int_{\Omega_t} \sigma\, dx = \int_{\Omega_t} \rho\, dx - \int_{\Omega_t} \rho_b(z)\, dx = \int_{\Omega_b} \rho_b(z)\, dx - \int_{\Omega_t} \rho_b(z)\, dx. \qquad (4.5.10)$$

Employing the expression (4.4.3) for ρ_b, we develop the integrals in (4.5.10) to obtain

$$\int_{\Omega_t} \sigma\, dx = \frac{k}{g} \int_{\Sigma} \Big(\rho_b(\zeta(x', t)) - \rho_b(h) \Big) dx' \neq 0! \qquad (4.5.11)$$

Finally, the energy equation (4.5.4) reduces to

$$\frac{d}{dt} E + D_u(t) = 0. \qquad (4.5.12)$$

where

$$E = \int_{\Omega_t} \Big\{ \rho \frac{1}{2} \mathbf{u}^2 + k\rho(\ln \rho - \ln \rho_b) \Big\} dx + k\rho_b(h) \int_{\Sigma} \eta\, dx'$$

$$+ \kappa \int_{\Sigma} (\sqrt{1 + |\nabla' \eta|^2} - 1) dx'. \qquad (4.5.13)$$

Remark 4.5.1 *In the last term in (4.5.13) we have added the constant $-\kappa|\Sigma|$ that doesn't change the derivative of E. The functional E defined in (4.5.13) represents the difference between the total energies in the two motions, the unsteady and the steady one. Each total energy is given by the sum of kinetic and Helmholtz free, potential, and surface energies, which are the classical energy terms, plus a new potential energy term $k\rho_b(h) \int_{\Omega_t} dx$, which we shall denote by $\mathcal{E}_{p_e}(\Omega_t)$. We observe that this new energy term is given, up to a constant by $E_{p_e} := \mathcal{E}_{p_e}(\Omega_t) - \mathcal{E}_{p_e}(\Omega_b)$.*

Remark 4.5.2 *Notice that, in general, the energy identity* (4.5.12) *doesn't furnish any "a priori" estimate for the solution* $\mathbf{u}$, ρ, ζ, *because it is yet to be proven the equivalence between* E *and some norm of solutions. Such equivalence holds only for suitable external forces.*

4.5.2 Nonlinear Stability

We are now in the position to prove a stability result. Indeed it is enough to prove that the total energy E is a Lyapunov functional.

Here we wish to prove that E is a Lyapunov functional. To this end, we add and subtract in E the quantity $k \int_\Sigma \int_0^\zeta \rho_b dx'$, and we recall that there holds the conservation of total mass $k \int_{\Omega_t} \rho dx = k\,M$. It is easy to prove that E up to a constant is equivalent to the form

$$E = \int_{\Omega_t} \rho \frac{1}{2} \mathbf{u}^2 dx + \int_{\Omega_t} \left\{ k\rho(\ln\rho - \ln\rho_b) - k(\rho - \rho_b) \right\} dx \tag{4.5.14}$$

$$+ kM + \int_\Sigma \left\{ k\rho_b(h)\eta - \kappa \int_0^\zeta \rho_b(z)dz \right\} dx' + \kappa \int_\Sigma \left\{ \left(\sqrt{1 + |\nabla'\eta|^2} - 1 \right) \right\} dx'$$

$$=: E_u + E_\sigma + E_{\rho_b,\eta} + E_\eta.$$

Let us make the following remarks

(I) The term

$$E_u = \int_{\Omega_t} \frac{1}{2} \rho \mathbf{u}^2 dx = E_v - E_{v_b},$$

represents a weighted L^2 norm of perturbation to velocity.

(II) Take the Taylor polynomial of grade 2 for $E_\sigma = E_\rho - E_{\rho_b}$, and the term

$$E_\sigma = k \int_{\Omega_t} \left\{ \rho(\ln\rho - \ln\rho_b) - (\rho - \rho_b) \right\} dx dz = \frac{k}{2} \int_{\Omega_t} \frac{\sigma^2}{\bar\rho} dx,$$

where $\bar\rho$ is a value between ρ and ρ_b. Equation (4.5.2) provides a weighted L^2-norm for the perturbation $\sigma = \rho - \rho_b$ to density ρ_b, cf. (4.5.2).

(III)

$$E_{\rho_b,\eta} := k\rho_b(h) \int_\Sigma \eta dx' + kM - k \int_\Sigma \int_0^\zeta \rho_b(z)dzdx'$$

$$= k \int_\Sigma \int_h^\zeta \rho_b(h) dz dx' + k \int_\Sigma \int_0^h \rho_b(z) dz dx' - k \int_\Sigma \int_0^\zeta \rho_b(z) dz dx'$$

$$= k \int_\Sigma \int_h^\zeta \rho_b(h) dz dx' - k \int_\Sigma \int_h^\zeta \rho_b(z) dz dx' = k \int_\Sigma \int_h^\zeta (\rho_b(h) - \rho_b(z)) dz dx'$$

$$= -k \int_\Sigma \int_h^\zeta \rho_b'(\overline{\zeta})(z - h)\, dz\, dx' = \int_\Sigma \frac{\overline{\wp}}{2} \eta^2 dx',$$

where $\overline{\zeta}$ is a scalar function in Σ with values between h and ζ and

$$\overline{\wp} = g\rho_b(\overline{\zeta}),$$

Hence,

$$E_{\rho_b,\eta} := \int_\Sigma \frac{\overline{\wp}}{2} \eta^2 dx'. \tag{4.5.15}$$

This term is new in literature, it is responsible of the simultaneous variation of density and boundary; cf. [21, 110].

(IV) By the Taylor expansion we have

$$E_{\Gamma_t} - E_{\Gamma_b} = \kappa \int_\Sigma (\sqrt{1 + |\nabla'\eta|^2} - 1) dx' = \kappa \int_\Sigma \frac{|\nabla'\eta|^2}{2\sqrt{1 + |\nabla'\overline{\eta}|^2}^3} dx', \tag{4.5.16}$$

where $\overline{\eta}$ is a value of η between 0 and η.

Finally, by (4.5.15) and (4.5.16) we deduce

$$E_\eta = E_{\Gamma_t} - E_{\Gamma_b} + E_{\rho,\Gamma_t} - E_{\rho_b,\Gamma_b} = \int_\Sigma \frac{\kappa|\nabla'\eta|^2}{2\sqrt{1 + |\nabla'\overline{\eta}|^2}^3} dx' + \int_\Sigma \frac{\overline{\wp}}{2} \eta^2 dx'. \tag{4.5.17}$$

Provided $g > 0$, i.e. the rigid plane is below the layer, E_η furnishes a norm in $W^{1,2}(\Sigma)$ for the perturbation η to the height $\zeta_b = h$.

We may conclude by stating that E equals the L^2 norms of all perturbations, and therefore (4.5.2) is a good Lyapunov functional. Specifically, we have

$$E = \frac{1}{2}\|\sqrt{\rho}\mathbf{u}\|_{L^2(\Omega_t)}^2 + \frac{k}{2}\left\|\frac{\sigma}{\sqrt{\rho}}\right\|_{L^2(\Omega_t)}^2 + \kappa\left\|\frac{\nabla'\eta}{(1 + |\nabla'\overline{\eta}|^2)^{3/4}}\right\|_{L^2(\Sigma)}^2 + \frac{1}{2}\left\|\sqrt{\overline{\wp}}\eta\right\|_{L^2(\Sigma)}^2, \tag{4.5.18}$$

and we call it an *energy norm*. In this way, the stability theorem is completely proven.

Theorem 4.5.2 *Let* $(\mathbf{u}, \sigma, \eta)$ *be solutions to problem* (4.5.3) *in* $\mathcal{W}$, *corresponding to the large data. Then, integrating in time* (4.5.12) *we find that the norm* (4.5.18) *is increased by its value at initial time, and the rest state is nonlinearly stable in the energy norm.*

Remark 4.5.3 *Under the same hypotheses as Theorem 4.5.2, from Korn's inequality proved in cf. [9] Lemma 9.7, again integrating* (4.5.12) *we deduce an estimate for the* L^2*-norm, in space and time, of* $\nabla \mathbf{u}$; *see also* [10, 11, 14].

Remark 4.5.4 *We remark that even in absence of surface tension* $\kappa = 0$, *we have a norm for the perturbation to the height* η *if gravity is present. In this case, the gravity has a stabilizing effect.*

Remark 4.5.5 *Notice that* (4.5.18) *represents a norm if*

$$k > 0, \qquad \kappa > 0, \qquad g > 0.$$

There are physical examples where each of these coefficients may be negative! Of course in this section all coefficients are assumed to have positive sign.

4.5.3 Nonlinear Exponential Stability

We must still prove exponential decay to the rest state for the L^2 norm of perturbations, which will solve the full system (4.5.3), under the action of **large potential forces without smallness conditions on initial data**. To this end, we use again the free work identity.

Theorem 4.5.3 *Let* $(\mathbf{u}, \sigma, \eta)$ *be solutions to problem* (4.5.3) *in* $\mathcal{W}$, *corresponding to the large data. Then the following **modified energy** equation holds*

$$\frac{d}{dt}\mathbb{E} + \mathcal{D} = \nu \mathcal{I}, \tag{4.5.19}$$

where ν *is an arbitrary, positive real parameter, having dimension one over time, and where*

$$\mathbb{E} = E_u + E_\sigma + E_\eta - \nu(\rho \mathbf{u}, \mathbf{V}),$$

*is the **modified energy functional**, and*

$$\mathcal{D} = D_u + \nu\, D_\sigma + \nu\, D_\eta,$$

denotes the sum of dissipation terms given by

$$D_\sigma = k\left\|\frac{\sigma}{\sqrt{\rho_b}}\right\|^2_{L^2(\Omega_t)},$$

$$D_\eta = \int_\Sigma \left[k\Big(\rho_b(\zeta) - \rho_b(h)\Big) V(\zeta) - \kappa \frac{|\nabla'\eta|^2}{\sqrt{1+|\nabla'\eta|^2}} V'(\zeta) \right] dx',$$

with

$$\mathbf{V}(\zeta) = \frac{k(\rho_b(\zeta) - \rho_b(h))}{g\rho_b(\zeta)} \mathbf{n} =: V(\zeta)\mathbf{n}, ,$$

$$V'(\zeta) = -\exp\left(\frac{g}{k}(\zeta - h)\right),$$

$$\mathcal{I} = \int_{\Omega_t} \left[\Big(\partial_t \mathbf{V}, \rho\mathbf{u}\Big) + \Big(\rho\mathbf{u}\cdot\nabla\mathbf{V}, \mathbf{u}\Big) + \mu S(\mathbf{u})\cdot\nabla\mathbf{V}\right] dx,$$

and $\mathbf{V}$ is the vector field defined in Lemma 4.8.2 of the Appendix.

Remark 4.5.6 It is worth noting that D_η defines a norm in $W^{1,2}(\Omega_t) \times L^2(\Omega_t) \times W^{1,2}(\Sigma)$ because V' is negative. However, the modified energy function $\mathbb{E}$ is not always positive definite, actually it depends on the value of ν.

Proof. We observe that it holds

$$-k\nabla\rho - \rho g\nabla z = -k\left(\frac{\rho_0\nabla\rho}{\rho_b} - \frac{\rho}{\rho_b}\nabla\rho_b\right) = -k\rho_b\nabla\left(\frac{\rho}{\rho_b}\right),$$

thus we may rewrite $(4.5.3)_2$ in the equivalent form

$$-k\rho_b\nabla\left(\frac{\sigma}{\rho_b}\right) = \rho\frac{d\mathbf{u}}{dt} - \mu\nabla\cdot S(\mathbf{u}). \tag{4.5.20}$$

Remember that the auxiliary function $\mathbf{V}$, given in Lemma 4.8.2 of the Appendix, satisfies the conditions

$$\mathbf{V}\cdot\mathbf{n}\Big|_{\Gamma_t} = V(\zeta)n_z := \frac{k(\rho_b(\zeta) - \rho_b(h))}{g\rho_b(\zeta)}n_z.$$

Let us multiply (4.5.20) by $\mathbf{V}$ and integrate over Ω_t; integrating by parts we obtain the Free Work Equation

$$-k\int_{\Gamma_t} \sigma V(\zeta)\cdot\mathbf{n}dS + k\int_{\Omega_t} \frac{\sigma}{\rho_b}\nabla\cdot(\rho_b\mathbf{V})dx \tag{4.5.21}$$

$$= \int_{\Omega_t} \rho\frac{d\mathbf{u}}{dt}\cdot\mathbf{V}\,dx - \int_{\Omega_t} \mu\nabla\cdot S(\mathbf{u})\cdot\mathbf{V}dx.$$

We analyze each term separately. The Reynolds transport theorem yields

$$\int_{\Omega_t} \rho\frac{d\mathbf{u}}{dt}\cdot\mathbf{V}\,dx = \frac{d}{dt}\int_{\Omega_t} \rho\mathbf{u}\cdot\mathbf{V}\,dx - \int_{\Omega_t} \rho\mathbf{u}\cdot\Big(\partial_t\mathbf{V} + \mathbf{u}\cdot\nabla\mathbf{V}\Big)dx. \tag{4.5.22}$$

Due to the boundary conditions $\mathbf{n} \cdot S(\mathbf{u}) \cdot \boldsymbol{\tau} = 0$, where $\boldsymbol{\tau}$ is an arbitrary tangent vector it holds

$$-\int_{\Omega_t} \mu \nabla \cdot S(\mathbf{u}) \cdot \mathbf{V} dx = -\int_{\Gamma_t} \mathbf{V}(\zeta) \cdot S(\mathbf{u}) \cdot \mathbf{n} dS + \mu \int_{\Omega_t} S(\mathbf{u}) \cdot \nabla \mathbf{V} dx$$

$$\text{(4.5.23)}$$

$$= -\int_{\Gamma_t} \mathbf{V}(\zeta) \cdot \mathbf{nn} \cdot S(\mathbf{u}) \cdot \mathbf{n} dS + \mu \int_{\Omega_t} S(\mathbf{u}) \cdot \nabla \mathbf{V} dx.$$

Substituting these identities in (4.5.21), we get

$$k \int_{\Omega_t} \frac{\sigma}{\rho_b} \nabla \cdot (\rho_b \mathbf{V}) dx = \frac{d}{dt} \int_{\Omega_t} \rho \mathbf{u} \cdot \mathbf{V} \, dx - \int_{\Omega_t} \rho \mathbf{u} \cdot \left(\partial_t \mathbf{V} + \mathbf{u} \cdot \nabla \mathbf{V} \right) dx$$

$$\text{(4.5.24)}$$

$$+ \mu \int_{\Omega_t} S(\mathbf{u}) \cdot \nabla \mathbf{V} dx - \int_{\Gamma_t} \left[\mu \mathbf{n} \cdot S(\mathbf{u}) \cdot \mathbf{n} - k\sigma \right] V(\zeta) n_z dS.$$

Furthermore, the boundary conditions yield

$$-k\rho(\zeta) + \mu \mathbf{n} \cdot S(u)\mathbf{n} = \kappa \mathcal{H}(\eta) - p_e,$$

and recalling that

$$-k\rho_b(h) = -p_e,$$

we deduce

$$k\left(-\rho(\zeta) + \rho_b(\zeta) \right) + \mu \mathbf{n} \cdot S(u)\mathbf{n} = \kappa \mathcal{H}(\eta) + k(\rho_b(\zeta) - \rho_b(h)),$$

$$k\sigma(\zeta) - \mu \mathbf{n} \cdot S(u)\mathbf{n} = -\kappa \mathcal{H}(\eta) - k(\rho_b(\zeta) - \rho_b(h)). \qquad \text{(4.5.25)}$$

Multiply (4.5.25) by $V(\zeta)$ and integrate over Σ, recalling that

$$\mathbf{V} \cdot \mathbf{n} = n_z V(\zeta), \qquad V(\zeta) = \frac{k(\rho_b(\zeta) - \rho_b(h))}{g\rho_b(\zeta)},$$

we deduce

$$\int_{\Sigma} \left[k\sigma(\zeta) - \mu \mathbf{n} \cdot S(\mathbf{u}) \cdot \mathbf{n} \right] V(\zeta) dx' = \qquad \text{(4.5.26)}$$

$$-\kappa \int_{\Sigma} \nabla' \cdot \left(\frac{\nabla' \eta}{\sqrt{1 + |\nabla' \eta|^2}} \right) V(\zeta) dx' - k \int_{\Sigma} \left(\rho_b(\zeta) - \rho_b(h) \right) V(\zeta) dx',$$

and we have

$$\int_{\Sigma}\Big[k\sigma - \mathbf{n}\cdot S(\mathbf{u})\cdot \mathbf{n}\Big]V(\zeta)dx' = \tag{4.5.27}$$

$$-\kappa\int_{\Sigma}\nabla'\cdot\Big(\frac{\nabla'\eta}{\sqrt{1+|\nabla'\eta|^2}}\Big)V(\zeta)dx' - k^2\int_{\Sigma}\frac{(\rho_b(\zeta)-\rho_b(h))^2}{g\rho_b(\zeta)}dx'.$$

Therefore, after integration by parts we arrive at

$$\int_{\Gamma_t}\Big[k\sigma V(\zeta) - \mathbf{V}(\zeta)\cdot S(\mathbf{u})\cdot\mathbf{n}\Big]dS = \tag{4.5.28}$$

$$+\int_{\Sigma}\Big\{\kappa\frac{|\nabla'\eta|^2}{\sqrt{1+|\nabla'\eta|^2}}V'(\zeta) - \frac{k^2}{g\rho_b(\zeta)}\Big(\rho_b(\zeta)-\rho_b(h)\Big)^2\Big\}dx'.$$

Substituting (4.5.28) into (4.5.21) we get the Free Work Equation

$$-\frac{d}{dt}\int_{\Omega_t}\rho\mathbf{u}\cdot\mathbf{V}\,dx + \int_{\Sigma}\Big\{-\kappa\frac{|\nabla'\eta|^2}{\sqrt{1+|\nabla'\eta|^2}}\Big)V'(\zeta) + \frac{k^2}{g\rho_b(\zeta)}\Big(\rho_b(\zeta)-\rho_b(h)\Big)^2\Big\}dx'$$

$$\tag{4.5.29}$$

$$+k\int_{\Omega_t}\frac{\sigma^2}{\rho_b}\,dx' = \mathcal{I},$$

with $\mathcal{F}$ defined in Theorem 4.5.3. We add (4.5.29) multiplied by a constant ν to (4.5.2) and obtain the desired Modified Energy Equation (4.5.19).

We are in the position to prove the stability result. □

Theorem 4.5.4 *Assume there exist solutions to* (4.5.3) $(\mathbf{u}(x',z,t),\rho(x',z,t),$ $\zeta(x',t))$ *in the regularity class* $\mathcal{W}$, *corresponding to initial data*

$$\mathbf{u}_0(x',z), \quad \rho_0(x',z) = \rho_b(x',z) + \sigma_0(x',z), \quad \zeta_0(x') = h + \eta_0(x').$$

Then, for any data $(\mathbf{u}_0,\rho_0,\eta_0)$ *in* $L^2_\sharp(\Omega_0)\times C^0_\sharp(\Omega_0)\times W^{1,\infty}_\sharp(\Sigma)$, *the rest state* S_b *is nonlinearly exponentially stable in the energy norm.*

Proof. The proof of Theorem 4.5.4 will be achieved once one proves

$$\frac{d}{dt}\mathbb{E}(t) + \delta\mathbb{E}(t) \le 0, \tag{4.5.30}$$

with $\mathbb{E}$, the *modified energy* given in Theorem 4.5.3, and δ a positive constant.

Let m, d, and d_1 denote the essential supremum in space and time of ρ, ζ and $\nabla\eta$, respectively.

We begin with a very simple equation (4.5.19), therefore we shall prove the theorem using only two steps:

(1) $\mathbb{E}$ is equivalent to a norm of perturbation
(2) there exists a δ such that $\mathcal{D} - \nu\mathcal{I} \ge \delta\mathbb{E}$

(1) Notice that, by using Schwartz inequality, Lemma 4.8.2 of the appendix and the Poincaré inequality, we have

$$|(\rho\mathbf{u}, \mathbf{V})| \leq \sqrt{m}\, \|\sqrt{\rho}\mathbf{u}\|_{L^2(\Omega_t)} \|\mathbf{V}\|_{L^2(\Omega_t)} \leq b_0 \|\nabla\mathbf{u}\|_{L^2(\Omega_t)} \Big(\|\sigma\|_{L^2(\Omega_t)} + \|\nabla'\eta\|_{L^2(\Sigma)}\Big),$$
$$(4.5.31)$$

with

$$b_0 = p_2 \sqrt{m}\, d,$$

d the upper bound of ζ, and p_2 given in (4.8.3) of Lemma 4.8.2 in the Appendix. Hence it holds

$$\mathbb{E} \geq \frac{1}{2}\|\sqrt{\rho}\,\mathbf{u}\|^2_{L^2(\Omega_t)} + \frac{k}{2\sqrt{m}}\|\sigma\|^2_{L^2(\Omega_t)} + \frac{C_0}{2}\Big(\|\nabla'\eta\|^2_{L^2(\Sigma)} + \|\eta\|^2_{L^2(\Sigma)}\Big)$$
$$(4.5.32)$$

$$- \nu b_0 \|\sqrt{\rho}\,\mathbf{u}\|_{L^2(\Omega_t)} \Big(\|\sigma\|_{L^2(\Omega_t)} + \|\nabla'\eta\|_{L^2(\Sigma)}\Big),$$

with

$$C_0 = \min\Big\{\frac{\kappa}{\sqrt{1+d_1^2}}, \frac{k}{g}\exp\Big(-\frac{2gh}{k}\Big)\Big\}.$$

(2) By the Schwartz, Korn, and Poincaré inequalities, and by Lemma 4.9.2, it follows

$$\Big(\partial_t \mathbf{V}, \rho\mathbf{u}\Big) + \Big(\rho\mathbf{u}\cdot\nabla\mathbf{V}, \mathbf{u}\Big) \leq m\|\mathbf{u}\|^2_{L^4(\Omega_t)}\|\nabla\mathbf{V}\|_{L^2(\Omega_t)} + \|\mathbf{u}\|_{L^2(\Omega_t)}\|\partial_t\mathbf{V}\|_{L^2(\Omega_t)}$$
$$(4.5.33)$$

$$\leq b_1 \|S(\mathbf{u})\|^2_{L^2(\Omega_t)}(\|\sigma\|_{L^2(\Omega_t)} + \|\nabla'\eta\|_{L^2(\Sigma)}) + b_2\|S(\mathbf{u})\|^2_{L^2(\Omega_t)},$$

where

$$b_1 = m\, d^2 p_1\, c_k,$$
$$b_2 = mC_P^2 p_1\, c_k^2 \|\sigma\|_{L^3(\Omega_t)},$$

with c_k and C_P Korn and Poincaré constants, respectively.

Remark 4.5.7 Notice that in general the Poincaré constant is a function of Ω_t and therefore also of time. However, by the hypothesis that we work with regular functions in the class $\mathcal{W}$, the Poincaré constant becomes less than a constant in time.

Furthermore, it holds

$$\nu\mu\Big(S(\mathbf{u}), \nabla\mathbf{V}\Big) \leq p_2\nu\mu\|S(\mathbf{u})\|_{L^2(\Omega_t)}\Big(\|\sigma\|_{L^2(\Omega_t)} + \|\eta\|_{W^{1,2}(\Sigma)}\Big). \quad (4.5.34)$$

We observe that from the definition of dissipation in Theorem 4.5.3 we deduce

$$\mathcal{D} \geq k\nu \exp\left(-\frac{4gh}{k}\right) \int_{\Sigma} \eta^2 \, dx' + \frac{\kappa\nu}{\sqrt{1+d_1^2}} \exp\left(-\frac{2gh}{k}\right) \int_{\Sigma} |\nabla'\eta|^2 dx'$$

$$(4.5.35)$$

$$+ \mu\|S(\mathbf{u})\|_{L^2(\Omega_t)}^2 + \frac{k\nu}{\sqrt{m}}\|\sigma\|_{L^2(\Omega_t)}^2$$

$$\geq \mu\|S(\mathbf{u})\|_{L^2(\Omega_t)}^2 + \frac{k\nu}{\sqrt{m}}\|\sigma\|_{L^2(\Omega_t)}^2 + d_b\nu\|\eta\|_{W^{1,2}(\Sigma)}^2,$$

with

$$d_b = \min\left\{k\exp\left(-\frac{4gh}{k}\right), \frac{\kappa}{\sqrt{1+d_1^2}}\exp\left(-\frac{2gh}{k}\right)\right\}.$$

Therefore, inequalities (4.5.31), (4.5.33) and (4.5.35) yield

$$\mathcal{D} - \nu\mathcal{I} \geq (\mu - \nu b_2)\|S(\mathbf{u})\|_{L^2(\Omega_t)}^2 + \nu\left(\frac{k}{\sqrt{m}} - 3p_2\frac{\lambda}{4}\right)\|\sigma\|_{L^2(\Omega_t)}^2 \quad (4.5.36)$$

$$+ \nu\left(d_b - 3p_2\frac{\lambda}{4}\right)\|\eta\|_{W^{1,2}(\Sigma)}^2 - \nu\left(b_1 + 2p_2\mu\right)\|S(\mathbf{u})\|_{L^2(\Omega_t)}$$

$$\times \left(\|\sigma\|_{L^2(\Omega_t)} + \|\zeta\|_{W^{1,2}(\Sigma)}\right)$$

$$\geq \frac{\mu}{2}\|S(\mathbf{u})\|_{L^2(\Omega_t)}^2 + \frac{k\nu}{2\sqrt{m}}\|\sigma\|_{L^2(\Omega_t)}^2 + \frac{\nu d_b}{2}\|\eta\|_{W^{1,2}(\Sigma)}^2,$$

Finally, $\mathcal{D} - \nu\mathcal{I}$ is a positive definite form if

$$\nu \leq \min\left\{\frac{\mu}{2b_2}, \frac{k}{3p_2\lambda m^{3/2}}, \frac{2k\mu}{m^{3/2}(b_1 + 2p_2\mu)^2}, \frac{d_b}{3p_2\lambda}, \frac{2d_b\mu}{(b_1 + 2p_2\mu)^2}\right\}.$$

Since ρ is uniformly bounded in space and time, the definition of E_σ implies

$$E_\sigma \geq \frac{k}{2m_1}\|\sigma\|_{L^2(\Omega_t)}^2, \quad (4.5.37)$$

where m_1 is the maximum of ρ.
We develop the energy term E_η as a function of η at point zero to get

$$E_\eta = \kappa \int_{\Sigma} \left(\sqrt{1+|\nabla'\eta|^2} - 1\right) dx' + k\rho_b \exp\left(-\frac{gh}{k}\right) \quad (4.5.38)$$

$$\int_{\Sigma} \left\{\eta + \frac{k}{g}\left[\exp\left(-\frac{g\eta}{k}\right) - 1\right]\right\} dx'$$

$$\geq \frac{\kappa}{2\sqrt{1+d_1^2}} \int_{\Sigma} |\nabla'\eta|^2 dx' + \frac{k\rho_b}{g} \int_{\Sigma} \eta^2 \, dx'.$$

Hence, there exists a positive number

$$s_0 = \max\{\frac{\kappa}{2\sqrt{(1+d_1^2)}}, \frac{k\rho_b}{g}\} \qquad (4.5.39)$$

such that

$$E_\eta \geq s_0 \|\eta\|_{W^{1,2}(\Sigma)}^2.$$

Then (4.5.32) yields

$$\mathbb{E} \geq \frac{m}{2}\|\mathbf{u}\|_{L^2(\Omega_t)}^2 + \frac{k}{2m_1}\|\sigma\|_{L^2(\Omega_t)}^2 + s_0\|\eta\|_{W^{1,2}(\Sigma)}^2 \qquad (4.5.40)$$

$$+ b_1\nu\|\mathbf{u}\|_{L^2(\Omega_t)}\Big(\|\sigma\|_{L^2(\Omega_t)} + \|\eta\|_{W^{1,2}(\Sigma)}\Big)$$

$$\geq m\|\mathbf{u}\|_{L^2(\Omega_t)}^2 + \frac{k}{m_1}\|\sigma\|_{L^2(\Omega_t)}^2 + s_0\|\eta\|_{W^{1,2}(\Sigma)}^2,$$

provided that

$$b_1\nu < \max\left\{m, \frac{k}{m_1}\right\}.$$

We recall that ν is an arbitrary auxiliary parameter, so for ν small enough, the above condition is always satisfied. From (4.5.36) we can finally conclude that

$$\mathbb{E} \leq \frac{1}{\delta}\Big(\mathcal{D} - \nu\mathcal{I}\Big), \qquad (4.5.41)$$

with

$$4\delta = \max\left\{\frac{m\,c_k}{\mu}, \frac{1}{m\nu}, \frac{s_0}{d_b\nu}\right\}.$$

and step two of Theorem 4.5.3 is completely proved.

Equation (4.5.30) is called the *modified energy inequality*, and $\mathbb{E}$ represents the natural Lyapunov function because $\mathbb{E}$ is equivalent to the L^2-norms of the perturbation $\mathbf{u}$, σ, η, only for regular solutions.

Using Gronwall's lemma in (4.5.30), we prove that the perturbation

$$\mathbf{u}, \qquad \sigma = \rho - \rho_b, \qquad \eta = \zeta - \zeta_b$$

decays to zero at an exponential rate, and theorem is proved.   □

Remark 4.5.8 *Notice that ν depends on ρ_b, k, μ, λ, and Ω_0. Indeed, our theorem demonstrates the exponential decay of perturbation to zero, however the decay rate is not optimal.*

Remark 4.5.9 *We leave as an open problem the proof of uniqueness of the rest state under general large potential forces. It would be interesting to consider self gravitating forces in domains, either bounded or exterior to B_0, where B_0 may be or not the empty set. In case of self gravitating forces* *there is the possibility for fixing the zero boundary condition on density, this represents the case of a star atmosphere; cf. [74].*

Remark 4.5.10 *We conjecture that all proof and considerations of this section continue to hold in the case of an infinite layer. We leave it as an open problem.*

4.6 Loss of Stability

This section focuses on the nonlinear aspect of the physical world. We set our attention to one of the central problems of fluid mechanics: when and how do laminar flows break down?

On one side, we suppose the fluid at rest and that by increasing a given non-dimensional number G the fluid loses tension at its free surface. We wish to study the phenomenon of nonlinear stability or instability for **initial data sufficiently small**, say in B_δ by varying G. We expect that the radius δ goes to zero as G increases toward G_L, furthermore we have already observed that initial data become no more controllable if δ is 'too small'.

On the other side, we suppose the fluid at rest and we wish to study the phenomenon of loss of initial data control for **sufficiently large initial data**.

Summarizing the above problems, it may be interesting to have information directly about the control or loss of control of a solution in terms of its initial data.

4.6.1 *Energy Inequality*

Because we aim for an instability result, we begin by choosing suitable external forces. In Sect. 4.4 we have proved that, in the parallelepiped $\Omega_b = \Sigma \times (0, h)$. There exists only the rest state

$$S_b = \left\{ \mathbf{u}_b = 0, \quad \rho_b = \rho_* \exp\left(\frac{gz}{k}\right), \quad \zeta_b = h = \frac{k}{g} \ln\left(1 - \frac{M}{p_e |\Sigma|}\right) \right\}, \quad (4.6.1)$$

with ρ_* given by

$$\rho_* = \frac{Mg}{k|\Sigma|\left(\exp\left(\frac{gh}{k}\right) - 1\right)}. \quad (4.6.2)$$

Recalling the definition of $\overline{\wp}$ in (4.5.2) we set

$$\wp \|\eta\|_{L^2(\Sigma)}^2 := g\rho_b(\overline{\zeta}(\widehat{x})) \int_\Sigma \eta^2 dx := \int_\Sigma g\rho_b(\overline{\zeta}(x))\eta^2 dx. \quad (4.6.3)$$

Assume there exists a regular, global motion in time $\mathbf{u}(x,t)$, $\rho = \rho_b(z) + \sigma(x,t)$, $\zeta = h + \eta(x,t)$ for any given initial data in a neighborhood of the basic flow. In this subsection, we prove the following nonlinear instability result, B_1 is defined in (4.2.25).

Theorem 4.6.1 *Assume that the form $\int_\Sigma \eta(y)B_1\eta(y)dS$ can take negative values for some η. Moreover, let $\mathrm{Ker}B_1 = \emptyset$. Then there exist arbitrarily small initial values $(\mathbf{u}_0, \sigma_0, \eta_0)$, and a value $\epsilon > 0$ such that the solution of (4.5.3) leaves, sooner or later, a certain neighborhood of zero, i.e. there exists a certain $t > 0$ such that the following inequality holds true*

$$\frac{1}{2}\|\sqrt{\rho}\mathbf{u}\|^2_{L^2(\Omega_t)} + \frac{k}{2}\left\|\frac{\sigma}{\sqrt{\rho}}\right\|^2_{L^2(\Omega_t)} + \kappa\left\|\frac{\nabla'\eta}{(1+|\nabla'\bar\eta|^2)^{3/4}}\right\|^2_{L^2(\Sigma)} + \frac{\wp}{2}\|\eta\|^2_{L^2(\Sigma)} \geq \epsilon > 0.$$

$$(4.6.4)$$

Proof. We follow the lines of [68,107,115] proposed for incompressible fluids.

The proof is achieved by contradiction; we assume that problem FBP has a solution defined for $t \geq 0$ that satisfies the property:

For all $\varepsilon > 0$ there exists a $\epsilon_0 > 0$, such that if $\|(\mathbf{u}_0, \sigma_0, \eta_0)\|_W \leq \epsilon_0$ then

$$\|\eta(t)\|^2_{C^1(\Sigma)} \leq \varepsilon, \qquad \forall t > 0. \qquad (4.6.5)$$

We aim to show that this infers the contradiction that given a small but fixed ε, there exists always a special (arbitrarily small) initial data such that for a particular time T (large) it holds

$$\|\eta(T)\|^2_{C^1(\Sigma)} > \varepsilon.$$

We remark that Theorem 4.5.2, and point (IV) of the previous subsection still hold. By noticing that our instability hypothesis requires a change of sign in the gravity, thus developing the volume integrals, keeping in mind we write

$$E_{\rho,\Gamma_t} - E_{\rho_b,\Gamma_b} = k\rho_b(h)\int_\Sigma \eta dx + kM - k\int_\Sigma\int_0^\zeta \rho_b(z)dzdx = -\frac{\wp}{2}\int_\Sigma \eta^2\,dx',$$

$$(4.6.6)$$

with $\wp > 0$ given in (4.6.3). Next, the Taylor polynomial of grade 2, at $\eta = 0$, yields

$$\kappa\int_\Sigma(\sqrt{1+|\nabla'\eta|^2} - 1)dx' = \frac{\kappa}{2}\int_\Sigma \frac{|\nabla'\eta|^2}{(1+|\nabla'\bar\eta|^2)^{3/2}}dx'. \qquad (4.6.7)$$

Thus the function

$$G[\eta] = \int_\Sigma \left[\kappa \left(\sqrt{1 + |\nabla' \eta|^2} - 1 \right) - \frac{\wp}{2} \eta^2 \right] dx', \qquad (4.6.8)$$

with $\wp$ defined in (4.6.3), is equivalent to the quadratic form in η of no definite sign

$$G[\eta] = \int_\Sigma \left[\kappa \frac{|\nabla' \eta|^2}{(1 + |\nabla' \bar\eta|^2)^{3/2}} - \frac{\wp}{2} \eta^2 \right] dx'.$$

The functional G defines the following nonlinear function

$$G[\eta] = -\int_\Sigma \left[\kappa \nabla' \cdot \left(\frac{\nabla' \eta}{(1 + |\nabla' \bar\eta|^2)^{3/2}} \right) + \frac{\wp}{2} \eta \right] \eta \, dx' =: \Big(\eta, B_1(\eta) \Big). \quad (4.6.9)$$

Under our assumptions the rigid plane is above, and the free surface is below, thus the total energy is not positive definite, because of the negative term in the L^2 norm of η. Therefore, the energy identity will not allow the control of the norm of perturbations. In order to once again deduce information from the energy identity, it is useful to change sign in the energy identity (4.5.2) to write it as follows

$$\frac{d}{dt} K(t) = \mu \|S(\mathbf{u})\|_{L^2(\Omega_t)}^2, \qquad (4.6.10)$$

$$K(t) := \left[\frac{\wp}{2} \|\eta\|_{L^2(\Sigma)}^2 - \kappa \left\| \frac{\nabla' \eta}{(1 + |\nabla' \bar\eta|^2)^{3/4}} \right\|_{L^2(\Sigma)}^2 - \frac{1}{2} \|\sqrt{\rho} \mathbf{u}\|_{L^2(\Omega_t)}^2 - \frac{k}{2} \left\| \frac{\sigma}{\sqrt{\rho}} \right\|_{L^2(\Omega_t)}^2 \right],$$

or equivalently

$$\frac{d}{dt} \left[\wp \|\eta\|_{L^2(\Sigma)}^2 - 2\kappa \|\nabla' \eta\|_{L^2(\Sigma)}^2 + 2\kappa (\mathcal{H}_{nl} \nabla' \eta, \nabla' \eta)_\Sigma - \|\sqrt{\rho} \mathbf{u}\|_{L^2(\Omega_t)}^2 - k \left\| \frac{\sigma}{\sqrt{\rho}} \right\|_{L^2(\Omega_t)}^2 \right]$$

$$= \mu \|S(\mathbf{u})\|_{L^2(\Omega_t)}^2. \qquad (4.6.11)$$

holds, where $\mathcal{H}_{nl}$ is the nonlinear part of $\mathcal{H}$ given by

$$(\mathcal{H}_{nl} \nabla' \eta, \nabla' \eta)_\Sigma = \int_\Sigma \frac{\sqrt{1 + |\nabla' \eta|^2}}{\sqrt{1 + |\nabla' \eta|^2} + 1} |\nabla' \eta|^2 \, dx'. \qquad (4.6.12)$$

$\square$

4.6.2 Equivalence of Norms

We assume that for every $\epsilon > 0$ there exists a $\delta > 0$ such that for all initial data less than δ the corresponding solution of (4.3.4) defined for all $t > 0$ satisfies

$$\|(\mathbf{u}, \sigma, \eta)\|_X^2(t) = \|\sqrt{\rho} \mathbf{u}\|_{L^2(\Omega_t)}^2 + k \left\| \frac{\sigma}{\sqrt{\rho}} \right\|_{L^2(\Omega_t)}^2 + \|\eta\|_{W^{1,2}(\Sigma)}^2 < \epsilon, \qquad \forall t > 0. \qquad (4.6.13)$$

Set

$$K_\eta(t) = \wp\|\eta\|^2_{L^2(\Sigma)}(t) - 2\kappa D_\eta(t). \qquad (4.6.14)$$

Integrating (4.6.11) in $(0, t)$ yields

$$K_\eta(t) = \|\sqrt{\rho}\mathbf{u}\|^2_{L^2(\Omega_t)} + k\left\|\frac{\sigma}{\sqrt{\rho}}\right\|^2_{L^2(\Omega_t)} + 2\mu\int_0^t \|S(\mathbf{u})\|^2_{L^2(\Sigma)}d\tau + K_0, \quad (4.6.15)$$

where $K_0 = K(0)$ is given by $(4.6.10)_2$:

$$K_0 = \wp\|\eta_0\|^2_{L^2(\Sigma)} - 2\kappa\int_\Sigma\left(\sqrt{1 + |\nabla'\eta_0|^2} - 1\right)dx' - \|\sqrt{\rho_0}\mathbf{u}_0\|^2_{L^2(\Omega_t)} - k\left\|\frac{\sigma_0}{\sqrt{\rho_0}}\right\|^2_{L^2(\Omega_t)},$$

which is positive by the assumption. We take $\mathbf{u}_0 = 0$ thus $K_\eta(0) = K_0$. Since η satisfies the average zero condition $(4.5.3)_7$, the Poincaré inequality

$$\|\eta\|^2_{L^2(\Sigma)} \le c_P\|\nabla'\eta\|^2_{L^2(\Sigma)} \qquad (4.6.16)$$

holds (cf. (4.2.1)). Furthermore it holds the inequality

$$\sqrt{1 + p^2} \le 1 + |p|. \qquad (4.6.17)$$

From the identity

$$D_\eta(t) = \int_\Sigma \frac{\left[\sqrt{1 + |\nabla'\eta|^2} - 1\right]\left[\sqrt{1 + |\nabla'\eta|^2} + 1\right]}{1 + \sqrt{1 + |\nabla'\eta|^2}}dx' = \int_\Sigma \frac{|\nabla'\eta|^2}{1 + \sqrt{1 + |\nabla'\eta|^2}}dx',$$

$$(4.6.18)$$

(4.6.16), and the hypothesis (4.6.13), (4.6.17) with $p = |\nabla'\eta| < \epsilon$ we obtain

$$\|\eta\|^2_{L^2(\Sigma)} \le c_P\|\nabla'\eta\|^2_{L^2(\Sigma)} \qquad (4.6.19)$$

$$= c_P\int_\Sigma \frac{|\nabla'\eta|^2}{1 + \sqrt{1 + |\nabla'\eta|^2}}[1 + \sqrt{1 + |\nabla'\eta|^2}]dx' \le c_P(2 + \epsilon)D_\eta.$$

Furthermore, from (4.6.15), and (4.6.14) we have

$$D_\eta(t) < \frac{\wp}{2\kappa}\|\eta\|^2_{L^2(\Sigma)}(t). \qquad (4.6.20)$$

Inequalities (4.6.19) and (4.6.20) imply

$$\frac{1}{c_P(2 + \epsilon)}\|\eta\|^2_{L^2(\Sigma)}(t) \le D_\eta(t) < \frac{\wp}{2\kappa}\|\eta\|^2_{L^2(\Sigma)}(t), \qquad (4.6.21)$$

namely, they imply the equivalence between the two norms of η, one in $L^2(\Sigma)$ and one weighted in $W^{1,2}(\Sigma)$. Therefore we claim that for any t there exists a number $l(t)$, function of t only, such that

$$l(t)\|\eta\|^2_{L^2(\Sigma)}(t) = D_\eta(t), \qquad \frac{1}{c_P(2+\epsilon)} < l(t) < \frac{\wp}{2\kappa}. \qquad (4.6.22)$$

We note that $\wp > \kappa$ from the assumption (4.6.13). Combining (4.6.14) and (4.6.22), we obtain

$$K_\eta(t) = (\wp - 2\kappa l(t))\|\eta\|^2_{L^2(\Sigma)}(t). \qquad (4.6.23)$$

Substituting (4.6.23) into (4.6.15), and using (4.6.13), we obtain

$$\wp - 2\kappa l(t) = \frac{K_\eta(t)}{\|\eta\|^2_{L^2(\Sigma)}(t)} > \frac{K_0}{C\epsilon}. \qquad (4.6.24)$$

4.6.3 Instability

To prove instability we use the free work equation. To this end we multiply $(4.5.3)_2$ by $\mathbf{V}$, constructed in Lemma 4.8.2 where we set $\nabla \cdot (\rho_b \mathbf{V}) = -\sigma$, and integrate over Ω_t. This gives

$$\frac{d}{dt}(\rho\mathbf{u}, \mathbf{V})_{\Omega_t} + \int_{\Gamma_t} (-\kappa\mathcal{H}(\eta) - \wp\eta)\mathbf{V} \cdot \mathbf{n}\, dS = +k \int_{\Omega_t} \frac{\sigma}{\rho_b} \nabla \cdot (\rho_b \mathbf{V}) dx = \mathcal{I}$$

$$(4.6.25)$$

where

$$\mathcal{I} = (\rho\mathbf{u}, \partial_t\mathbf{V})_{\Omega_t} + ((\rho\mathbf{u} \cdot \nabla)\mathbf{V}, \mathbf{u})_{\Omega_t} - \mu(S(\mathbf{u}), \nabla\mathbf{V})_{\Omega_t}. \qquad (4.6.26)$$

Integrating by parts in the surface integral in (4.6.25), and recalling the boundary value of $\mathbf{V}$ it yields

$$\frac{d}{dt}(\rho\mathbf{u}, \mathbf{V})_{\Omega_t} + \int_\Sigma \left(\frac{\kappa}{\sqrt{1+|\nabla'\eta|^2}}|\nabla'\eta|^2 - \wp|\eta|^2 \right) dx' - k \int_{\Omega_t} \frac{\sigma^2}{\rho_b} dx = \mathcal{I}.$$

$$(4.6.27)$$

It holds that

$$\int_\Sigma \frac{\kappa}{\sqrt{1+|\nabla'\eta|^2}}|\nabla'\eta|^2 dx' < \int_\Sigma \frac{2\kappa}{1+\sqrt{1+|\nabla'\eta|^2}}|\nabla'\eta|^2 dx' = 2\,\kappa D_\eta(t),$$

and therefore

$$\frac{d}{dt}(\rho\mathbf{u}, \mathbf{V})_{\Omega_t} \geq \wp\|\eta\|^2_{L^2(\Sigma)}(t) - 2\kappa D_\eta(t) + k \int_{\Omega_t} \frac{\sigma^2}{\rho_b} dx + \mathcal{I}$$

$$= K_\eta(t) + k \int_{\Omega_t} \frac{\sigma^2}{\rho_b} dx + \mathcal{I}. \qquad (4.6.28)$$

Let us multiply by an arbitrary positive number ν (4.6.28) and add it to (4.6.11) to have

$$\frac{d}{dt}z(t) \geq \mu\|S(\mathbf{u})\|^2_{L^2(\Omega_t)}(t) + \nu K_\eta(t) + \nu k \int_{\Omega_t} \frac{\sigma^2}{\rho_b} dx + \nu\mathcal{I}, \qquad (4.6.29)$$

where

$$z(t) = K_\eta(t) - \frac{1}{2}\|\sqrt{\rho}\mathbf{u}\|^2_{L^2(\Omega_t)} - \frac{k}{2}\left\|\frac{\sigma}{\sqrt{\rho}}\right\|^2_{L^2(\Omega_t)} + \nu(\rho\mathbf{u}, \mathbf{V})_{\Omega_t}(t). \quad (4.6.30)$$

Now we estimate $\mathcal{F}(\mathbf{u}, \mathbf{V})$. Since $\mathbf{u} = 0$ on Γ_B, the Poincaré inequality

$$\|\mathbf{u}\|_{L^2(\Omega_t)} \leq c_P\|\nabla\mathbf{u}\|_{L^2(\Omega_t)} \qquad (4.6.31)$$

holds. From (4.6.31) and Korn's inequality, we obtain

$$\|\mathbf{u}\|_{L^2(\Omega_t)} \leq C\|S(\mathbf{u})\|_{L^2(\Omega_t)}, \qquad C = c_P c_K. \qquad (4.6.32)$$

By Lemma 4.8.2, we obtain

$$\|\partial_t \mathbf{V}\|_{L^2(\Omega_t)} \leq p_1(\|\nabla\mathbf{u}\|_{L^2(\Omega_t)} + \|\nabla'\eta\|_{L^2(\Sigma)})$$
$$\leq p_1(c_K\|S(\mathbf{u})\|_{L^2(\Omega_t)} + \|\nabla'\eta\|_{L^2(\Sigma)}). \qquad (4.6.33)$$

By Sobolev embedding theorem $W^{1,2}(\Omega_t) \subset L^6(\Omega_t)$, and Korn's inequality it follows that there exist a limit embedding constant c_e, and a Korn constant c_K such that

$$\|\mathbf{u}\|_{L^6(\Omega_t)} \leq c_e\|\mathbf{u}\|_{W^{1,2}(\Omega_t)} \leq c_e c_K\|S(\mathbf{u})\|_{L^2(\Omega_t)}. \qquad (4.6.34)$$

Remark 4.6.1 Notice that in general the Poincaré constant is a function of Ω_t and therefore of time. However, by hypothesis (4.6.13) the Poincaré constant becomes less than a constant in time.

By using Schwartz and the limit Sobolev (4.6.34) inequalities, it is easy to see that $((\rho\mathbf{u} \cdot \nabla)\mathbf{V}, \mathbf{u})$ is estimated by

$$|((\rho\mathbf{u} \cdot \nabla)\mathbf{V}, \mathbf{u})_{\Omega_t}| \leq \|\rho\|_{L^\infty}\left(\int_{\Omega_t} |\mathbf{u}|^4 dx\right)^{\frac{1}{2}}\left(\int_{\Omega_t} |\nabla\mathbf{V}|^2 dx\right)^{\frac{1}{2}}$$
$$\leq C_1\|S(\mathbf{u})\|_{L^2(\Omega_t)}\|\nabla\mathbf{V}\|_{L^2(\Omega_t)}, \qquad (4.6.35)$$

with

$$C_1 := c_K\|\rho\|_{L^\infty}\|\mathbf{u}\|^{1/2}_{L^2(\Omega_t)}\|\nabla\mathbf{u}\|^{1/2}_{L^2(\Omega_t)}.$$

By (4.6.32), (4.6.33), (4.6.35), (4.5.33), (4.6.13), Lemma 4.8.2, and (4.6.34), we obtain

$$|\mathcal{I}| \leq |(\rho\mathbf{u}, \partial_t\mathbf{V})_{\Omega_t} + |((\rho\mathbf{u}\cdot\nabla)\mathbf{V}, \mathbf{u})_{\Omega_t}| + \mu|(S(\mathbf{u}), \nabla\mathbf{V})_{\Omega_t}|$$

$$\leq Cp_1\|S(\mathbf{u})\|_{L^2(\Omega_t)}(c_K\|S(\mathbf{u})\|_{L^2(\Omega_t)} + \|\nabla'\eta\|_{L^2(\Sigma)}) \tag{4.6.36}$$

$$+ p_2(C_1+\mu)\|S(\mathbf{u})\|_{L^2(\Omega_t)}(\|\sigma\|_{L^2(\Omega_t)} + \|\nabla'\eta\|_{L^2(\Sigma)})$$

$$\leq C_2\|S(\mathbf{u})\|^2_{L^2(\Omega_t)} + C_3\|S(\mathbf{u})\|_{L^2(\Omega_t)}\|\nabla'\eta\|_{L^2(\Sigma)} + C_4\|S(\mathbf{u})\|_{L^2(\Omega_t)}\|\sigma\|_{L^2(\Omega_t)},$$

with

$$C_2 := Cp_1c_K, \qquad C_3 := Cp_1 + p_2(C_1+\mu), \qquad C_4 := p_2(C_1+\mu).$$

We recall that by (4.6.14) and (4.6.24) it follows

$$K_\eta(t) = (\wp - 2\kappa l(t))\|\eta\|^2_{L^2(\Sigma)}(t) > \frac{K_0}{C\epsilon}\|\eta\|^2_{L^2(\Sigma)}(t). \tag{4.6.37}$$

Combining (4.6.24), (4.6.20) and (4.6.37), it holds that

$$\|\nabla'\eta\|^2_{L^2(\Sigma)}(t) \leq (2+\epsilon)D_\eta(t) < (2+\epsilon)\frac{\wp}{2\kappa}\|\eta\|^2_{L^2(\Sigma)}(t)$$

$$< (2+\epsilon)\frac{\wp}{2\kappa}\frac{C\epsilon}{K_0}K_\eta(t). \tag{4.6.38}$$

Applying Cauchy's inequality to (4.6.36) and using (4.6.38), we get

$$\nu\mathcal{I} \leq \left(\nu C_2 + \frac{\mu}{2}\right)\|S(\mathbf{u})\|^2_{L^2(\Omega_t)} + \nu^2\frac{C_1^2}{\mu}\|\sigma\|^2_{L^2(\Omega_t)} + \nu^2\frac{C_3^2}{2\mu}\|\nabla'\eta\|^2_{L^2(\Sigma)}$$

$$\leq \left(\nu C_2 + \frac{\mu}{2}\right)\|S(\mathbf{u})\|^2_{L^2(\Omega_t)} + \nu^2\frac{C_1^2}{\mu}\|\sigma\|^2_{L^2(\Omega_t)} + \nu^2\frac{C_3^2}{\mu}(2+\epsilon)\frac{\wp}{2\kappa}\frac{C\epsilon}{K_0}K_\eta,$$

hence we deduce

$$-\nu\mathcal{I} \geq -\left(\frac{\mu}{2}+\nu C_2\right)\|S(\mathbf{u})\|^2_{L^2(\Sigma)}(t) - \nu^2\frac{C_1^2}{\mu}\|\sigma\|^2_{L^2(\Omega_t)} - \nu^2\frac{C_3^2}{2\mu}(2+\epsilon)\frac{\wp}{2\sigma}\frac{C\epsilon}{K_0}K_\eta(t). \tag{4.6.39}$$

The free work identity (4.6.29) implies

$$\frac{dz}{dt}(t) > \left(\frac{\mu}{2} - \nu C_2\right)\|S(\mathbf{u})\|^2_{L^2(\Omega_t)}(t) \tag{4.6.40}$$

$$+ \nu\left(k - \nu\frac{C_1^2}{\mu}\right)\|\sigma\|^2_{L^2(\Omega_t)} + \nu\left(1 - \nu\frac{C_3^2}{2\mu}(2+\epsilon)\frac{\wp}{2\sigma}\frac{C\epsilon}{K_0}\right)K_\eta(t).$$

If we choose ν so small that $\nu C < \mu/2$, and

$$\frac{1}{2} > \nu \frac{(C_1\epsilon + \mu)^2}{2\mu}(2+\epsilon)\frac{\wp}{2\kappa}\frac{C\epsilon}{K_0},$$

then we get

$$\frac{d}{dt}z(t) > \frac{\nu}{2}K_\eta(t). \tag{4.6.41}$$

By (4.6.30), Cauchy's inequality and (4.6.37), we obtain

$$z(t) = K_\eta(t) - \frac{1}{2}\|\sqrt{\rho}\mathbf{u}\|_{L^2(\Omega)}^2 - \frac{k}{2}\left\|\frac{\sigma}{\sqrt{\rho}}\right\|_{L^2(\Omega)}^2 + \nu(\rho\mathbf{u},\mathbf{V})_{\Omega_t}(t) \tag{4.6.42}$$

$$\leq K_\eta(t) + \frac{\nu^2}{4}\|\eta\|_{L^2(\Sigma)}^2(t) < \left(1 + \frac{\nu^2 C\epsilon}{4K_0}\right)K_\eta(t).$$

Combining (4.6.41) with (4.6.42), we obtain

$$\frac{d}{dt}z(t) > \frac{\nu}{2}\frac{1}{1 + \frac{\nu^2 C\epsilon}{4K_0}}z(t). \tag{4.6.43}$$

Moreover, by (4.6.30) and (4.6.23), it holds that

$$z(t) = K_\eta(t) - \frac{1}{2}\|\sqrt{\rho}\mathbf{u}\|_{L^2(\Omega_t)}^2 - \frac{k}{2}\left\|\frac{\sigma}{\sqrt{\rho}}\right\|_{L^2(\Omega_t)}^2 + \nu(\mathbf{u},\mathbf{V})_{\Omega_t}(t)$$

$$\leq K_\eta(t) + \nu(\rho\mathbf{u},\mathbf{V})_{\Omega_t}(t)$$

$$\leq (\wp - 2\sigma l(t))\|\eta\|_{L^2(\Sigma)}^2(t) + \frac{\nu}{2}\|\mathbf{u}\|_{L^2(\Omega_t)}^2(t) + \frac{\nu}{2}\|\eta\|_{L^2(\Sigma)}^2(t)$$

$$\leq \left(\wp + \frac{\nu}{2}\right)(\|\mathbf{u}\|_{L^2(\Omega_t)}^2 + \|\eta\|_{L^2(\Sigma)}^2)(t)$$

$$\leq \left(\wp + \frac{\nu}{2}\right)(\|\mathbf{u}\|_{L^2(\Omega_t)}^2 + c\|\sigma\|_{L^2(\Omega_t)}^2 + \|\eta\|_{W^{1,2}(\Sigma)}^2)(t).$$

We set

$$\beta = \frac{\nu}{2}\frac{1}{1 + \frac{\nu^2 C\epsilon}{4K_0}}.$$

Applying Gronwall's Lemma to (4.6.43) we obtain an absurdum. Indeed for any given time $t \in (0,\infty)$ we get

$$e^{\beta t}z(0) \leq z(t) \leq \left(\wp + \frac{\nu}{-2}\right)(\|\mathbf{u}\|_{L^2(\Omega_t)}^2 + c\|\sigma\|_{L^2(\Omega_t)}^2 + \|\eta\|_{W^{1,2}(\Sigma)}^2)(t). \tag{4.6.44}$$

This contradicts (4.6.13), which completes the proof of the theorem.

4.6.4 Loss of Initial Data Control

In this subsection Theorem 4.3.4 is proven following upon the proof of the
Theorem 4.3.3, and the proof is given by absurdum. We assume that a
solution $(\mathbf{u}, \sigma, \eta)$ of (4.3.4) defined for all $t > 0$ has a norm uniformly bounded
in time, i.e. that satisfies

$$\|\mathbf{u}\|^2_{W^{1,2}(\Omega_t)}(t) + \|\sigma\|^2_{W^{1,\infty}(\Omega_t)}(t) + \|\eta\|^2_{W^{1,\infty}(\Sigma)}(t) < 2\,\alpha, \qquad t \in (0, \infty).$$
$$(4.6.45)$$

By carefully repeating all steps of previous Sect. 4.3 by changing ϵ with α it
is not difficult to prove that for any given $\alpha > 0$, for all times $t > 0$ such that

$$e^{\beta t} z(0) \leq z(t) \leq \left(\wp + \frac{\gamma}{2}\right)(\|\mathbf{u}\|^2_{L^2(\Omega_t)} + c\|\sigma\|^2_{L^2(\Omega_t)} + \|\eta\|^2_{W^{1,2}(\Sigma)})(t).$$
$$(4.6.46)$$

This contradicts (4.6.45), which completes the proof of the theorem.

4.7 Sign of E_0 in Terms of Size of Perturbation

In this part, we furnish explicit characteristic parameters κ, $\wp$, d, h, ensuring
that the linearized initial energy E_0 is positive for any initial data. This
furnishes an example of rest state with characteristic parameters κ, $\wp$, d, h
ensuring linear stability. Next, for the same set of characteristic parameters
κ, $\wp$, d and h, we construct sufficiently large initial data to ensure the total
energy at initial time is negative. This latter result is reached provided the
volume of the fluid is sufficiently large; cf. (4.7.23).

Add a subscript c to a Sobolev spaces X; X_c means the same space
quotiented with respect to constants. Here we consider a plane rigid section
$\Sigma \times 0$, and we prove the following Theorems

Theorem 4.7.1 *It is possible to determine precise values of $\wp$, κ, d, h such
that:*

(i) *There exists a positive constant $\delta(\wp, \kappa, h) > 0$ such that all solutions to
(4.5.3) with $\tau \in L^\infty(0, \infty); W^{1,\infty}(\Sigma))$ satisfy*

$$\int_\Sigma \tau \, dx = 0, \qquad h + \tau > \frac{h}{2}, \qquad \sup_{t \in (0,\infty)} \|\tau\|_{W^{1,\infty}(\Sigma)} < \delta;$$

(ii) *For all τ satisfying (i), the energy at initial time with $\mathbf{u}_0 = 0$, $\sigma_0 = 0$,
$\eta_0 = \tau$, verifies also*

$$E(0, 0, \tau) > 0.$$
$$(4.7.1)$$

Theorem 4.7.2 *Under hypotheses of Theorem 4.7.1 there exist at least one initial data $\eta_0 = \tau$ sufficiently large*

$$\|\tau\|_{W^{1,\infty}(\Sigma)} > \alpha,$$

such that the nonlinear total energy computed for initial data ($\mathbf{u}_0 = 0, \sigma_0 = 0$, $\eta_0 = \tau$) *is negative, $E(0,0,\tau) < 0$.*

4.7.1   Proof of Theorem 4.7.1

We begin by observing that it is possible to find in $L_0^2(\Sigma)$ an orthonormal basis of functions τ_k and a sequence $\lambda_k \in R$ in such a way that it yields

$$- \Delta'\tau_k - G\tau_k = \lambda_k\tau_k, \qquad \int_{\Sigma} \tau_k \, dx' = 0. \tag{4.7.2}$$

In particular, we choose τ_k, as product of periodic functions $f(2\,i\pi\,x_1)\,g(\frac{2\,j\,\pi\,x_2}{d})$, $f, g \in \{sin, cos\}$, $i,j \in \mathbb{N}_0$, $k = k(i,j)$, $(i,j) \neq (0,0)$, multiplied by coefficients α_{ij} that provide unitary norms. Also the smallest of the corresponding eigenvalues is given by

$$\lambda_1 = G_d - G, \tag{4.7.3}$$

where

$$G_d := 4\,\pi^2 \min\Big\{1, \frac{1}{d^2}\Big\}. \tag{4.7.4}$$

Hence we have the following equivalent inequalities

$$G < G_d := 4\pi^2 \min\Big\{1, \frac{1}{d^2}\Big\}, \quad \Longleftrightarrow \quad \lambda_1 = G_d - G > 0.$$

This yields $\lambda_k \geq \lambda_1 > 0, \forall k \in N.$[4] Our proofs require the hypothesis

$$\lambda_1 > 0,$$

this explains the request that eigenfunctions must have zero mean values, say we work in the quotient space of $L^2(\Sigma)$.

---

[4]This result can be read physically that for a liquid layer below a rigid plane $(4.7.1)$ is certainly true under the linear stability $(4.3.13)$ hypothesis:

$$\frac{\wp|\Sigma|}{\kappa} < 4\pi^2 \min\Big\{1, \frac{1}{d^2}\Big\}.$$

Namely, the linearized total energy of perturbation is positive for every regular functions.

We notice that multiplying (4.7.2) times τ_k, integrating over Σ, and integrating by parts it furnishes

$$(\lambda_k + G) \int_\Sigma \tau_k^2 \, dx' = -\int_\Sigma \tau_k \Delta' \tau_k \, dx'$$

$$= \int_\Sigma |\nabla' \tau_k|^2 \, dx' - \int_{\partial\Sigma} \tau_k \nabla' \tau_k \cdot \nu \, dl = \int_\Sigma |\nabla' \tau_k|^2 \, dx'.$$

Notice that the line integral over $\partial\Sigma$ becomes zero either when τ_k vanishes, say the Dirichlet condition holds, or when the normal derivative of τ_k is zero, say the Neumann condition holds, or in our case, for periodic functions.

Setting $\mathbf{u}_0 = 0$, $\sigma_0 = 0$, the total energy reduces to

$$E(\eta_0) = \left(2 \int_\Sigma \left(\sqrt{1 + |\nabla' \eta_0|^2} - 1\right) dx' - G \int_\Sigma \eta_0^2 \, dx'\right),$$

$$G = \frac{\wp|\Sigma|}{\kappa} < 4\pi^2 \min\left\{1, \frac{1}{d^2}\right\} = G_d. \tag{4.7.5}$$

Then the initial energy is given by $\mathcal{E}(0, \eta_0) = \kappa E(\eta_0)$. Of course the sign of $E(\eta_0)$ depends on the parameters κ, $\wp$, h, d. In the sequel we set $\eta_0 = \tau$.

Let $\mathbf{u}_0 = 0$, $\sigma_0 = 0$. The total energy reduces to

$$E(\eta_0) = \kappa\left(2 \int_\Sigma \left(\sqrt{1 + |\nabla' \eta_0|^2} - 1\right) dx' - G \int_\Sigma \eta_0^2 \, dx'\right),$$

$$G = \frac{\wp|\Sigma|}{\kappa} < 2\pi^2\left(1 + \frac{1}{a^2}\right). \tag{4.7.6}$$

Then the initial energy is given by $E(\eta_0) = \kappa E_0$. Of course the sign of E_0 depends on the parameters κ, $\wp$, h

The aim of these calculations are twofold:

(a) To prove that there exists $\varepsilon > 0$ such that if $\tau \in C^1(\Sigma)$ satisfies the conditions

$$\tau|_{\partial\Sigma} = 0, \quad \int_\Sigma \tau \, dx' = 0, \quad \|\nabla' \tau\|_\infty = \sup_\Sigma |\nabla' \tau| < \varepsilon \tag{4.7.7}$$

then we have $E(\tau) \geq 0$

(b) To prove that there exists $\alpha > 0$ such that if $\tau \in C^1(\Sigma)$ satisfies the conditions

$$\tau|_{\partial\Sigma} = 0, \quad \int_\Sigma \tau \, dx' = 0, \quad \|\nabla' \tau\|_\infty = \sup_\Sigma |\nabla' \tau| > \alpha, \tag{4.7.8}$$

then we have $E(\tau) < 0$.

(a) Positiveness of $E(\eta_0)$ for small initial data

We notice that it holds

$$\sqrt{1+p^2} - 1 = \frac{p^2}{\sqrt{1+p^2}+1} = p^2\left(\frac{1}{2} + \frac{1}{\sqrt{1+p^2}+1} - \frac{1}{2}\right)$$

$$= \frac{p^2}{2} + p^2\frac{1-\sqrt{1+p^2}}{2(1+\sqrt{1+p^2})} = \frac{p^2}{2} - \frac{p^4}{2(1+\sqrt{1+p^2})^2} = \frac{p^2}{2} - R(p).$$

Since $1 + \sqrt{1+p^2} \geq 2$, the following estimate is true:

$$R(p) = |R(p)| \leq \frac{|p|^4}{8}.$$

Hence, for all $\tau \in W^{1,\infty}(\Sigma)$ it holds

$$E(\tau) = \kappa\left(2\int_\Sigma \left(\sqrt{1+|\nabla'\tau|^2} - 1\right) dx' - G\int_\Sigma \tau^2\, dx'\right) \qquad (4.7.9)$$

$$\geq \kappa\left(\int_\Sigma |\nabla'\tau|^2\, dx' - G\int_\Sigma \tau^2\, dx' - \frac{1}{4}\int_\Sigma |\nabla'\tau|^4\, dx'\right).$$

Notice that from (4.7.9) it becomes clear that $E(\tau)$ is positive for suitably small $\|\nabla'\eta\|_{L^\infty(\Sigma)}$ if and only if its linearized part $E_l(\tau)$ is positive. Therefore, provided the linearized energy $E_l(\tau)$, namely $\int_\Sigma |\nabla'\tau|^2\, dx' - G\int_\Sigma \tau^2\, dx'$ is positive, the total energy $E(\tau)$ will also be positive for $\nabla'\tau$ sufficiently small in the norm of L^∞. Thus to completely prove Theorem 4.7.1 we show that $E(\tau)$ for small $\|\tau\|$ is still positive. For an arbitrary function τ we consider the Fourier expansion and we get the same result. By using the eigenfunction expansion in L_0^2, with $\gamma_j = (\tau, \tau_j)$, τ is expressed by,

$$\tau = \sum_{k=1}^\infty \gamma_j \tau_j, \quad \nabla'\tau = \sum_{k=1}^\infty \gamma_j \nabla'\tau_j.$$

It follows that

$$\int_\Sigma |\nabla'\tau|^2\, dx' - G\int_\Sigma \tau^2\, dx' = \sum_{k=1}^\infty \gamma_j^2 \int_\Sigma \left(|\nabla'\tau_j|^2\, dx' - G\tau_j^2\right) dx'$$

$$= \sum_{k=1}^\infty \gamma_j^2 \lambda_j \int_\Sigma \tau_j{}^2\, dx' = \sum_{k=1}^\infty \gamma_j^2 \lambda_j.$$

On the other hand, by the assumption : $\|\nabla' \tau\|_\infty \leq \delta$, we get

$$\int_\Sigma |\nabla' \tau|^4 \, dx' \leq \delta^2 \int_\Sigma |\nabla' \tau|^2 \, dx'$$

$$= \delta^2 \sum_{k=1}^\infty \gamma_j^2 \int_\Sigma |\nabla' \tau_j|^2 \, dx' = \delta^2 \sum_{k=1}^\infty \gamma_j^2 \, (G + \lambda_j) \, .$$

As a consequence it yields:

$$E(\tau) \geq \sum_{k=1}^\infty \left(\gamma_j^2 \lambda_j - \frac{\delta^2}{4} \gamma_j^2 (G + \lambda_j) \right) \int_\Sigma \tau_j{}^2 \, dx'$$

$$= \sum_{k=1}^\infty \gamma_j^2 \left[\lambda_j \left(1 - \frac{\delta^2}{4} \right) - \frac{\delta^2}{4} G \right] \int_\Sigma \tau_j{}^2 \, dx'.$$

Hence if

$$\lambda_1 \left(1 - \frac{\delta^2}{4} \right) - \frac{\delta^2}{4} G \geq 0,$$

we obtain

$$E(\tau) \geq 0.$$

Therefore, it is enough to let:

$$\lambda_1 (4 - \delta^2) - \delta^2 \, G \geq 0$$

or equivalently

$$0 < \delta \leq \sqrt{\frac{4\,\lambda_1}{\lambda_1 + G}} = \sqrt{\frac{4\,\lambda_1}{\lambda_1 + \frac{\varrho}{\kappa}}}. \tag{4.7.10}$$

Therefore we have proven that if τ satisfies (4.7.7) with δ bounded above by (4.7.10), for δ sufficiently small, the initial value $(\mathbf{u}_0, \sigma_0, \tau) = (\mathbf{0}, 0, \tau)$ lets the initial energy (4.3.10) be positive.

4.7.2 Proof of Theorem 4.7.2

Here we prove that in our example it is possible to construct large initial values which let the initial energy be negative. We look for a function τ with $E(\tau) < 0$. We choose $\eta_0 = L\,\tau_1$ for sufficiently large number L. Using the inequality (4.6.17) we deduce that recalling (4.7.9) it holds

$$E(\tau) = k \int_\Sigma \left(2(\sqrt{1 + |\nabla' \tau|^2} - 1) - G\,\tau^2 \right) dx'.$$

We must construct initial data such that $E(\tau) < 0$. Take $\tau = L\tau_1$, where

$$\|\tau_1\|_{L^2(\Sigma)}^2 = \frac{1}{\lambda_1 + G}\|\nabla'\tau_1\|_{L^2(\Sigma)}^2 = \frac{1}{G_d}\|\nabla'\tau_1\|_{L^2(\Sigma)}^2$$

is the first eigenvalue and L is the intensity of initial data in L^2. To simplify calculations, in this computation we set

$$L|\nabla'\tau_1| = p,$$

and substitute the value of first eigenvalue, $\lambda_1 = G_d - G$,

$$E(\tau) = k \int_\Sigma \left(2(\sqrt{1+p^2} - 1) - \frac{G}{G_d} p^2 \right) dx'. \tag{4.7.11}$$

The sign of this function is reduced to the sign of the integrand, hence we compute the sign of the integrand. Simple calculations lead to

$$\left(1 - \frac{G}{G_d}\right)p^2 - \frac{p^2}{(1+\sqrt{1+p^2})^2}p^2 = p^2\left\{ \frac{G_d - G}{G_d} - \frac{p^2}{(1+\sqrt{1+p^2})^2} \right\}. \tag{4.7.12}$$

If

$$\frac{G_d - G}{G_d} \int_\Sigma p^2\, dx' - \int_\Sigma p^4 \frac{1}{(1+\sqrt{1+p^2})^2} dx' < 0, \tag{4.7.13}$$

then $E(\tau) < 0$. To give a bound in terms of the intensity L of the L^2 norm of initial data, we consider the case $d < 1$ which furnishes $\nabla'\tau_1 = \sqrt{\frac{2}{d}}\, 2\pi \sin(2\pi x)$. Notice that the coefficient $\sqrt{\frac{2}{d}}$ is needed in order the L^2 norm if τ_1 is one. Therefore it holds

$$0 < |\cos^2(2\pi x_1)| < 1 \qquad \Longleftrightarrow \qquad 0 < p^2 < \frac{8\pi^2 L^2}{d}. \tag{4.7.14}$$

(4.7.14) infers

$$-\int_\Sigma p^4 \frac{1}{(1+\sqrt{1+p^2})^2} dx' \leq -\frac{1}{(1+\sqrt{1+(8\pi^2 L^2/d)})^2} \int_\Sigma p^4\, dx', \tag{4.7.15}$$

$$\int_\Sigma \sin(2\pi x)^4 dx' = \int_\Sigma \cos(2\pi x_1)^4 dx' = \frac{3\,d}{8}.$$

Furthermore, it holds

$$\int_\Sigma p^2 dx' = 4\pi^2 L^2. \tag{4.7.16}$$

Of course the energy is negative if

$$\frac{G_d - G}{G_d} \int_\Sigma p^2 \, dx' - \int_\Sigma p^4 \frac{1}{(1 + \sqrt{1 + p^2})^2} \, dx' < 0. \tag{4.7.17}$$

From the above calculations we deduce that E is negative if

$$\frac{G_d - G}{G_d} 4\pi^2 L^2 - \frac{24\pi^4 L^4}{d(1 + \sqrt{1 + (8\pi^2 L^2/d)})^2} < 0. \tag{4.7.18}$$

We set $\gamma := (G_d - G)/G_d$ and write

$$\gamma - \frac{6\pi^2 L^2}{d(1 + \sqrt{1 + \frac{8\pi^2 L^2}{d}})^2} < 0.$$

Developing the square we have

$$2\gamma d + \gamma d \frac{8\pi^2 L^2}{d} + 2\gamma d \sqrt{1 + \frac{8\pi^2 L^2}{d}} - \frac{3 d}{4} \frac{8\pi^2 L^2}{d} < 0.$$

Equivalently, we write

$$\left(1 + \frac{8\pi^2 L^2}{d}\right)\left(\gamma - \frac{3}{4}\right) + 2\gamma \sqrt{1 + \frac{8\pi^2 L^2}{d}} + 2\gamma - \left(\gamma - \frac{3}{4}\right) < 0.$$

Set $x := \sqrt{1 + \frac{8\pi^2 L^2}{d}}$, we get

$$\left(\gamma - \frac{3}{4}\right)x^2 + 2\gamma x + \left(\gamma + \frac{3}{4}\right) < 0.$$

This algebraic inequality admits two resolutions in function of the coefficient of x^2. We assume

$$\left(\gamma - \frac{3}{4}\right) < 0, \quad \Longleftrightarrow \quad 4G > G_d,$$

in order to find a condition on initial data. Precisely, we first change sign to all inequality

$$x^2 \left(\frac{3}{4} - \gamma\right) - 2\gamma x - \left(\gamma + \frac{3}{4}\right) > 0.$$

Taking the positive root:

$$x \geq \frac{4\gamma + 3}{(3 - 4\gamma)}. \tag{4.7.19}$$

Recall that $4\gamma - 3 < 0$. Substituting the value of x we have

$$\frac{8\pi^2}{d}L^2 > \left(\frac{7G_d - 4G}{4G - G_d}\right)^2 - 1. \tag{4.7.20}$$

Therefore, we have

$$L^2 > \frac{d}{8\pi^2}\frac{48\,G_d}{(4G - G_d)^2}(G_d - G), \tag{4.7.21}$$

that is

$$L^2 > 24\,d\frac{G_d - G}{(4G - G_d)^2}. \tag{4.7.22}$$

For such L, we obtain $E(\tau) < 0$. Therefore we know that for initial data $\eta_0 = L\tau_1$ with L satisfying condition (4.7.22), the initial value $(\mathbf{u}_0, \eta_0) = (\mathbf{0}, L\tau_1)$ has an initial negative energy E_0.

Remark 4.7.1 *We read condition (4.7.22) as a condition on the Grashoff number G, recalling that G must belong to the interval $(\frac{G_d}{4}, G_d)$. The smaller G is, the more stable the system is. As G goes to $G_d/4$, L^2 must go to ∞, and there is instability only for initial data extremely large, which means never! This result seems interesting. Moreover, if G goes to G_d then initial data may become extremely small and there is instability also for small initial data. Another factor to be analyzed is the size of the wet area namely the area of contact section $|\Sigma|$ between liquid and plane. If d goes to zero also L^2 goes to zero, thus instability may occur for thin contact sections.*

Remark 4.7.2 *The conditions on $\wp$, κ, h, d need a physical comment. In our example, we require the following conditions*

$$\frac{G_d}{4} < G < G_d, \tag{4.7.23}$$

$$h + L\tau_1 > \frac{h}{2},$$

$$volV = hd = 1.$$

From the first two inequalities, and (4.7.22) we get

$$\frac{h}{2} + L\tau_1 > \frac{h}{2} - L|\tau_1| \geq \frac{h}{2} - L\sqrt{\frac{2}{d}}. \tag{4.7.24}$$

Hence the left hand side of (4.7.24) will be positive if

$$\frac{h}{2} \geq L\sqrt{\frac{2}{d}} = 2\sqrt{6}\,d\frac{\sqrt{G_d - G}}{(4G - G_d)}\sqrt{\frac{2}{d}} = 4\sqrt{3}\frac{\sqrt{G_d - G}}{(4G - G_d)}. \tag{4.7.25}$$

We may read (4.7.25) as a compatibility condition on the depth of the layer in function of G: for instability to occur the depth must be sufficiently large. On the other side, from (4.7.23)$_3$ we obtain a compatibility condition on the cross section in function G, specifically

$$(4G - G_d) > 8\sqrt{3}d\sqrt{(G_d - G)}. \tag{4.7.26}$$

We read condition (4.7.26) as a condition of smallness on the contact area $|\Sigma|$, the wet area. For instability to occur the wet area must be sufficiently small.

Remark 4.7.3 *The hypothesis of upwardly directed gravity may appear artificial. In this regard we notice that a rigid plane which is wet below represents just a customary example of a thin layer of fluid with a rigid surface above and a free surface below. In this example the motion occurs due to capillarity phenomena. Another example that we propose as an open problem is that of a fluid on an inclined plane. In this case it is important to find the critical inclination angle.*

4.8   Auxiliary Lemmas

We need two lemmas in order to consider the free boundary problem. We denote by ρ_b the extension of the rest state to

$$\Omega_\zeta = \{(x', z) : x' \in \Sigma, 0 < z < \zeta(x')\},$$

and we set
$$\Gamma_\zeta = \{x \in \bar{\Omega}_\zeta : x' \in \Sigma, z = \zeta(x')\}.$$

In order to state our theorem we introduce the following regularity class

$$\mathcal{V} = \{\sigma(x', z), \mathbf{u}(x', z), \zeta(x') : \mathbf{u} \in W_\sharp^{1,2}(\Omega_\zeta); \ \sigma \in C_\sharp^0(\bar{\Omega}_\zeta, \inf |\zeta| > \frac{h}{4}, \zeta \in W_\sharp^{1,\infty}(\Sigma)\}.$$

4.8.1   Function V for Uniqueness

Lemma 4.8.1 *Let $(\sigma(x', z), \mathbf{u}(x', z), \zeta(x')) \in \mathcal{V}$ be given, and set*

$$V(\zeta(x')) = \frac{k}{g}(\rho_b(\zeta(x')) - \rho_b(h)).$$

Then there exists a vector field $\mathbf{V} \in W_0^{1,2}(\Omega_\zeta))$ solution to the following problem

$$\nabla \cdot \mathbf{V} = \sigma, \quad x \in \Omega_\zeta$$

$$\mathbf{V}(x', 0) = 0,$$

$$\mathbf{V} \cdot \mathbf{n}(x', \zeta(x')) = n_z(\nabla'\zeta)\frac{V(\zeta(x'))}{\rho_b(\zeta(x'))}. \tag{4.8.1}$$

Moreover, there exist two constants l_1, l_2 depending on ρ_b, ρ and ζ, such that the following estimates hold true:

$$\|\mathbf{V}\|_{L^6(\Omega_\zeta)} \leq l_1 \left[\|\sigma\|_{L^2(\Omega_\zeta)} + \|\zeta\|_{W^{1,2}(\Sigma)}\right],$$

$$\|\nabla \mathbf{V}\|_{L^2(\Omega)} \leq l_2 \left[\|\sigma\|_{L^2(\Omega_\zeta)} + \|\zeta\|_{W^{1,2}(\Sigma)}\right]. \tag{4.8.2}$$

Proof. First of all verify that the compatibility condition

$$\int_{\Omega_\zeta} \sigma(x', z)dx = \int_{\Gamma_\zeta} V(\zeta(x'))n_z(\nabla'\zeta)dx',$$

requested by integration by parts of $(4.8.1)_1$ over Ω_ζ is satisfied. We have

$$\int_{\Omega_\zeta} \sigma(x)dx = \int_{\Omega_\zeta} (\rho(x) - \rho_b(x))dx = M - \rho_b \int_\Sigma dx' \int_0^{\zeta(x')} \exp\left(-\frac{gz}{k}\right)dz$$

$$= M - \rho_b\frac{k}{g} \int_\Sigma \left(1 - \exp\left(-\frac{g\zeta(x')}{k}\right)\right)dx' = \frac{k}{g} \int_\Sigma \left(\rho_b(\zeta(x')) - \rho_b(h)\right)dx'$$

$$= \int_\Sigma V(\zeta(x'))dx', \tag{4.8.3}$$

that provides the wanted compatibility condition.

We look for the vector field in the form

$$\rho_b\mathbf{V} = \rho_b\mathbf{U} + \mathbf{W},$$

where $\mathbf{U}$, $\mathbf{W}$ solve the simpler problems

$$\begin{aligned}
\nabla \cdot (\rho_b\mathbf{U}) &= \sigma(x) - \bar{\sigma}, & \Omega_\zeta, \\
\mathbf{U}(x) &= 0, & \Gamma_\zeta, \\
\nabla \cdot (\rho_b\mathbf{W}) &= \bar{\sigma}, & \Omega_\zeta, \\
\mathbf{W}(x', 0) &= 0, & \Sigma, \\
\mathbf{W}(x', \zeta(x')) \cdot \mathbf{n}(\nabla'\zeta) &= V(\zeta(x'))n_z(\nabla'\zeta), & \Sigma,
\end{aligned} \tag{4.8.4}$$

where $\bar{\sigma} = \frac{1}{|\Omega_\zeta|} \int_{\Omega_\zeta} \sigma(x)dx$ denotes the mean value of σ. We ask that the following estimates be satisfied

$$
\|\nabla U\|_{L^2(\Omega_\zeta)} \le c \left[\|\sigma - \bar{\sigma}\|_{L^2(\Omega_\zeta)}\right],
$$
$$
\|\nabla W\|_{L^2(\Omega_\zeta)} \le c \left[\|\bar{\sigma}\|_{L^2(\Omega_\zeta)} + \|\zeta\|_{W^{1,2}(\Sigma)}\right].
$$
(4.8.5)

We solve such problems by constructing explicit solutions. We observe that the domain Ω_ζ is the union of domains $\Omega_i = \Sigma_i \times (0, \zeta(x'))$, $\Sigma_i \subseteq \Sigma$, $i = 1, \ldots, N$, star-shaped with respect to certain balls B_i, because $\zeta > h/4$, and the gradient of ζ is uniformly bounded. We make the reasoning for $N = 1$, but our arguments hold for arbitrary N. We define $\rho_b U$ according to the formula due to Bogowski [16],

$$
\rho_b(x)U(x) = \int_{\Omega_\zeta} (\sigma(y) - \bar{\sigma}) \left[\frac{x - y}{|x - y|^n} \int_{|x-y|}^\infty \omega \left(y + \xi \frac{x - y}{|x - y|}\right) \xi^2 d\xi\right] dy =
$$

$$
\int_{\Omega_\zeta} (\sigma - \bar{\sigma}) N(x, y) dy,
$$
(4.8.6)

where n is the dimension. Furthermore, we may extend by periodicity U onto the whole domain $x' \in R^2$, $0 < z < \zeta$. Inequality $(5.4.13)_1$ builds upon the Calderon–Zygmund theorem because the kernel $N(x, y)| \le c|x - y|^{-n+1}$. Furthermore, we take the vector W in the form

$$
W(x) = \nabla \times (A\chi(z)) + \bar{\sigma}z e_3, \qquad if \quad n = 3,
$$
$$
W \equiv \left(-\frac{\partial}{\partial x_2}[b(x')\chi(x_2)], \chi(x_2)\frac{\partial b(x')}{\partial x_2} + \bar{\sigma}x_2 e_2\right), \qquad if \quad n = 2,
$$
(4.8.7)

where $\chi(x_i)$, $i = 2, 3$ is a regular cut-off function vanishing for $x_i < (h/4)$, and equal to one in a neighborhood of $x_i(x' \ge \min \zeta(x') > (h/4)$. In addition, $A = (-b_2(x'), b_1(x'), 0)$, and

$$
\frac{\partial b_1}{\partial x_1} + \frac{\partial b_2}{\partial x_2} = (V - \bar{\sigma}\zeta)(x'), \qquad if \quad n = 3,
$$
$$
\frac{\partial b}{\partial x_1} = (V - \bar{\sigma}\zeta)(x_1), \qquad if \quad n = 2.
$$
(4.8.8)

Since the compatibility condition

$$
\int_\Sigma (V - \bar{\sigma}\zeta)dx' = \int_{\Gamma_\zeta} (V - \bar{\sigma}\zeta)n_z dS = |\Gamma_\zeta|\bar{V} - \bar{\sigma}\int_{\Gamma_\zeta} \zeta n_z dS,
$$

is satisfied, there exists a periodic vector field $\mathbf{b} \equiv (b_1, b_2)$, or a scalar function b satisfying (4.8.8), such that

$$\|b\|_{W^{2,2}(\Sigma)} \leq c \left[\|V - \bar{\sigma}\zeta\|_{W^{1,2}(\Sigma)} \right],$$
$$\|\mathbf{b}\|_{W^{2,2}(\Sigma)} \leq c \left[\|V - \bar{\sigma}\zeta\|_{W^{1,2}(\Sigma)} \right]. \tag{4.8.9}$$

These imply (4.8.2), and the Lemma is proved.   □

4.8.2 Functions V for Stability and Instability

We need two further lemmas in order to consider the free boundary problem. We shall continue to denote by ρ_b the extension of the rest state to

$$\Omega_t = \{(x', z) : x' \in \Sigma, \, 0 < z < \zeta(x', t)\},$$

and we set

$$\Gamma_t = \{x \in \bar{\Omega}_t : x' \in \Sigma, \, z = \zeta(x*, t)\}.$$

In order to state our theorem we introduce the following regularity class

$$\mathcal{V} = \{\mathbf{u}(x', z, t), \sigma(x', z, t), \eta(x', t) : \, \mathbf{u} \in L^2(0, \infty; W^{1,2}_\sharp(\Omega_t));$$

$$\sigma \in C^0(0, \infty; C^0_\sharp(\bar{\Omega}_t)), \, \inf |\zeta| > \frac{h}{4}, \, \zeta \in L^\infty(0, \infty; W^{1,\infty}_\sharp(\Sigma))\}.$$

Lemma 4.8.2 Let the fields $(\mathbf{u}(x', z, t), \sigma(x', z, t)\zeta(x', t)) \in \mathcal{V}$ be given, and let

$$\partial_t \sigma = -\nabla \cdot (\rho\mathbf{u}) \in L^2(0, \infty; L^2(\Omega)). \tag{4.8.10}$$

Then there exists a vector field $\mathbf{V} \in L^\infty(0, \infty; W^{1,2}_0(\Omega_t))$ solution to the following problem

$$\nabla \cdot (\rho_b \mathbf{V}) = \sigma, \qquad\qquad\qquad\qquad x \in \Omega_t$$

$$\mathbf{V}(x', 0, t) = 0, \qquad\qquad\qquad\qquad x \in \Sigma,$$

$$\mathbf{V} \cdot \mathbf{n}(x', \zeta(x', t)) = n_z(\nabla'\zeta)\frac{V(\zeta(x', t))}{\rho_b(\zeta(x', t))}, \qquad x \in \Sigma, \tag{4.8.11}$$

$$V(\zeta(x', t)) = \frac{k}{g}(\rho_b(\zeta(x', t) - \rho_b(h)) \qquad\qquad x \in \Sigma.$$

Moreover, there exist two constants p_1 depending on ρ_b, ρ and ζ, p_2 function of ζ, such that when setting $\zeta = h + \eta$ the following estimates hold true:

$$\|\partial_t \mathbf{V}\|_{L^2(\Omega)} \leq p_1 \left[\|S(\mathbf{u})\|_{L^2(\Omega_t)} + \|\eta\|_{W^{1,2}(\Sigma)} \right],$$
$$\|\nabla \mathbf{V}\|_{L^2(\Omega)} \leq p_2 \left[\|\sigma\|_{L^2(\Omega_t)} + \|\eta\|_{W^{1,2}(\Sigma)} \right]. \tag{4.8.12}$$

Proof. First of all, we verify that compatibility condition

$$\int_{\Omega_t} \sigma(x', z, t) dx = \int_{\Gamma_t} V(\zeta(x', t)) n_z(\nabla' \eta) dx',$$

requested by integration by parts of $(4.8.11)_1$ over Ω_t is satisfied. We have

$$\int_{\Omega_t} \sigma(x, t) dx = \int_{\Omega_t} (\rho(x, t) - \rho_b(x)) dx = M - \rho_b \int_\Sigma dx' \int_0^{\zeta(x', t)} \exp\left(-\frac{gz}{k}\right) dz$$

$$= M - \rho_b \frac{k}{g} \int_\Sigma \left(1 - \exp\left(-\frac{g\zeta(x', t)}{k}\right)\right) dx' = \frac{k}{g} \int_\Sigma \left(\rho_b(\zeta(x', t)) - \rho_b(h)\right) dx'$$

$$= \int_\Sigma V(\zeta(x', t)) dx',$$

that provides the wanted compatibility condition.

We look for the vector field in the form

$$\rho_b \mathbf{V} = \rho_b \mathbf{U} + \mathbf{W},$$

where $\mathbf{U}$, $\mathbf{W}$ solve the simpler problems

$$\begin{aligned}
\nabla \cdot (\rho_b \mathbf{U}) &= \sigma(x) - \bar{\sigma}, & \Omega_t, \\
\mathbf{U}(x, t) &= 0, & \partial\Omega_t, \\
\nabla \cdot (\rho_b \mathbf{W}) &= \bar{\sigma}, & \Omega_t, \\
\mathbf{W}(x', 0, t) &= 0, & \Sigma, \\
\mathbf{W}(x', \zeta(x', t)) \cdot \mathbf{n}(\nabla' \eta) &= V(\zeta(x', t)) n_z(\nabla' \eta(x', t)), & \Sigma,
\end{aligned}$$

$$(4.8.13)$$

where $\bar{\sigma} = \frac{1}{|\Omega_t|} \int_{\Omega_t} \sigma(x) dx$ denotes the mean value of σ. We ask the following estimates to be satisfied

$$\|\nabla \mathbf{U}\|_{L^2(\Omega_t)} \le c \left[\|\sigma - \bar{\sigma}\|_{L^2(\Omega_t)}\right],$$

$$\|\partial_t \mathbf{U}\|_{L^2(\Omega_t)} \le c \left[\|\sigma_t - \partial_t \bar{\sigma}\|_{L^2(\Omega_t)} + \|\mathbf{u} \cdot \mathbf{n}\|_{L^2(\Gamma_t)}\right],$$

$$\|\partial_t \mathbf{W}\|_{L^2(\Omega_t)} \le c \left[\|\bar{\sigma}\|_{L^2(\Omega_t)} + \|\mathbf{u} \cdot \mathbf{n}\|_{L^2(\Gamma_t)}\right],$$

$$\|\nabla \mathbf{W}\|_{L^2(\Omega_t)} \le c \left[\|\bar{\sigma}\|_{L^2(\Omega_t)} + \|\zeta\|_{W^{1,2}(\Sigma)}\right].$$

We solve such problems by constructing explicit solutions. We observe that the domain Ω_t is the union of domains $\Omega_i = \Sigma_i \times (0, \zeta(x', t))$, $\Sigma_i \subseteq \Sigma$, $i = 1, \ldots, N$, star-shaped with respect to certain balls B_i, because $\zeta > h/4$, and the gradient of ζ is uniformly bounded. We make the reasoning for $N = 1$, but our arguments hold for arbitrary N. We define $\rho_b \mathbf{U}$ according to the formula attributed to Bogowski, cf. [16],

$$\rho_b(x)\mathbf{U})(x,t) = \int_{\Omega_t} (\sigma(y,t) - \bar{\sigma})\Big[\frac{\mathbf{x} - \mathbf{y}}{|x - y|}\int_{|x-y|}^{\infty} \omega\Big(y + \xi\frac{\mathbf{x} - \mathbf{y}}{|x - y|}\Big)\xi^2 d\xi\Big] dy =$$

$$\int_{\Omega_t}(\sigma(y,t) - \bar{\sigma})\mathbf{N}(\mathbf{x},\mathbf{y})dy,$$

(4.8.14)

where ω is a compact support function in Ω_t. Thus, by periodicity we extend $\mathbf{U}$ onto the whole domain $x' \in R^2$, $0 < z < \zeta$. Differentiation of (4.8.6) with respect to time gives

$$\rho_b(x)\partial_t\mathbf{U})(x,t) = \int_{\Omega_t} \partial_t(\sigma(y,t) - \bar{\sigma})\Big[\frac{\mathbf{x} - \mathbf{y}}{|x - y|^n}\int_{|x-y|}^{\infty} \omega\Big(y + \xi\frac{\mathbf{x} - \mathbf{y}}{|x - y|}\Big)\xi^2 d\xi\Big] dy +$$

$$\int_{\Gamma_t}(\sigma(y,t) - \bar{\sigma})\mathbf{N}(\mathbf{x},\mathbf{y})\mathbf{u}\cdot\mathbf{n}dS = \mathbf{I}_1 + \mathbf{I}_b.$$

(4.8.15)

Now we use (4.8.10) to estimate differentiation in time of σ $\partial_t\sigma(y,t)$ that coincides with $\nabla\cdot(\rho\mathbf{u})$ and can be estimated in terms of $\nabla\mathbf{u}$. Next, we compute the time derivative of $\bar{\sigma}$. It holds

$$\partial_t\bar{\sigma} = \frac{d}{dt}\Big(\frac{M - \int_{\Omega_t}\rho_b(z)dzdx}{|\Omega_t|}\Big) =$$

$$-\Big(\frac{M - \int_{\Omega_t}\rho_b(z)dzdx}{|\Omega_t|^2}\Big)\frac{d|\Omega_t|}{dt} + \frac{1}{|\Omega_t|}\frac{d}{dt}\int_{\Sigma}\int_0^{\zeta(x',t)}\rho_b(z)dzdx'$$

$$\frac{d|\Omega_t|}{dt} = \frac{d}{dt}\int_{\Sigma}\zeta(x',t)dx' = \int_{\Sigma}\partial_t\zeta dx' = \int_{\Sigma}\mathbf{u}\cdot\tilde{\mathbf{n}}dx',$$

$$\frac{d}{dt}\int_{\Sigma}\int_0^{\zeta(x',t)}\rho_b(z)dzdx' = \int_{\Sigma}\rho_b(\zeta(x',t))\partial_t\zeta(x',t)dx' = \int_{\Sigma}\rho_b(\zeta(x',t))\mathbf{u}\cdot\tilde{\mathbf{n}}dx'.$$

Hence, the time derivative of $\bar{\sigma}$ furnishes

$$\partial_t\bar{\sigma} = -\frac{M - \int_{\Omega_t}\rho_b(z)dzdx}{|\Omega_t|^2}\int_{\Sigma}\mathbf{u}\cdot\tilde{\mathbf{n}}dx' + \frac{1}{|\Omega_t|}\int_{\Sigma}\rho_b(\zeta(x',t))\mathbf{u}\cdot\tilde{\mathbf{n}}dx'.$$

(4.8.16)

Both the integrals on the left side of (4.8.16) can be bounded through the L^2 norm of $\nabla\mathbf{u}$, and by Korn's inequality by $S(\mathbf{u})$. Inequality (5.4.13)$_1$ for $\rho_b\mathbf{U}$ follows by the Calderon–Zygmund theorem.

Furthermore, we take the vector $\mathbf{W}$ in the form

$$\mathbf{W})(x) = \nabla \times (\mathbf{A}\chi(z)) + \bar{\sigma}z\mathbf{k}, \qquad if \quad n = 3,$$

$$\mathbf{W} = \left(-\frac{\partial}{\partial x}[b(x',t)\chi(z)], \chi(z)\frac{\partial b(x',t)}{\partial x} \right) + \bar{\sigma}z\mathbf{k}, \qquad if \quad n = 2, \tag{4.8.17}$$

where $\chi(z)$, is a regular cut-off function vanishing for $z < (h/4)$, and equal to one in a neighborhood of $\zeta(x',t) \geq \min_{(x',t)} \zeta(x',t) > (h/4)$. Also it is $\mathbf{A} = (-b_2(x',t), b_1(x',t), 0)$, and

$$\frac{\partial b_1}{\partial x_1} + \frac{\partial b_2}{\partial x_b} = \left(V - \bar{\sigma}\eta \right)(x',t), \qquad if \quad n = 3,$$

$$\frac{\partial b}{\partial x_1} = \left(V - \bar{\sigma}\eta \right)(x_1, t), \qquad if \quad n = 2. \tag{4.8.18}$$

Since the compatibility condition

$$\int_{\Sigma}(V - \bar{\sigma}\eta)dx' = \int_{\Gamma_t}(V - \bar{\sigma}\eta)n_z dS = |\Gamma_t|\bar{V} - \bar{\sigma}\int_{\Gamma_t}\eta n_z dS,$$

is satisfied, there exists a periodic vector $\mathbf{b} \equiv (b_1, b_2)$, respectively scalar b, fields satisfying (4.8.18), and such that

$$\|\mathbf{b}\|_{W^{2,2}(\Sigma)} \leq c\|V - \bar{\sigma}\eta\|_{W^{1,2}(\Sigma)},$$

$$\|b\|_{W^{2,2}(\Sigma)} \leq c\|V - \bar{\sigma}\eta\|_{W^{1,2}(\Sigma)}. \tag{4.8.19}$$

Moreover, it holds. These imply (4.8.12) and the Lemma is proved. □

Some inequalities

The following inequalities hold true.

Lemma 4.8.3 *Let $\mathbf{u}$ be a solenoidal vector field in $W^{1,2}(\Omega)$, with Ω a bounded domain. If $\mathbf{u}$ is orthogonal to rigid motions, then the following inequalities hold true*

$$\|\mathbf{u}\|_{L^2(\Omega)} \leq c_P\|\nabla\mathbf{u}\|_{L^2(\Omega)}, \tag{4.8.20}$$

$$\|\mathbf{u}\|_{L^6(\Omega)} \leq c_S\|\nabla\mathbf{u}\|_{L^2(\Omega)}, \tag{4.8.21}$$

$$\|\nabla\mathbf{u}\|_{L^2(\Omega)} \leq c_K\|S(\mathbf{u})\|_{L^2(\Omega)}, \tag{4.8.22}$$

where c_P, c_S and c_K are the Poincare', Sobolev, and Korns constants.

Proof. The first two inequalities $(4.8.20)_{1,2}$ are true in a domain, bounded at least in one direction, when $\nabla\mathbf{u} = 0$ implies $\mathbf{u} = 0$, cf. [5]. This statement is true thanks to the hypothesis that $\mathbf{u}$ vanishes on $\Sigma \times \{0\}$.

The third inequality $(4.8.20)_2$ is the Korn inequality, again true because $\mathbf{u}$ is zero at bottom; cf. [29, 35, 41, 58].   □

4.8.3 Bibliographical Notes

For compressible fluids with free boundaries we quote few existence theorems of steady flows, such as Zajazkowsky Pileckas [52, 119], and some results can be quoted for the existence for unsteady flows; cf. [50, 51, 74–76, 126, 126, 127, 133, 137, 138, 142], stability and uniqueness results have been furnished for large initial data, cf. [45, 46, 113]. For viscous isothermal fluids having upper free surface uniqueness and stability results have been furnished by the author and Solonnikov in [113], however no results are known in the general barotropic case or when the upper membrane is elastic. Here we have proven stability of the rest state when outside the fluid there is the vacuum. The proof of uniqueness of a compressible fluid bounded by an elastic membrane, and the study of stability of the rest state is in preparation. The study of non-linear instability has recently been started by the author; cf. [107], the author and Solonnikov [113–118].

Below we quote some earlier results of the 'well-posedness' theory.

Existence results, incompressible fluids – The motion of a horizontal layer of viscous, incompressible fluid bounded below by a rigid plane and above by a free surface, taking surface tension into account, has been studied by several authors, we quote only the pioneering results by Beale [9, 10], and Beale and Nishida [11]. For incompressible fluid drops existence theorems of motions have been started by Solonnikov [1, 64, 132, 133], and for equilibrium configurations, see [32]. Other existence results in incompressible fluid motions with free boundary problems or with fluid-elastic interaction can be found in [19, 48, 49], and [46] for continuous dependence, if $K(\zeta)$ satisfies either (1.6.10), or (1.6.11). Recently it has been studied the stability of a capillary Poiseuille flow moving over a inclined plane with upper free boundary [79, 81, 117].

Stability results, incompressible fluids – The Rayleigh (1842-1919)-Taylor (1886-1975) stability for incompressible fluids presents a wide literature and we omit it. We just quote some books for references Batchelor (1920-2000) [6], Drazin and Reid [22], Chandrasekhar (1910-1995) [18], Kopachevsky and Krein [59], Yih [150].

Existence results, compressible fluids – For a section of a layer of compressible fluid with upper free surface we quote the results about the existence of non-steady motions by Secchi and Valli [126], by Solonnikov [131], by Solonnikov and Tani [137], [13], by Tani [142], and by Jin and the author [51] and about stationary motions by Jin and the author [53].

Besides the stability results for inviscid fluids by Sedenko and Yudovich [127], stability has been mainly been studied for viscous fluids, we quote two typical examples. In one case we study the stability of a rotating fluid drop, quoting the paper by Solonnikov and the author; cf. [114]. In the second case it is demonstrated that the absence of the lower rigid boundary allows vertical disturbance in the velocity in the vertical direction z that may produce novel features of instability, so called Rayleigh–Taylor instability. The Rayleigh–Taylor stability of the problem has been studied by Whitehead and Chen [149], Helffer and Lafitte [55], Iooss and Rossi [62], and by Solonnikov and the author [113]. All the above results are proved in using small initial data.

Here we wish to quote previous results about stability of a fluid with free surface. We notice that such results are available in the two cases of a layer and of a drop. Nonlinear stability results have been proved separately for a layer [97, 106, 113], and for a liquid drop [115, 132, 134]. A first nonlinear instability result under a technical assumption was proved in [107]. Several linear instability results can be quoted [68, 114, 134–136] for a liquid drop and analogous proof can be developed true also for a layer with rigid surface upward. All results but [113] hold for incompressible fluids.

Chapter 5
Polytropic Fluids with Rigid Boundary

E vidi lume in forma di riviera
fulgido di fulgore, in tra due rive
dipinte di mirabil primavera.
61, XXX, Paradiso, A. Dante

Experience up to now justifies our faith that nature actualizes the simplest
mathematically conceivable ideas.
A. Einstein

5.1 Introduction

The process of heat transfer in a fluid is quite complex, because it combines
with the motion of the fluid. Think to a heated body immersed in a fluid,
it cools more rapidly in a moving fluid than in a fluid at rest. Such cooling
phenomenon due typically to the motion of the fluid is called **convection**.

In this chapter we study the uniqueness and stability properties of rest
state S_b of an horizontal layer of a viscous, heat conducting fluid, in the class
of regular steady flows, such a problem is known as *Benard Problem* [18].

The most studied problem is given by the Boussinesq approximation,
and deals with incompressible fluids. Not always experiments fits with the
Boussinesq approximation, when this happens the more complex and rigorous
compressible problem has to be studied.

In this chapter we study the full Benard problem for polytropic viscous
fluids.

Results in this chapter represent a modified version of the results proved
by the author in [15, 20, 94, 105].

M. Padula, *Asymptotic Stability of Steady Compressible Fluids*,
Lecture Notes in Mathematics 2024, DOI 10.1007/978-3-642-21137-9_5,
© Springer-Verlag Berlin Heidelberg 2011

Other boundary conditions for velocity and temperature have been discussed in Sect. 1.6, and the same algorithm should apply also to these boundary conditions. However, until now there have been no results in this direction.

The plane of the chapter is as follows:

Section 5.1 The **boundary value problem**, (BVP), and the **initial boundary value problem**, (IBVP), are set for steady and unsteady motions, respectively of a portion of heavy, polytropic viscous gas filling a section of a horizontal layer, with upper and lower rigid horizontal planes, of thickness h, heated from below. In case of non homogeneous thermal boundary condition, the rest state $S_b \equiv (0, \rho_b, \Theta_b)$ with non uniform temperature and density fields, constitutes an example of exact solution.

Section 5.2 Under periodicity assumptions on the horizontal variables, and under smallness hypothesis on density and temperature gradients, $\nabla \rho_b$, $\nabla \Theta_b$ of the basic flow, say hypothesis A, in a given class of steady regular flows, it is proved uniqueness of rest state S_b of a viscous polytropic gas.

Section 5.3 Under hypothesis A nonlinear exponential stability of rest state S_b is proved. As physically expected, the stability result depends on temperature gradient of the basic flow. If the temperature gradient is zero then no restrictions are requested for the proof of exponential stability.

Section 5.4 Auxiliary Lemmas are proven, there suitable functions are constructed, which are used as test functions both in the proofs of uniqueness and asymptotic stability as well.

We conjecture that all proof and considerations of this section continue to hold in the case of a infinite layer. We leave it as open problem. Another open problem is the study of uniqueness and stability of a basic flow of a fluid filling either a closed fixed region whose perfectly conducting walls are not uniformly heated, or a layer with upper free boundary.

5.1.1 *Equations of Motion*

In the horizontal layer we reduce the problem to the study of uniqueness in a parallelepiped Ω, with height h, with four vertical walls S_l, and with two horizontal rectangular bases Σ, Σ_h. We assume periodicity conditions on vertical lateral opposite walls. Introduce a orthonormal reference frame $\mathcal{R} = \{O, \mathbf{i}, \mathbf{j}, \mathbf{k}\}$ with $O \in \Sigma$, $\mathbf{k}$ orthogonal to Σ, and upward directed, with (x, y, z) the coordinates of a point in $\mathcal{R}$. We set $x' \equiv (x, y)$ to denote a point on Σ, in general for a three-dimensional vector $\mathbf{v}$, by $\mathbf{v}'$ we denote the two dimensional vector $\mathbf{v}' \equiv (v_x, v_y)$. Thus, the bottom Σ is described by the equation $z = 0$, while the upper plane Σ_h is described in Cartesian coordinates by $z = h$, and the domain Ω occupied by the fluid is $\Omega = \{\mathbf{x} = (x', z) : x' \in \Sigma, 0 < z < h\}$. We denote by $\nabla' = (\partial_x, \partial_y)$, the derivatives along horizontal variables, and it holds $\nabla' \cdot \mathbf{u}' = \partial_x u_x + \partial_y u_y$. The periodicity cell is denoted by Σ, the fluid volume by V. As usual $\nabla = (\partial_x, \partial_y, \partial_z) = (\nabla', \partial_z)$ denotes the cartesian gradient operator, $\Delta = \partial_x^2 + \partial_y^2 + \partial_z^2$ the Laplacian operator.

At horizontal plane boundaries Σ, and Σ_h, for velocity and temperature fields we assume Dirichlet boundary conditions. Precisely, for the velocity we prescribe the no-slip condition given by the following boundary condition

$$\mathbf{u} = \mathbf{0}, \quad \text{nosplip.} \tag{5.1.1}$$

For the temperature we prescribe Dirichlet condition corresponding to perfectly conducting walls.

$$\Theta = \Theta_b \quad \text{perfectly heat conducting.} \tag{5.1.2}$$

In order to study the Benard problem we recall that unsteady motions of polytropic fluids satisfy the following Initial Boundary Value Problem

$$\rho_t + \nabla \cdot (\rho \mathbf{u}) = 0, \qquad\qquad \Omega \times (0, T),$$

$$\rho(\mathbf{u}_t + \mathbf{u} \cdot \nabla)\mathbf{u} = -R_* \nabla(\rho \Theta) + \mu \Delta \mathbf{u} + (\lambda + \mu)\nabla \nabla \cdot \mathbf{u} - \rho g \mathbf{k}, \qquad\qquad \Omega \times (0, T),$$

$$\rho c_v(\Theta_t + \mathbf{u} \cdot \nabla \Theta) = \chi \Delta \Theta - R_* \rho \Theta \nabla \cdot \mathbf{u} + 2\mu D^2(\mathbf{u}) + \lambda(\nabla \cdot \mathbf{u})^2, \qquad\qquad \Omega \times (0, T), \tag{5.1.3}$$

$$\rho(x, 0) = \rho_0(x), \quad \mathbf{u}(x, 0) = \mathbf{u}_0(x), \quad \Theta(x, 0) = \Theta_0(x), \qquad\qquad \Omega,$$

$$\mathbf{u}(x', 0, t) = \mathbf{u}(x', h, t) = 0, \qquad\qquad \Sigma \cup \Sigma_h \times (0, T),$$

$$\Theta(x', 0, t) = \Theta_h + \beta h, \quad \Theta(x', h, t) = \Theta_h, \qquad\qquad \Sigma \cup \Sigma_h \times (0, T),$$

where μ is the shear viscosity, λ the bulk viscosity, $D(\mathbf{u}) = (\nabla \mathbf{u} + \nabla^T \mathbf{u})/2$ is the rate-of-strain tensor, c_v the specific heat at constant volume, χ the heat conduction.

We assume for the pressure the Boyle–Mariot law $p = R_* \rho \Theta$, R_* the universal gas constant, c_v is the specific heat at constant volume, χ is the coefficient of thermal conductivity. Moreover, β is the basic temperature gradient, Θ_h is the given temperature at level $z = h$.

System (5.1.3) is complete.

Remark 5.1.1 *We remark that the total mass*

$$\int_\Omega \rho \, dx = M,$$

is determined by the value of density ρ_0 at initial time. In Sect. 5.3 we shall study the stability of a steady motion S_b perturbing initial data. The perturbation ρ_0 to the density must satisfy the condition that the total $\int_\Omega \rho_0\, dx$ must coincide with the total mass of S_b.

We also recall that steady motions of heat conducting viscous polytropic fluids satisfy the following Boundary Value Problem

$$\nabla \cdot (\rho \mathbf{u}) = 0, \qquad\qquad\qquad\qquad\qquad \Omega, \quad (5.1.4)$$

$$\rho(\mathbf{u} \cdot \nabla)\mathbf{u} = -R_*\nabla(\rho\Theta) + \mu\Delta\mathbf{u} + (\lambda + \mu)\nabla\nabla \cdot \mathbf{u} - \rho g\mathbf{k}, \qquad\qquad \Omega,$$

$$\rho c_v(\mathbf{u} \cdot \nabla)\Theta = \chi\Delta\Theta - R_*\rho\Theta\nabla \cdot \mathbf{u} + 2\mu D^2(\mathbf{u}) + \lambda(\nabla \cdot \mathbf{u})^2, \qquad\qquad \Omega,$$

$$\mathbf{u}(x',0) = \mathbf{u}(x',h) = 0, \qquad \Theta(x',0) = \Theta_h + \beta h, \quad \Theta(x',h) = \Theta_h, \quad \Sigma,$$

$$\int_\Omega \rho = M.$$

The periodicity cell Σ, the total mass M, the data h Θ_h, β, g, and all the physical coefficients μ, λ, c_v, χ, are given positive constants.

Exact solution: rest state S_b

The rest state satisfies the following Boundary Value Problem

$$0 = -R_*\nabla(\rho_b, \Theta_b) - \rho_b g\mathbf{k}, \qquad\qquad \Omega,$$

$$0 = \chi\Delta\Theta_b, \qquad\qquad \Omega, \qquad (5.1.5)$$

$$\Theta_b(x',0) = \Theta_h + \beta h, \quad \Theta_b(x',h) = \Theta_h,, \qquad \Sigma,$$

$$\int_\Omega \rho_b dx = M.$$

We begin by construct an exact solution called the rest state $S_b = (\mathbf{u}_b, \Theta_b, \rho_b)$, where

$$\mathbf{u}_b = 0, \qquad\qquad\qquad\qquad \Theta_b = \beta(h - z) + \Theta_h, \qquad (5.1.6)$$

$$\rho_b = \rho_*(\beta(h - z) + \Theta_h)^m, \qquad\qquad m := \frac{g}{R_*\beta} - 1,$$

satisfy (5.1.5) in $\Omega = \Sigma \times (0, h)$.

Moreover, $(5.1.5)_5$ implies

$$\rho_* = \frac{M\beta(m + 1)}{|\Sigma|[(\beta h + \Theta_h)^{m+1} - \Theta_h^{m+1}]},$$

$$\frac{M}{h|\Sigma|}\frac{1}{(\beta h + \Theta_h)^m} < \rho_* < \frac{M}{h|\Sigma|}\frac{1}{\Theta_h^m}.$$

Remark 5.1.2 *Notice that β small implies that $\nabla\rho_b$ is small. Thus it is enough to suppose β small in our theorems.*

5.2 Uniqueness of the Rest State

In this section we prove that under suitable conditions on the physical parameters the total mass M, $\|\rho\|_{L^\infty}$ the data h, β, Θ_h, g, and μ, λ, c_v, χ, the rest state S_b given by (5.1.6) is the unique solution to the Boundary Value Problem (5.1.4) in a suitable regularity class of steady solutions.

In order to present the main result we introduce the following regularity class $(\mathbf{u}(x', z), \rho(x', z), \Theta(x')) \in \mathcal{V}$

$$\mathcal{V} = W_\sharp^{1,2}(\Omega) \times C_\sharp^0(\Omega) \cap W^{1,3}(\Omega) \times W_\sharp^{1,2}(\Omega), \qquad l > 0.$$

Notice that this space contains densities ρ in $L^\infty(\Omega)$.

Aim of this section is the proof of the following uniqueness theorem.

Theorem 5.2.1 *The rest state S_b given in (5.1.6) is the unique solution to the Boundary Value Problem (5.1.4), in the class of solutions $\mathcal{V}$, corresponding to the same boundary data and the same total mass, provided conditions (5.4.5), (5.2.21) below are satisfied.*

Proof. Assume by absurdum that there exists another solution $\mathbf{u}(x), \rho(x),$ $\Theta(x)$ defined in Ω, satisfying the system (5.1.4), with the same boundary data $(5.1.4)_{4-7}$, and the same total mass, we wish to prove a contradiction.

It is useful to observe that from the property of the rest state

$$-g\mathbf{k} = R_* \frac{\nabla(\Theta_b\rho_b)}{\rho_b}, \tag{5.2.1}$$

the following identity yields

$$-R_*\nabla(\Theta_b\rho) - g\rho\nabla z = -R_*\left(\nabla(\Theta_b\rho) - \rho\frac{\nabla(\Theta_b\rho_b)}{\rho_b}\right) = \tag{5.2.2}$$

$$-R_*\Theta_b\rho\left[\frac{\nabla(\Theta_b\rho)}{\Theta_b\rho} - \frac{\nabla(\Theta_b\rho_b)}{\Theta_b\rho_b}\right] = -R_*\rho\Theta_b\nabla\ln\left(\frac{\rho}{\rho_b}\right).$$

Therefore it holds

$$-R_*\nabla(\rho\theta + \rho\Theta_b) - g\rho\nabla z = -R_*\nabla(\rho\theta) - R_*\rho\Theta_b\nabla\ln\left(\frac{\rho}{\rho_b}\right).$$

Substituting this identity in (5.1.4) we find that the difference $\mathbf{u}(x)$, $\sigma = \rho(x) - \rho_b(x)$, $\theta = \Theta(x) - \Theta_b(x)$ satisfies the perturbed Boundary Value Problem

$$\nabla \cdot ((\rho_b + \sigma)\mathbf{u}) = 0, \qquad\qquad\qquad\qquad\qquad \Omega \quad (5.2.3)$$

$$\rho(\mathbf{u} \cdot \nabla)\mathbf{u} - \mu\Delta\mathbf{u} - (\lambda + \mu)\nabla\nabla \cdot \mathbf{u} = -R_*\nabla\rho\theta - R_*\rho\Theta_b\nabla\ln\left(\frac{\rho}{\rho_b}\right), \quad \Omega$$

$$\rho c_v(\mathbf{u} \cdot \nabla)\Theta = \chi\Delta\theta - R_*\rho\Theta\nabla \cdot \mathbf{u} + 2\mu D^2(\mathbf{u}) + \lambda(\nabla \cdot \mathbf{u})^2, \qquad\qquad \Omega$$

$$\mathbf{u}(x',0) = \mathbf{u}(x',h) = 0, \qquad \theta(x',0) = \theta(x',h) = 0, \qquad\qquad \Sigma$$

$$\int_\Omega \sigma\,dx = 0.$$

We set

$$\|(\mathbf{u}, \theta, \sigma)\|_X^2 := \|\mathbf{u}\|_{L^2}^2 + \|\theta\|_{L^2}^2 + \|\sigma\|_{L^2}^2.$$

The line of proof consists in the construction of a functional $\mathbb{G}$ quadratic in the L^2 norms of perturbations, that along the solution of (5.2.3) satisfies the inequality

$$\mathbb{G} \le 0.$$

If $\mathbb{G}$ is positive definite, inequality $\mathbb{G} \le 0$ can be satisified only if the variables are intenticaly zero, which contradicts the starting hypothesis of the theorem and achieves the thesis.

Therefore, sufficient conditions to guarantee the positiveness of the form $\mathbb{G}$ are also sufficient to ensure uniqueness of the rest state. □

Energy Equation of Perturbation

Set $T = \theta/\Theta$. Sometime in the same formula both symbols T and θ may appear to indicate the perturbation to the temperature.

Let us multiply equations $(5.2.3)_2$ by $\mathbf{u}$, $(5.2.3)_3$ by $T/(1+T)$, and integrate over Ω. Integrating by parts, taking into account $(5.2.3)_1$ and boundary conditions $(5.2.3)_{4-7}$ we obtain

$$\int_\Omega \left(\mu(\nabla\mathbf{u})^2 + (\lambda + \mu)(\nabla \cdot \mathbf{u})^2\right)dx = -R_*\int_\Omega \mathbf{u} \cdot \nabla(\rho\theta) - R_*\int_\Omega \rho\Theta_b\mathbf{u} \cdot \nabla\ln\left(\frac{\rho}{\rho_b}\right)dx,$$
$$(5.2.4)$$

$$\chi\int_\Omega \nabla\theta \cdot \nabla\left(\frac{T}{1+T}\right)dx = -c_v\int_\Omega \rho\mathbf{u} \cdot \nabla\left(\Theta_b(1+T)\right)\frac{T}{1+T}\,dx$$
$$- R_*\int_\Omega \rho\theta\nabla \cdot \mathbf{u}\,dx + \int_\Omega \frac{T}{1+T}\left(\mu(\nabla\mathbf{u})^2 + (\lambda + \mu)(\nabla \cdot \mathbf{u})^2\right)dx.$$

We now develop separately each term in $(5.2.4)_2$.

For the integral at left hand side of $(5.2.4)_2$, since Θ_b is linear in z, and T is zero at boundaries Σ, Σ_h, by periodicity on lateral surface, it is easy to check that the following two identities hold

$$\Delta\Theta_b = 0, \qquad \int_{\partial\Omega} \mathbf{k}\cdot\mathbf{n}\left(\ln(1+T) + \frac{1}{1+T}\right)dS = 0.$$

The integral $\int_\Omega \nabla\theta\cdot\nabla(T/(1+T))dx$ apparently has no definite sign, however integrating by parts, and taking into account that $T = 0$ at horizontal planes Σ, Σ_h of the boundary $\partial\Omega$, and using the periodicity conditions on the lateral walls one verifies that it holds

$$\int_\Omega \nabla(\Theta_b T)\cdot\nabla\left(\frac{T}{1+T}\right)dx = \int_\Omega \left(\Theta_b\nabla T + T\nabla\Theta_b\right)\cdot\frac{\nabla T}{(1+T)^2}dx$$

$$= \int_\Omega \Theta_b\frac{|\nabla T|^2}{(1+T)^2}dx + \int_\Omega \nabla\Theta_b\cdot\nabla T\frac{T}{(1+T)^2}dx$$

$$= \int_\Omega \Theta_b\frac{|\nabla T|^2}{(1+T)^2}dx - \beta\int_\Omega \mathbf{k}\cdot\nabla\left(\ln(1+T) + \frac{1}{1+T}\right)dx \qquad (5.2.5)$$

$$= \int_\Omega \Theta_b\frac{|\nabla T|^2}{(1+T)^2}dx - \beta\int_{\partial\Omega} \mathbf{k}\cdot\mathbf{n}\left(\ln(1+T) + \frac{1}{1+T}\right)dx$$

$$= \int_\Omega \Theta_b\frac{|\nabla T|^2}{(1+T)^2}dx = \int_\Omega \Theta_b|\nabla \ln(1+T)|^2.$$

Concerning the first integral at right hand side we observe what follows

$$\nabla\left(\Theta_b(1+T)\right)\frac{T}{1+T} = T\nabla\Theta_b + \Theta_b\left(1 - \frac{1}{1+T}\right)\nabla T = \nabla(\Theta_b T) - \Theta_b\nabla\ln(1+T).$$
$$(5.2.6)$$

Notice that $\int_\Omega \rho\mathbf{u}\cdot\nabla(\Theta_b T) = 0$. Substituting (5.2.6) in $(5.2.4)_2$, and integrating by parts in $(5.2.4)_2$ yields

$$\chi\int_\Omega \Theta_b|\nabla \ln(1+T)|^2 dx = -c_v\beta\int_\Omega \rho\mathbf{u}\cdot\mathbf{k}\ln(1+T)\,dx - R_*\int_\Omega \rho\theta\nabla\cdot\mathbf{u}\,dx +$$

$$\int_\Omega \frac{T}{1+T}\left(\mu(\nabla\mathbf{u})^2 + (\lambda+\mu)(\nabla\cdot\mathbf{u})^2\right)dx. \qquad (5.2.7)$$

Adding $(5.2.4)_1$ to (5.2.7) yields

$$\int_\Omega \frac{\Theta_b}{\Theta}\left(\mu(\nabla\mathbf{u})^2 + (\lambda+\mu)(\nabla\cdot\mathbf{u})^2\right)dx + \chi\int_\Omega \Theta_b|\nabla \ln(1+T)|^2\,dx \qquad (5.2.8)$$

$$= -c_v\beta\int_\Omega \rho\ln(1+T)\mathbf{u}\cdot\mathbf{k}\,dx - R_*\beta\int_\Omega \rho\ln\left(\frac{\rho}{\rho_b}\right)\mathbf{u}\cdot\mathbf{k}\,dx.$$

We observe that by the Lagrange mean value theorem it holds

$$\ln\left(\frac{\rho}{\rho_b}\right) = \frac{1}{\bar{\rho}}\sigma, \tag{5.2.9}$$

with $\bar{p}$ a point between ρ and ρ_b. Therefore the r.h.s. of (5.2.8) can be increased by the norm $\|\cdot\|_X$ of disturbances, thus it is bounded above by $\|\nabla\mathbf{u}\|_{L^2}^2$, $\|\sigma\|_{L^2}^2$, $\|\nabla\ln(1+T)\|_{L^2}^2$. Specifically we have

$$-c_v\beta\int_\Omega \rho\ln(1+T)\mathbf{u}\cdot\mathbf{k}\,dx - R_*\beta\int_\Omega \rho\ln\left(\frac{\rho}{\rho_b}\right)\mathbf{u}\cdot\mathbf{k}\,dx$$

$$\le c_v\beta\|\rho\|_{L^{3/2}}\|\mathbf{u}\|_{L^6}\|\ln(1+T)\|_{L^6} + R_*\beta\left\|\frac{\rho}{\bar{\rho}}\right\|_{L^3}\|\mathbf{u}\|_{L^6}\|\sigma\|_{L^2} \tag{5.2.10}$$

$$= \beta\,a_1\|\nabla\mathbf{u}\|_{L^2}\|\nabla\ln(1+T)\|_{L^2} + \beta\,a_2\|\nabla\mathbf{u}\|_{L^2}\|\sigma\|_{L^2},$$

and

$$a_1 := c_v\,c_S^2\|\rho\|_{L^{3/2}}$$

$$a_2 := R_*\,c_S\left\|\frac{\rho}{\bar{\rho}}\right\|_{L^3}.$$

In this way we deduce

$$\mu\inf_\Omega\left|\frac{\Theta_b}{\Theta}\right|\|\nabla\mathbf{u}\|_{L^2}^2 + \chi\inf_\Omega\Theta_b\|\nabla\ln(1+T)\|_{L^2}^2 \tag{5.2.11}$$

$$\le \beta\,a_1\|\nabla\mathbf{u}\|_{L^2}\|\nabla\ln(1+T)\|_{L^2} + \beta\,a_2\|\nabla\mathbf{u}\|_{L^2}\|\sigma\|_{L^2}.$$

Inequality (5.2.11) by Cauchy inequality can be simplified as

$$b_1\|\nabla\mathbf{u}\|_{L^2}^2 + b_2\|\nabla T\|_{L^2}^2 \le \beta\,a_2\|\nabla\mathbf{u}\|_{L^2}\|\sigma\|_{L^2}. \tag{5.2.12}$$

where we have assumed

$$b_1 := \mu\inf_\Omega\frac{\Theta_b}{\Theta} - \beta\frac{a_1}{2} > 0, \tag{5.2.13}$$

$$b_2 := \chi\inf_\Omega\Theta_b - \beta\frac{a_1}{2} > 0, \tag{5.2.14}$$

and α is an arbitrary positive number.

Unfortunately at the left hand side of (5.2.12) there appear only the norms $\|\nabla\mathbf{u}\|_{L^2}$, and $\|\nabla T\|_{L^2}$, while the norm $\|\sigma\|_{L^2}$ is missing. Hence the right hand side of (5.2.12) **cannot** be increased by the left hand side of (5.2.12)!

The key point of the proof of our uniqueness theorem consists in finding an additional inequality that can provide these terms. In the spirit of the papers [104] and [113], we compute the free work identity, using again the equation of momentum $(5.2.3)_2$.

Free Work Equation

Our goal is to estimate $\|\sigma\|_{L^2}$ in terms of $\|\nabla \mathbf{u}\|_{L^2}$, and $\|\nabla T\|_{L^2}$.

We multiply $(5.2.3)_2$ by a suitable vector field $\mathbf{V}$, solution of the problem $(5.4.4)$ introduced in Lemma 5.4.2 of Sect. 5.4, and integrate over Ω. We obtain the free work equation:

$$\int_\Omega \rho \mathbf{u} \cdot \nabla \mathbf{u} \cdot \mathbf{V} dx - \int_\Omega (\mu \Delta \mathbf{u} + (\lambda + \mu) \nabla \cdot \mathbf{u}) \cdot \mathbf{V} dx + R_* \int_\Omega \mathbf{V} \cdot \nabla (\rho \theta) dx$$

$$= -R_* \int_\Omega \rho \Theta_b \mathbf{V} \cdot \nabla \ln \left(\frac{\rho}{\rho_b} \right) dx. \tag{5.2.15}$$

Integrating by parts the left hand side of (5.2.15) we obtain the nonlinear functional

$$J_1 = \int_\Omega \rho \mathbf{u} \cdot \nabla \mathbf{u} \cdot \mathbf{V} dx + \int_\Omega \left(\mu \nabla \mathbf{u} : \nabla \mathbf{V} + (\lambda + \mu) \nabla \cdot \mathbf{u} \nabla \cdot \mathbf{V} \right) dx - R_* \int_\Omega \rho \theta \nabla \cdot \mathbf{V} \, dx.$$

To find a dissipative term on σ we choose as test function $\mathbf{V}$ the one constructed in Lemma 5.4.2. Employing (5.2.9), and (5.4.4) it yields

$$R_* \int_\Omega \nabla \cdot (\rho \Theta_b \mathbf{V}) \ln \left(\frac{\rho}{\rho_b} \right) dx = R_* \int_\Omega \frac{\sigma^2}{\bar\rho} dx,$$

where $\bar\rho$ is a point between ρ and ρ_b. Above calculations, exchanging the sides in equation (5.2.15), yield the free work equation

$$R_* \int_\Omega \frac{\sigma^2}{\bar\rho} dx = -J_1, \tag{5.2.16}$$

where $R_* \int_\Omega \frac{\sigma^2}{\bar\rho} dx$ represents the squared natural norm where the problem should be studied. In Lemma 5.4.2, under hypothesis (5.4.5) it is proved the estimate

$$\|\nabla \mathbf{V}\|_{L^2} \le f_3 \|\sigma\|_{L^2}. \tag{5.2.17}$$

Remark 5.2.1 *We remark that (5.4.5) implies existence of a minimum for the density ρ. Also notice that estimates for the auxiliary function $\mathbf{V}$ exist only if it exists a minimum for the density ρ.*

Finally, employing the property (5.2.17) satisfied by $\mathbf{V}$, and using the Poincare' inequality we estimate J_1 as follows

$$J_1 \le \left(\|\rho \mathbf{u}\|_{L^3} \|\mathbf{u}\|_{L^6} + (\lambda + 2\mu) \|\nabla \mathbf{u}\|_{L^2} + R_* \|\rho \Theta\|_{L^3} \|T\|_{L^6} \right) \|\nabla \mathbf{V}\|_{L^2}$$

$$\le \left(c_1 \|\nabla \mathbf{u}\|_{L^2} + c_2 \|\nabla T\|_{L^2} \right) f_3 \|\sigma\|_{L^2}, \tag{5.2.18}$$

where

$$c_1 := c_S \|\rho \mathbf{u}\|_{L^3} + (\lambda + 2\mu),$$

$$c_2 := R_* c_S \|\rho \Theta\|_{L^3}.$$

Thus (5.2.16) together with inequality

$$\frac{R_*}{\|\rho\|_{L^\infty}} \|\sigma\|_{L^2}^2 \leq J_1,$$

yields the wanted estimate for σ,

$$\|\sigma\|_{L^2} \leq \frac{f_3}{R_*} \|\rho\|_{L^\infty} \Big(c_1 \|\nabla \mathbf{u}\|_{L^2} + c_2 \|\nabla T\|_{L^2} \Big). \tag{5.2.19}$$

We remark that the coefficient $\lambda + 2\mu$ cannot be taken small. Substituting (5.2.19) into (5.2.12) we deduce the algebraic inequality

$$b_1 \|\nabla \mathbf{u}\|_{L^2}^2 + b_2 \|\nabla T\|_{L^2}^2 \leq \beta \, a_2 \frac{f_3}{R_*} \|\rho\|_{L^\infty} \|\nabla \mathbf{u}\|_{L^2} \Big(c_1 \|\nabla \mathbf{u}\|_{L^2} + c_2 \|\nabla T\|_{L^2} \Big)$$

$$\leq \beta \, d_1 \|\nabla \mathbf{u}\|_{L^2}^2 + \beta \, d_2 \|\nabla \mathbf{u}\|_{L^2} \|\nabla T\|_{L^2}, \tag{5.2.20}$$

with

$$d_1 = c_1 a_2 \frac{f_3}{R_*} \|\rho\|_{L^\infty},$$

$$d_2 = c_2 a_2 \frac{f_3}{R_*} \|\rho\|_{L^\infty}.$$

We consider the inequality

$$\mathbb{G} := (b_1 - \beta \, d_1) \|\nabla \mathbf{u}\|_{L^2}^2 + b_2 \|\nabla T\|_{L^2}^2 - \beta \, d_2 \|\nabla \mathbf{u}\|_{L^2} \|\nabla T\|_{L^2} \leq 0,$$

which states that the quadratic form $\mathbb{G}$ in the variables $\|\nabla \mathbf{u}\|_{L^2}$, $\|\nabla T\|_{L^2}$ is less than zero. If

$$(b_1 - \beta \, d_1) b_2 \geq \left(\frac{\beta \, d_2}{2} \right)^2, \tag{5.2.21}$$

the quadratic form $\mathbb{G}$ is positive definite, and we obtain contradiction. Hence uniqueness has been proved.

Remark 5.2.2 *It is obvious that assumption of homogeneous temperature at boundary implies $\beta = 0$, then condition (5.2.21) is always satisfied. In general it is $\beta \neq 0$, thus uniqueness holds only for large kinematic and heat viscosity coefficients (small Rayleigh number). The reasoning could be straightforward extended to general potential forces, and Theorem 5.2.1 is recovered.*

Notice that to get uniform temperature it is enough in our case to prescribe uniform temperature at boundaries. The same is still true for a gas in a rigid vessel.

We also observe that the r.h.s. of (5.2.8) is naturally increased in terms of the L^2 norms of logarithmic functions of temperature and density perturbations. For temperature we find the same norm as dissipative term at l.h.s. of (5.2.8), for the density we do not have an analogous dissipative term at l.h.s. of (5.2.8) therefore we should derive artificially it. Here we aim to construct a dissipative term for the density through the auxiliary test function $\mathbf{V}$ satisfying a problem of kind

$$\nabla \cdot (\rho \Theta_b \mathbf{V}) = \ln \left(\frac{\rho}{\rho_b} \right) - c$$

$$\mathbf{V}|_{\partial \Omega} = 0,$$

(5.2.22)

with c suitable constant

$$c = \int_v \ln \left(\frac{\rho}{\rho_b} \right) dv.$$

The problem to derive the L^2 norm of the logarithmic function of density perturbation remains till now an open problem.

5.3 Stability Problem

In this Section we study the stability problem of the rest state of compressible heat conducting fluids, when the domain is a section of a horizontal layer Ω already defined in the previous section. We conjecture that all proof and considerations of this Section continue to hold in the case of a infinite layer. We leave it as open problem.

In order to present the main result we introduce the following regularity class $\mathcal{V}$ defined by

$$\mathcal{V} = \{\mathbf{u}(x', z, t), \rho(x', z, t), \Theta(x', t)$$

$$\in L^2(0, \infty; W_\sharp^{1,2}(\Omega)) \times C^0(0, \infty; C_\sharp^0(\Omega) \cap W^{1,3}(\Omega)) \times L^2(0, \infty; W_\sharp^{1,2}(\Omega))\}.$$

For all reasonings and proofs done in this section we shall suppose that there exists a regular, global solution $\left(\mathbf{u}(x, t), \rho(x, t), \Theta(x, t) \right)$ defined in Ω, satisfying system (5.1.3), corresponding to given initial data $\left(\mathbf{u}_0(x), \rho_0(x), \Theta_0(x) \right)$.

5.3.1 Energy Equation

As known in the presence of potential forces, a balance equation of total energy, still holds true for solutions to (5.1.3), namely it holds

$$\frac{dE}{dt} = \Phi, \tag{5.3.1}$$

where

$$\Phi = \chi \int_{\partial\Omega} \mathbf{n} \cdot \nabla\Theta \, dx$$

is the heat flux, and

$$E(t) = \int_{\Omega} \rho\left(\frac{u^2}{2} + \Pi + c_v\Theta\right) dx$$

is the **total energy** represented by the sum of specific **kinetic, potential** $\Pi = g \int_{\Omega} \rho z \, dx$ and **internal energy** $c_v\Theta$ see (1.7.10) [24]. Below we give the proof of equation (5.3.1). To this end we multiply $(5.1.3)_2$ times $\mathbf{u}$ and integrate over Ω, by use of $(5.1.3)_1$ and transport theorem, since $\mathbf{u}$ vanishes at boundary, we have

$$\frac{d}{dt}\int_{\Omega}\rho\left(\frac{u^2}{2} + \Pi\right)dx = -\mu\, D_u - \int_{\Omega}\mathbf{u}\cdot\nabla p(\rho,\Theta)dx, \tag{5.3.2}$$

where $D(\mathbf{u}) = \frac{\nabla\mathbf{u} + \nabla\mathbf{u}^T}{2}$ is the strain rate tensor.

Integrating $(5.1.3)_3$ over Ω and again using the transport theorem we get

$$\frac{d}{dt}\int_{\Omega}\rho\, c_v\Theta \, dx = \chi\int_{\partial\Omega}\mathbf{n}\cdot\nabla\Theta \, dx - \int_{\Omega}\nabla\cdot\mathbf{u}\,p(\rho,\Theta)dx + \mu\, D_u. \tag{5.3.3}$$

Adding (5.3.3) to (5.3.2) we obtain the celebrated energy equation cf. [24] (5.3.1).

It is evident that, despite the barotropic case, the difference $E(t) - E_b$, between the energies $E(t)$ computed along the unsteady thermal process $\left(\mathbf{u}, \rho, \Theta\right)$ and E_b computed along the rest state $\left(0, \rho_b, \Theta_b\right)$ doesn't furnish any more a positive definite quadratic form in the L^2-norm of perturbations $\left(\mathbf{u}, \sigma, \theta\right)$. *Therefore from energy equation we cannot anymore deduce a stability result, even for the rest state.* Because of this difficulty, in this section we construct first the **energy of perturbation** $\mathcal{E}(t)$, thus we deduce a **energy equation of perturbation**. Finally we construct a modified energy $\mathbb{E}(t)$ which we choose as appropriate Lyapunov functional.

The **equation governing evolution in time for** $\mathbb{E}$ will furnish stability when the Dirichlet data on the temperature are homogeneous, without any further hypothesis. In fact such equation provides an a priori estimate for the solutions.

5.3.2 Energy Equation of Perturbation

Here we keep the notation $T = \theta/\Theta_b$ introduced in previous section, and we shall use both notations θ, T in the same relation.

Let us begin by deriving a modified energy equation, to this end we write the perturbation equations

$$\sigma_t + \nabla \cdot ((\rho_b + \sigma)\mathbf{u}) = 0, \qquad\qquad \Omega \times (0, T),$$

$$\rho(\mathbf{u}_t + \mathbf{u} \cdot \nabla)\mathbf{u}) = \mu\Delta\mathbf{u} + (\lambda + \mu)\nabla\nabla \cdot \mathbf{u} - R_*\nabla(\rho\theta) + R_*\rho\Theta_b\nabla\left(\frac{\rho}{\rho_b}\right), \qquad \Omega \times (0, T),$$

$$c_v\rho(\theta_t + \mathbf{u} \cdot \nabla\theta) = \chi\Delta\theta + c_v\,\beta\rho\mathbf{u} \cdot \mathbf{k} - R_*\rho\Theta\nabla \cdot \mathbf{u} + 2\mu\,D^2(\mathbf{u}) + \lambda(\nabla \cdot \mathbf{u})^2, \quad \Omega \times (0, T),$$
$$(5.3.4)$$

$$\rho(x, 0) = \rho_0(x), \qquad \mathbf{u}(x, 0) = \mathbf{u}_0(x), \qquad \theta(x, 0) = \theta_0(x), \qquad\qquad \Omega,$$

$$\mathbf{u}(x', 0, t) = \mathbf{u}(x', h, t) = 0, \qquad \theta(x', 0, t) = \theta(x', h, t) = 0, \qquad \Omega \times (0, T),$$

$$\int_\Omega \sigma\,dx = 0, \qquad\qquad (0, T).$$

Theorem 5.3.1 Energy Equation of Perturbations. *Let* $\mathbf{u}$, $\rho = \rho_b + \sigma$, $\Theta = \Theta_b + \theta$ *solve* (5.1.3) *with* $\mathbf{u}, \sigma, \theta \in V$. *Then setting* $\theta/\Theta = T/(1 + T)$ *the following* **equation for the energy of perturbations** $\mathcal{E}$ *holds*

$$\frac{d\mathcal{E}}{dt} = I_0 - \mu\,\mathcal{D}_u(t) - \chi\,\mathcal{D}_\theta(t); \qquad\qquad (5.3.5)$$

$$\mathcal{E} = E_u + E_\sigma + E_\theta;$$

$$E_u = \frac{1}{2}\int_\Omega \rho\mathbf{u}^2\,dx;$$

$$E_\sigma = \int_\Omega \Theta_b\rho\left(\ln\frac{\rho}{\rho_b} - 1\right)dx;$$

$$E_\theta = c_v\int_\Omega \rho\Theta_b\left(T - \ln\frac{T}{1 + T}\right)dx;$$

$$\mu \mathcal{D}_u = \int_\Omega \frac{\Theta_b}{\Theta}(2\mu D(\mathbf{u}) : D(\mathbf{u}) + \lambda |\nabla \cdot \mathbf{u}|^2)\, dx;$$

$$\chi \mathcal{D}_\theta = \chi \int_\Omega \frac{\Theta_b^3}{\Theta^2}|\nabla T|^2\, dx;$$

$$I_0 = R_* \beta \int_\Omega \rho \ln\left(\frac{\rho}{\rho_b}\right)\mathbf{u} \cdot \mathbf{k}\, dx + c_v \beta \int_\Omega \rho \ln\left(\frac{\Theta}{\Theta_b}\right)\mathbf{u} \cdot \mathbf{k}\, dx.$$

Proof. Let us multiply $(5.3.4)_2$ by $\mathbf{u}$. Integrating by parts over Ω, taking into account the boundary conditions $(5.3.4)_{7-10}$, we obtain the following integral equation:

$$\frac{d}{dt}\int_\Omega \rho \frac{\mathbf{u}^2}{2}\, dx + \mu \|\nabla \mathbf{u}\|_{L^2}^2 + (\lambda + \mu)\|\nabla \cdot \mathbf{u}\|_{L^2}^2 \qquad (5.3.6)$$

$$= -R_* \int_\Omega \mathbf{u} \cdot \nabla(\rho\theta)\, dx + R_* \int_\Omega \rho\Theta_b \mathbf{u} \cdot \nabla \ln\left(\frac{\rho}{\rho_b}\right) dx.$$

In a fixed domain Ω the following identity is true

$$\frac{d}{dt}\int_\Omega \Theta_b \rho\left(\ln \frac{\rho}{\rho_b} - 1\right) dx = \int_\Omega \Theta_b\left\{\partial_t \rho\left(\ln \frac{\rho}{\rho_b} - 1\right) + \rho\partial_t \ln \frac{\rho}{\rho_b}\right\} dx \quad (5.3.7)$$

$$= \int_\Omega \Theta_b \partial_t \rho \ln \frac{\rho}{\rho_b}\, dx.$$

Employing the continuity equation

$$\partial_t \rho = -\nabla \cdot (\rho\mathbf{u}),$$

we get

$$\int_\Omega \Theta_b \partial_t \rho \ln \frac{\rho}{\rho_b}\, dx = -\int_\Omega \Theta_b \ln \frac{\rho}{\rho_b}\nabla \cdot (\rho\mathbf{u})\, dx \qquad (5.3.8)$$

$$= \int_\Omega \rho\mathbf{u} \cdot \nabla\left(\Theta_b \ln \frac{\rho}{\rho_b}\right) dx = -\beta \int_\Omega \rho\mathbf{u} \cdot \mathbf{k} \ln \frac{\rho}{\rho_b}\, dx + \int_\Omega \Theta_b \rho\mathbf{u} \cdot \nabla \ln \frac{\rho}{\rho_b}\, dx.$$

From (5.3.8) and (5.3.7) it follows

$$\int_\Omega \Theta_b \rho\mathbf{u} \cdot \nabla \ln \frac{\rho}{\rho_b}\, dx = \frac{d}{dt}\int_\Omega \Theta_b \rho\left(\ln \frac{\rho}{\rho_b} - 1\right) dx + \beta \int_\Omega \rho\mathbf{u} \cdot \mathbf{k} \ln \frac{\rho}{\rho_b}\, dx.$$
$$(5.3.9)$$

Substituting (5.3.9) into (5.3.6) we deduce

$$\frac{d}{dt} \int_\Omega \left\{ \rho \frac{\mathbf{u}^2}{2} \, dx + R_* \Theta_b \rho \left(\ln \frac{\rho}{\rho_b} - 1 \right) \right\} dx + \mu \|\nabla \mathbf{u}\|_{L^2}^2 + (\lambda + \mu)\mu \|\nabla \cdot \mathbf{u}\|_{L^2}^2$$

$$\tag{5.3.10}$$

$$= -R_* \int_\Omega \mathbf{u} \cdot \nabla(\rho\theta) \, dx + R_*\beta \int_\Omega \rho \mathbf{u} \cdot \mathbf{k} \ln \left(\frac{\rho}{\rho_b} \right) dx.$$

Notice that the trivial identity holds true

$$\left[(\Theta_b T)_t + \mathbf{u} \cdot \nabla(\Theta_b T) \right] \frac{T}{1+T} = \Theta_b (T_t + \mathbf{u} \cdot \nabla T) \frac{T}{1+T} + \mathbf{u} \cdot \nabla \Theta_b \frac{T^2}{1+T}$$

$$= \Theta_b \frac{d(T - \ln(1+T))}{dt} - \beta \mathbf{u} \cdot \mathbf{k} \frac{T^2}{1+T}.$$

$$\tag{5.3.11}$$

Hence, since it is $\Theta_b T = \theta$, integrating (5.3.11) in Ω and integrating by parts, it yields

$$\int_\Omega \rho(\theta_t + \mathbf{u} \cdot \nabla\theta) \frac{T}{1+T} dx = \int_\Omega \rho \left(\Theta_b \frac{d(T - \ln(1+T))}{dt} - \beta \mathbf{u} \cdot \mathbf{k} \frac{T^2}{1+T} \right) dx$$

$$= \frac{d}{dt} \int_\Omega \rho\Theta_b (T - \ln(1+T)) dx + \beta \int_\Omega \rho \left(T - \ln(1+T) - \frac{T^2}{1+T} \right) \mathbf{u} \cdot \mathbf{k} \, dx$$

$$= \frac{d}{dt} \int_\Omega \rho\Theta_b (T - \ln(1+T)) dx + \beta \int_\Omega \rho \left(\frac{T}{1+T} - \ln(1+T) \right) \mathbf{u} \cdot \mathbf{k} \, dx.$$

$$\tag{5.3.12}$$

Moreover integrating over Ω equation $(5.3.4)_3$ multiplied by $\theta/\Theta = T/(1+T)$, and using previous identity (5.3.12) we obtain

$$c_v \frac{d}{dt} \int_\Omega \rho\Theta_b (T - \ln(1+T)) dx + \chi \int_\Omega \nabla\theta \cdot \nabla \left(\frac{T}{1+T} \right) dx \tag{5.3.13}$$

$$= c_v \beta \int_\Omega \rho \ln(1+T) \mathbf{u} \cdot \mathbf{k} \, dx - R_* \int_\Omega \rho\theta \nabla \cdot \mathbf{u} \, dx$$

$$+ \int_\Omega \frac{T}{1+T} \left(\mu(\nabla\mathbf{u})^2 + (\lambda + \mu)(\nabla \cdot \mathbf{u})^2 \right) dx.$$

Adding (5.3.6), (5.3.13), and recalling (5.2.5) we obtain

$$\frac{d}{dt} \int_\Omega \left\{ \rho \frac{\mathbf{u}^2}{2} + R_*\rho\Theta_b \left(\ln \left(\frac{\rho}{\rho_b} - 1 \right) \right) + c_v \rho\Theta_b (T - \ln(1+T)) \right\} dx \tag{5.3.14}$$

$$+ \mu \int_\Omega \frac{\Theta_b}{\Theta} (\nabla\mathbf{u})^2 dx + (\lambda + \mu) \int_\Omega \frac{\Theta_b}{\Theta} (\nabla \cdot \mathbf{u})^2 dx + \chi \int_\Omega \frac{\Theta_b^3}{\Theta^2} \left| \nabla T \right|^2 dx = I_0$$

with I_0 given in $(5.3.5)_7$.

$\square$

**The energy of perturbations $\mathcal{E}$ is equivalent to L^2 norm of pertur-
bations.**

From $(5.3.5)_2$ we see that $\mathcal{E}$ is the sum of three energy terms.

The first term E_u represents the kinetic energy and it is a weighted norm
of the velocity in L^2. Furthermore, if we assume that density ρ is bounded
from below for all times, then E_u can be decreased by the L^2 norm of the
velocity **u**.

For the second term E_σ, we notice that by Taylor expansion it holds

$$E_\sigma = R_* \int_\Omega \left\{ \Theta_b \rho \left(\ln \frac{\rho}{\rho_b} - 1 \right) dx = \frac{R_*}{2} \int_\Omega \Theta_b \frac{\sigma^2}{\bar{\rho}} \, dx,$$

and E_σ represents a weighted L^2 norm of σ. Furthermore, if we assume that
density is bounded from above for all times, and the basic temperature Θ_b is
bounded from below, then E_σ is decreased by the L^2 norm of the perturbation
to the density σ. Concerning the third and last term E_θ, again by Taylor
expansion we get

$$E_\theta = c_v \int_\Omega \rho \Theta_b \left(T - \ln(1 + T) \right) dx = \frac{c_v}{2} \int_\Omega \rho \frac{\Theta_b^3}{\bar{\Theta}^2} T^2 \, dx,$$

with $\bar{\rho}$ between ρ and ρ_b, $\bar{\Theta}$ between Θ and Θ_b. Thus E_θ represents a weighted
L^2 norm of θ. Finally, if we assume that temperature is bounded from above
for all times, and the basic temperature Θ_b is bounded from below, then E_θ
is decreased by the L^2 norm of the perturbation to the temperature θ.

Hence, *the modified energy $\mathcal{E}$ in (5.3.5) may represent a good Lyapunov
functional.*

**If the time derivative of the energy of perturbations $\mathcal{E}$ in (5.3.14) is
negative, then $\mathcal{E}$ is a good Lyapunov functional.**

Remark 5.3.1 *Finally, as expected if $\beta = 0$ that is for uniform temperature
at boundary, it results $I_0 = 0$, and from (5.3.5) we see that the energy $\mathcal{E}$
is decreased by a positive quadratic form in the L^2 norms of perturbations.
Hence (5.3.5)* **furnishes for $\beta = 0$ a stability result**, *namely a control for
all times of the L^2 norms of perturbations.
Stability remains true for $\mu = 0$, $\chi = 0$.*

Remark 5.3.2 *Let the layer be heated from above, i.e. let $\beta < 0$.
In this case, in the Boussinesq (1882-1929) approximation it follows directly
from the energy equation of perturbations that the rest is stable.
In compressible case (5.3.5) doesn't imply any more stability for $\beta < 0$, this
is a very challenging question, cf. [47].*

Differential equation (5.3.14) involves two functionals say $\mathcal{E}$, and

$$\mathcal{G} = I_0 - \mu \, \mathcal{D}_u(t) - \chi \, \mathcal{D}_\theta(t)$$

involving perturbations in the weighted L^2 space.

If it would hold $c\mathcal{E} < \mathcal{G}$, with c positive constant, one would obtain

$$\frac{d\mathcal{E}}{dt} + c\mathcal{E} \leq 0, \tag{5.3.15}$$

and in virtue of Gronwall's lemma, one would deduce

$$\mathcal{E}(t) \leq \mathcal{E}(0) \exp^{-ct}. \tag{5.3.16}$$

Since $\mathcal{E}$ is equivalent to a quadratic form positive definite in the L^2 norms of perturbations, thus (5.3.16) provides the exponential decay to zero of perturbations.

Therefore sufficient conditions that guarantee the positive definiteness of $\mathcal{E}$, and the validity of inequality $c\mathcal{E} \leq \mathcal{G}$ would be also sufficient conditions for exponential stability of the rest state.

We recall that by its definition, in $(5.3.5)_7$, I_0 contains certain norms in X of the disturbances and $\mathcal{G} = \mathcal{D}_u + \mathcal{D}_\theta - I_0$ may be increased by a quadratic form in X but it can never be positive definite because positive terms in σ are missing! Thus inequality $c\mathcal{E} \leq \mathcal{G}$ doesn't hold. The key point of the proof of our asymptotic stability theorem consists in finding an additional inequality that provides a dissipative term in σ. This will be achieved by use of the *Free work equation.*

5.3.3 *Nonlinear Exponential Stability*

In this subsection by direct Lyapunov method we prove an asymptotic stability result. The plane of the proof consists in two steps. First, we deduce the **free work equation** that provides an artificial dissipative L^2 norm for the perturbation σ to density. Next we derive a "Modified Energy Estimate" that furnishes a balance between the time derivative of the modified energy $\mathbb{E}$ which is expressed as L^2 norm of perturbations, and a functional $\mathcal{I}$ expressed as sum of the dissipative viscous $-\mathcal{D}_u$, heat diffusive $-\mathcal{D}_\theta$ terms, and an artificial dissipative term for the perturbation to density plus a quadratic functional in the perturbations. The modified energy method furnishes asymptotic exponential decay for the L^2 norms of all perturbations, provided the gradient of basic temperature is not too large, it this subsection we follow the lines of [15].

In the spirit of the paper by [104] and [15], we compute the free work equation using perturbations $\mathbf{u}(x,t)$, $\sigma(x,t)$, $\theta(x,t)$ that satisfy the Initial Boundary Value Problem (5.3.4).

Remark 5.3.3 *Hypothesis (5.4.5) in Lemma 5.4.1 of next section infers existence of minimum for the density ρ.*

Theorem 5.3.2 Free Work Equation. *Let* $\mathbf{u}$, $\rho = \rho_b + \sigma$, $\Theta = \Theta_h + \theta$ *solve* (5.3.4) *with* $(\mathbf{u}, \sigma, \theta) \in \mathcal{W}$. *Then, under assumption* (5.4.5) *the free work equation holds*

$$-\frac{d}{dt}\int_\Omega \rho \mathbf{u}\cdot\mathbf{V}dx + R_*\int_\Omega \frac{\sigma^2}{\bar{\rho}}\,dx = I_1, \qquad (5.3.17)$$

where

$$I_1 = -\int_\Omega \rho\Big(\partial_t\mathbf{V} + \mathbf{u}\cdot\nabla\mathbf{u}\cdot\mathbf{V}\Big)\cdot\mathbf{u}\,dx$$
$$-\int_\Omega \Big(\mu\nabla\mathbf{u}:\nabla\mathbf{V} + (\lambda+\mu)\nabla\cdot\mathbf{u}\nabla\cdot\mathbf{V}\Big)dx - R_*\int_\Omega \rho\theta\nabla\cdot\mathbf{V}\,dx.$$
$$(5.3.18)$$

We multiply $(5.3.4)_2$ by the auxiliary function $\mathbf{V}$, we integrate over Ω and take integrations by parts. We obtain:

$$\int_\Omega \rho\Big(\partial_t\mathbf{u} + \mathbf{u}\cdot\nabla\mathbf{u}\Big)\cdot\mathbf{V}\,dx + R_*\int_\Omega \rho\Theta_b\mathbf{V}\cdot\nabla\ln\left(\frac{\rho}{\rho_b}\right)dx = \qquad (5.3.19)$$
$$- R_*\int_\Omega \mathbf{V}\cdot\nabla(\rho\theta)\,dx + \int_\Omega \Big(\mu\Delta\mathbf{u} + (\lambda+\mu)\nabla\cdot\mathbf{u}\Big)\mathbf{V}\,dx.$$

Integrating by parts, and employing Reynolds transport theorem, and changing sign in the equation, we get

$$-\frac{d}{dt}\int_\Omega \rho\mathbf{u}\cdot\mathbf{V}\,dx + R_*\int_\Omega \ln\left(\frac{\rho}{\rho_b}\right)\nabla\cdot(\rho\Theta_b\mathbf{V})\,dx = I_1, \qquad (5.3.20)$$

$$I_1 := -R_*\int_\Omega \rho\theta\nabla\cdot\mathbf{V}\,dx + \int_\Omega \Big(\mu\nabla\mathbf{u}:\nabla\mathbf{V} + (\lambda+\mu)\nabla\cdot\mathbf{u}\nabla\cdot\mathbf{V}\Big)dx$$
$$+ \int_\Omega \rho\Big(\partial_t\mathbf{V} + \mathbf{u}\cdot\nabla\mathbf{V}\Big)\cdot\mathbf{u}\,dx.$$

By taking as $\mathbf{V}$ a suitable vector field, solution of the problem (5.4.12) introduced in Lemma 5.4.3, we have

$$R_*\int_\Omega \nabla\cdot(\rho\Theta_b\mathbf{V})\ln\left(\frac{\rho}{\rho_b}\right)dx = \frac{R_*}{2}\int_\Omega \frac{\sigma^2}{\bar{\rho}}\,dx, \qquad (5.3.21)$$

where $\bar{\rho}$ is a point between ρ and ρ_b. Hence, we arrive at the free work identity (5.3.17).

We denote by c_e any embedding constant.

Theorem 5.3.3 Nonlinear Exponential Stability *Let the conditions* (5.3.25), (5.4.5) *be verified. Then the rest state* S_b *is asymptotically stable* *in the class of motions $\mathcal{W}$ solutions to system (5.1.3), corresponding to the same data.*

Proof. Adding (5.3.20) multiplied by an arbitrary constant ν to $(5.3.5)_1$ we get

$$\frac{d}{dt}\mathbb{E} + \mathcal{D} = I_0 + \nu I_1,$$ (5.3.22)

where $\mathbb{E}$ is called the **Modified Energy Functional** and is given by

$$\mathbb{E} = \int_\Omega \left\{ \rho\frac{\mathbf{u}^2}{2} + R_*\frac{\sigma^2}{2\overline{\rho}} + c_v\rho\frac{\Theta_b}{2\overline{\Theta}^2}\theta^2 - \nu\rho\mathbf{u}\cdot\mathbf{V} \right\} dx.$$

Also it holds

$$\mathcal{D} = \int_\Omega \frac{\Theta_b}{\Theta}\left[\mu(\nabla\mathbf{u})^2 dx + (\lambda+\mu)(\nabla\cdot\mathbf{u})^2\right]dx + \chi\int_\Omega \frac{\Theta_b^3}{\Theta^2}\left|\nabla\left(\frac{\theta}{\Theta}\right)\right|^2 dx$$

$$+\nu R_*\left\|\frac{\sigma}{\sqrt{\overline{\rho}}}\right\|_{L^2}^2 \geq \min_\Omega\frac{\Theta_b}{\Theta}\left(\mu\|\nabla\mathbf{u}\|_{L^2}^2 + (\lambda+\mu)\|\nabla\cdot\mathbf{u}\|_{L^2}^2\right)$$

$$+\chi\min_\Omega\left(\frac{\Theta_b^3}{\Theta^2}\right)\left\|\nabla T\right\|_{L^2}^2 + \frac{\nu R_*}{\|\rho\|_{L^\infty}}\|\sigma\|_{L^2}^2$$

$$\geq c_d\left[\mu\|\nabla\mathbf{u}\|_{L^2}^2 + \chi\|\nabla T\|_{L^2}^2 + \nu\|\sigma\|_{L^2}^2\right],$$ (5.3.23)

where c_d is a constant function of Θ, and ρ

$$c_d := min\left\{\inf\frac{\Theta_b}{\Theta}, \inf\frac{\Theta_b^3}{\Theta^2}, \frac{R_*}{\|\rho\|_{L^\infty}}\right\}.$$

To prove Theorem 5.3.3, now we estimate I_0 and I_1 defined by the last of (5.3.5), and by (5.3.20), respectively. Concerning I_0 it holds

$$I_0 = R_*\int_\Omega \rho\ln\left(\frac{\rho}{\rho_b}\right)\mathbf{u}\cdot\nabla\Theta_b\,dx - c_v\int_\Omega \rho\ln\left(1+T\right)\mathbf{u}\cdot\nabla\Theta_b dx$$

$$\leq i_0\left(\|\sigma\|_{L^2} + \|\nabla T\|_{L^2}\right)\|\nabla\mathbf{u}\|_{L^2},$$ (5.3.24)

where the constant i_0 is given by

$$i_0 := c_P\|\rho\|_{L^\infty}\|\nabla\Theta_b\|_{L^\infty}max\left\{\left\|\frac{\rho_b}{\rho}\right\|_{L^\infty}, \left\|\frac{\Theta_b}{\Theta}\right\|_{L^\infty}\right\}.$$

*Notice that i_0 **can be taken small**. Actually becomes small for small temperature gradients!*

Concerning I_1 we have

$$I_1 \leq R_* c_P p_3 \|\rho\|_{L^\infty} \|\nabla T\|_{L^2} \|\sigma\|_{L^2} + \mu \|\nabla \mathbf{u}\|_{L^2} \|\nabla \mathbf{V}\|_{L^2}$$
$$+ (\lambda + \mu) \|\nabla \cdot \mathbf{u}\|_{L^2} \|\nabla \cdot \mathbf{V}\|_{L^2}$$
$$+ \|\rho\|_{L^\infty} c_P \Big(\|\partial_t \mathbf{V}\|_{L^2} \|\nabla \mathbf{u}\|_{L^2} + \|\mathbf{u}\|_{L^\infty} \|\nabla \mathbf{V}\|_{L^2} \|\nabla \mathbf{u}\|_{L^2} \Big)$$
$$\leq c_3 \Big(\|\nabla \mathbf{u}\|_{L^2} + \|\nabla T\|_{L^2} \Big) \|\sigma\|_{L^2} + c_4 \|\nabla \mathbf{u}\|_{L^2}^2.$$

with

$$c_3 = p_3 \, max \, \{ R_* c_P \|\rho\|_{L^\infty}, (\lambda + \mu), \mu, \|\rho\|_{L^\infty} c_P \|\mathbf{u}\|_{L^\infty} \},$$
$$c_4 := \|\rho\|_{L^\infty} c_P p_4.$$

*Also we notice that the coefficient $3\lambda + 2\mu$, hence c_3, c_4 **cannot be taken small**, while i_0 can be taken small provided $\nabla \Theta_b$ is small.*

Thus we have

$$\mathcal{D} - I_0 - \nu I_1 \geq c_d \left[\mu \|\nabla \mathbf{u}\|_{L^2}^2 + \chi \|\nabla T\|_{L^2}^2 + \nu \|\sigma\|_{L^2}^2 \right] - c_4 \|\nabla \mathbf{u}\|_{L^2}^2$$
$$- \left[i_0 \Big(\|\sigma\|_{L^2} + \|\nabla T\|_{L^2} \Big) \|\nabla \mathbf{u}\|_{L^2} + \nu c_3 \Big(\|\nabla \mathbf{u}\|_{L^2} + \|\nabla T\|_{L^2} \Big) \|\sigma\|_{L^2} \right] =: \mathcal{I}.$$

We observe that $\mathcal{I}$ is a quadratic form $\mathcal{I}$ in the variables $\|\nabla \mathbf{u}\|_{L^2}$, $\|\nabla T\|_{L^2}$, $\|\sigma\|_{L^2}$. Since ν is an arbitrary number, by taking ν, and i_0 suitably small we may let $\mathcal{I}$ be a positive definite quadratic form, say there exists a positive number d such that the following inequality is verified

$$\mathcal{I} \geq \frac{1}{d} \left[\|\nabla \mathbf{u}\|_{L^2}^2 + \|\nabla T\|_{L^2}^2 + \|\sigma\|_{L^2}^2 \right].$$

A sufficient condition in order to exist a ν small enough, is expressed by conditions

$$\nu c_4 < c_d \mu,$$
$$c_d^2 \mu \chi > i_0^2,$$
$$\nu < min \left\{ \frac{\mu}{12 c_3^2}, \frac{\mu}{4 c_4}, \frac{4\chi}{g c_3^2} \right\} =: \nu_m, \qquad (5.3.25)$$
$$\mu > \frac{12 i_0^2}{\nu_M}.$$

Finally we may infer

$$\mathcal{D} - I_0 - I_1 \geq d \mathcal{E}. \qquad (5.3.26)$$

This yields

$$\frac{d}{dt}\mathcal{E} + d\mathcal{E} \leq 0. \qquad (5.3.27)$$

Finally, applying Gronwall's Lemma we obtain the exponential decay, for any initial data, and the theorem is completely proved.   □

Remark 5.3.4 *Notice that smallness hypotheses on initial data will be needed in order to prove existence of global unsteady solutions. Furthermore, our theorem holds only for basic flow with small temperature gradient. This assumption is also needed in the stability proof of Bènard problem in the Boussinesq approximation.*

Condition $(5.3.25)_2$ *represents the analogous of the stability condition in the Boussinesq approximation.*

5.4 Auxiliary Lemmas

We conclude Chap. 5 by giving two generalizations of Lemmas proved in Sect. 3.7. Let us consider the fields ρ_b, $\Theta_b(x)$ given by (5.1.6) solutions to (5.1.5)

$$\inf(\rho_b\Theta_b) = \rho_*\Theta_h^{m+1} =: 2c_p$$

$$(m+1)\rho_*\beta(\beta h + \Theta_h)^m h = \frac{\ell}{R_*}. \qquad (5.4.1)$$

5.4.1 Some Inequalities

The following inequalities hold true.

Lemma 5.4.1 *Let* $\mathbf{u}$ *be a solenoidal vector field in* $W^{1,2}(\Omega)$, *with* Ω *a bounded domain. If* $\mathbf{u}$ *vanishes at boundary, then the following inequalities hold true*

$$\|\mathbf{u}\|_{L^2(\Omega)} \leq c_P\|\nabla\mathbf{u}\|_{L^2(\Omega)}, \qquad \textit{Poincare' inequality,} \quad (5.4.2)$$

$$\|\mathbf{u}\|_{L^6(\Omega)} \leq c_S\|\nabla\mathbf{u}\|_{L^2(\Omega)}, \qquad \textit{Sobolev inequality,} \quad (5.4.3)$$

where c_P, *and* c_S *are the Poincare', and Sobolev constants.*

Inequalities (5.4.2) are true because the Dirichlet boundary conditions on Σ, and Σ_h imply that $\nabla\mathbf{u} = 0$ infers $\mathbf{u} = 0$, cf. [5]. Furthermore, since $\mathbf{u}$ vanishes on $\Sigma \times \{0\}$, an explicit Poincare' constant can be computed.

5.4.2 Function V for Uniqueness

Lemma 5.4.2 Uniqueness of Rest State. *Let $\Omega \in C^1$ piecewise, and let be bounded. Let $\rho \in W^{1,2}$ and $\sigma \in L^2(\Omega)$, with $\int_\Omega \sigma \, dx = 0$. Then there exists a vector field $\mathbf{V} \in W^{1,2}(\Omega))$ which satisfies the following boundary value problem*

$$\nabla \cdot (\rho \Theta_b \mathbf{V}) = \sigma, \qquad\qquad x \in \Omega,$$
$$\mathbf{V}(x)|_{\partial\Omega} = 0. \tag{5.4.4}$$

Furthermore, provided

$$\inf(\rho \Theta_b) \geq \frac{\rho_* \Theta_h^{m+1}}{2} =: c_b > 0 \tag{5.4.5}$$

there exist positive constants f_i, $i = 1, 2, 3$, such that the following estimates hold true:

$$\|\mathbf{V}\|_{L^2} \leq f_1 \|\sigma\|_{L^2},$$
$$\|\mathbf{V}\|_{L^6} \leq f_2 \|\sigma\|_{L^2},$$
$$\|\nabla \mathbf{V}\|_{L^2} \leq f_3 \|\sigma\|_{L^2},$$
$$\|\nabla \mathbf{V}\|_{L^3} \leq f_4 \|\sigma\|_{L^3}.$$

Proof. The proof of existence of a function $\mathbf{W}$ solution to

$$div\,\mathbf{W} = \sigma, \qquad\qquad x \in \Omega,$$
$$\mathbf{W}(x)|_{\partial\Omega} = 0,$$
$$\|\mathbf{W}\|_{L^2} \leq c_P c_e \|\sigma\|_{L^2},$$
$$\|\mathbf{W}\|_{L^6} \leq c_S c_e \|\sigma\|_{L^2}, \tag{5.4.6}$$

where c_S, C_P denote the Sobolev and Poincare' constants, is similar to that of Lemma 3.7.2, cf. [36]. Moreover the problem

$$div(\rho \Theta_b \mathbf{V}) = div\,\mathbf{W}, \qquad\qquad x \in \Omega,$$
$$\mathbf{V}(x)|_{\partial\Omega} = 0, \tag{5.4.7}$$

admits a solution $\mathbf{V}$ satisfying the estimates

$$\|\rho\Theta_b\mathbf{V}\|_{L^2} \le c_e\,\|\mathbf{W}\|_{L^2}\,,$$
$$\|\rho\Theta_b\mathbf{V}\|_{L^6} \le c_e\,\|\mathbf{W}\|_{L^6}\,, \tag{5.4.8}$$
$$\|\nabla(\rho\Theta_b\mathbf{V})\|_{L^2(\Omega)} \le c_e\,\|\nabla\cdot\mathbf{W}\|_{L^2}\,.$$

Now we require hypothesis (5.4.5) of minimum for $\rho\Theta_b$. To develop estimates in L^p for $\mathbf{V}$ we employ estimates in (5.4.6), (5.4.8), to deduce

$$\|\mathbf{V}\|_{L^2} \le \frac{c_e}{c_b}\|\mathbf{W}\|_{L^2} \le \frac{c_P c_e^2}{c_b}\|\sigma\|_{L^2} =: f_1\|\sigma\|_{L^2},$$

$$\|\mathbf{V}\|_{L^6} \le \frac{c_e}{c_b}\|\mathbf{W}\|_{L^6} \le \frac{c_S c_e^2}{c_b}\|\sigma\|_{L^2} =: f_2\|\sigma\|_{L^2}.$$

Next developing the derivative at left hand side of $(5.4.9)_3$, using the triangular inequality, we obtain an estimate for the L^2 norm of $\nabla\mathbf{V}$

$$\|\nabla\mathbf{V}\|_{L^2} \le \frac{1}{\inf|\rho\Theta_b|}\Big(\|\mathbf{V}\|_{L^6}\|\nabla(\rho\Theta_b)\|_{L^3} + c_e\|\sigma\|_{L^2}\Big). \tag{5.4.9}$$

Also by $(5.4.9)_2$ we obtain

$$\|\nabla\mathbf{V}\|_{L^2} \le \Big(\frac{c_S c_e^2}{c_b^2}\|\nabla(\rho\Theta_b)\|_{L^3} + \frac{c_e}{c_b}\Big)\|\sigma\|_{L^2} =: f_3\|\sigma\|_{L^2}. \tag{5.4.10}$$

Starting from (5.4.10) we may compute an estimate for the L^3-norm of $\mathbf{V}$.

$\square$

5.4.3 Auxiliary Function for Stability

In this subsection we prove the existence of the auxiliary function $\mathbf{V}$.

Let there be given the fields $(\mathbf{u},\sigma) \in \mathcal{X}$, and let

$$\partial_t\sigma = -\nabla\cdot(\rho\mathbf{u}) \in L^2(0,\infty;L^2(\Omega)). \tag{5.4.11}$$

Lemma 5.4.3 *Provided (5.4.5), and (5.4.11) hold, there exists a vector field* $\mathbf{V} \in L^\infty(0,\infty;W_0^{1,2}(\Omega))$ *with* $\partial_t\mathbf{V} \in L^\infty(0,\infty;L^2(\Omega))$, *which satisfies the following problem*

$$\nabla\cdot(\rho_b\Theta_b\mathbf{V}) = \sigma, \quad x \in \Omega,$$
$$\mathbf{V}(x)|_{\partial\Omega} = 0. \tag{5.4.12}$$

Moreover, there exist three constants p_1, p_2, p_3, *depending on* ρ_b, ρ, *and* Ω *such that the following estimates hold true:*

$$\|\mathbf{V}\|_{L^2} \leq p_1 \|\sigma\|_{L^2} ,$$

$$\|\mathbf{V}\|_{L^6} \leq p_2 \|\sigma\|_{L^2} ,$$

$$\|\nabla\mathbf{V}\|_{L^2(\Omega)} \leq p_3\|\sigma\|_{L^2(\Omega)}, \qquad (5.4.13)$$

$$\|\partial_t\mathbf{V}\|_{L^2(\Omega)} \leq p_4 \|\nabla\mathbf{u}\|_{L^2(\Omega)} ,$$

$$\|\nabla\mathbf{V}\|_{L^2(\Omega)} \leq p_5 \|\sigma\|_{L^2(\Omega)} .$$

Proof. The lines of the proof are exactly the same as in previous subsection. To deduce the L^2 estimate for the time derivative we must make use of problem (5.4.7). Taking in (5.4.7) the partial derivative with respect to time furnishes

$$div(\rho\Theta_b\partial_t\mathbf{V}) = div\partial_t\mathbf{W} - div(\partial_t\rho\Theta_b\mathbf{V}). \qquad (5.4.14)$$

Thus using the continuity equation, and estimates (5.4.7) we get

$$\|\partial_t\mathbf{V}\|_{L^2} \leq \frac{1}{c_b}\Big(\|\partial_t\mathbf{W}\|_{L^2} + \|div(\rho\mathbf{u})\Theta_b\mathbf{V}\|_{L^2}\Big). \qquad (5.4.15)$$

To obtain an estimate for $\mathbf{W}$ we take in (5.4.7) the partial derivative with respect to time and recall the continuity equation to deduce

$$div\partial_t\mathbf{W} = -div(\rho\mathbf{u}). \qquad (5.4.16)$$

From (5.4.16) we have computed the estimate

$$\|\partial_t\mathbf{W}\|_{L^2} \leq c_e\|\rho\mathbf{u}\|_{L^2}. \qquad (5.4.17)$$

Substituting this estimate together with estimate for $\mathbf{V}$ in (5.4.15) it yields

$$\|\partial_t\mathbf{V}\|_{L^2} \leq \frac{1}{c_b}\Big(c_e\|\rho\mathbf{u}\|_{L^2} + \|div(\rho\mathbf{u})\Theta_b\|_{L^3}\|\mathbf{V}\|_{L^2}\Big)$$

$$\leq \frac{c_e}{c_b}\|\rho\|_{L^3}\|\mathbf{u}\|_{L^6} + \frac{1}{c_b}\Big(\|(\rho\nabla\mathbf{u} + \mathbf{u}\cdot\nabla\rho)\Theta_b\mathbf{V}\|_{L^2}\Big)$$

$$\leq \frac{c_S c_e}{c_b}\|\rho\|_{L^3}\|\nabla\mathbf{u}\|_{L^2} + \frac{1}{c_b}\|\Theta_b\|_{L^\infty}\|\mathbf{V}\|_{L^\infty}$$

$$\times\Big(\|\rho\|_{L^\infty}\|\nabla\mathbf{u}\|_{L^2} + c_S\|\nabla\mathbf{u}\|_{L^2}\|\nabla\rho\|_{L^3}\Big)$$

$$\leq p_4\|\nabla\mathbf{u}\|_{L^2}. \qquad (5.4.18)$$

$\square$

5.4.4 Bibliographical Notes

It is likely that the case of polytropic gases with free boundary could be solved with tools analogous to those employed for barotropic gases, however we have not studied the general problem here, just we remark that it is still an open problem. We have limited ourselves to the study of stability of the rest state of an heavy isothermal fluid in a horizontal layer, with rigid boundaries. It appears not difficult to deal with the case of barotropic, or polytropic gases, under the action of potential forces, with free boundary, actually it constitutes an open problem. Still, it is worth mentioning that the stability problem for incompressible fluids with free boundary, in the Boussinesq approximation has been just recently solved in [45].

Results in Chap. 5 represent a short version of the results proved by the author in [15, 20, 94, 105]. About uniqueness we quote [92, 93], existence theorems of unsteady solutions, and stability of zero solution we quote [27, 28, 64, 69–71, 131].

Concerning the stability of a basic flow of a fluid filling a fixed region whose perfectly conducting walls are not uniformly heated, since the works of [139, 144, 149], the literature on stability in this field is not too rich, we quote [15, 20, 30, 94, 112, 123]. It is also worth of mention the paper [45] where non linear stability has been studied for a layer of incompressible fluid with upper free boundary, in the presence of Marangoni effect.

Nonlinear stability results for heat conducting fluids in exterior domain has been proven in [60, 112].

Bibliography

1. Abels, H.: *The intial value problem for the Navier-Stokes equations with a free surface in L^q-Sobolev spaces*, Adv. Diff. Eq., **10**, 45–64, 2005.
2. R. Abeyaratne & J.K. Knowles, Elastic materials with two stress free configurations, J. Elasticity, **67**, 61–69, 2002.
3. Agmon, S., Douglis, A., and Niremberg, L., *Estimates near the boundary for solutions of elliptic partial differential equations satisfying general boundary conditions I*, Comm. Pure Appl. Math. **12**, 623–727, 1959.
4. V.I. Arnold, *Conditions for nonlinear stability of stationary plane curvilinear flows of an ideal fluid*, Doklady **162** 1965, 773–777.
5. H. Babovski, and Padula M. *A new contribution in the nonlinear stability of a discrete velocity model*, Comm. Math. Phys. **144** 1992, 87
6. Batchelor G.K., *An introduction to Fluid dynamics*, Cambridge Univ. Press, 1967
7. M. Bause, and J. G. Heywood and A. Novotny, and M. Padula, *An Iterative Scheme for Steady Compressible Viscous Flow, modified to treat numerical solutions*, Mathematical Fluid Mechanics: recent results and numerical methods, eds.J. Neustupa, P. Penel, Basel; Boston; Berlin: Birkhauser **5**, 2003.
8. M. Bause, and J. G. Heywood and A. Novotny, and M. Padula, *An Iterative Scheme for Steady Compressible Viscous Flow, modified to treat Large Potential Forces*, Mathematical Fluid Mechanics: recent results and open questions, eds.J. Neustupa, P. Penel, Basel; Boston; Berlin: Birkhauser 2003, pp.27–46.
9. J.T. Beale, *The initial value problem for the Navier-Stokes equations with a free surface.* Comm. Pure Appl. Math. **34** 1981, no. 3, 359–392.
10. J.T. Beale, *Large-time regularity of viscous surface waves.* Arch. Rat. Mech. Anal. **84** 1983/84, no. 4, 307–352
11. J.T. Beale & T. Nishida, *Large-time behavior of viscous surface waves.* Recent topics in nonlinear PDE, II (Sendai, 1984), 1–14, North-Holland Math. Stud., **128**, North-Holland, Amsterdam, 1985.
12. H. Beirão da Veiga, *An L^p-theory for the n-dimensional, stationary, compressible Navier-Stokes equations, and the incompressible limit for compressible fluids. The equilibrium solutions*, Commun. Math. Phys., **109** 1987, 229–248.
13. H. Beirao da Veiga, *Existence results in Sobolev spaces for a stationary transport equation*, Ricerche di Matematica suppl. **36** 1987, 173–184.
14. Bernstein, B., Toupin, R.A.: *Korn's inequalities for the sphere and circle*, Arch. Ratl. Mech. Anal., **6** (1960) 51–64.
15. R. Benabidallah & M. Padula, *Sulla stabilita' di un fluido politropico, viscoso, pesante in un contenitore limitato da pareti perfette conduttrici di calore*, Annali dell'Universita' di Ferrara, sez. VII, Sc. Mat.**45** 1998.

M. Padula, *Asymptotic Stability of Steady Compressible Fluids*,
Lecture Notes in Mathematics 2024, DOI 10.1007/978-3-642-21137-9,
© Springer-Verlag Berlin Heidelberg 2011

16. Bogovski, M.E., *Solutions of the first boundary value problem for the equation of continuity of an incompressible medium*, Soviet Math. Dokl. **20** 1979, 1094–1098.

17. A.L. Cauchy, Bulletin de la Societe de Philo-mathique, 1823.

18. S. Chandrasekhar, *Hydrodynamic and hydromagnetic stability*, Dover Pub. inc. New York, 1981.

19. Chambolle, A.; Desjardin, B.; Esteban, M.J.; Grandmont, C.: *Existence of weak solutions for unsteady fluid-plate interaction problem.* J. Math. Fluid Mech. **7** 2005, no. 3, 368–404.

20. Coscia V. & M. Padula, *Nonlinear energy stability in compressible atmosphere*, Geophys. Astrophys. Fluid Dynamics **54** 1990, 49

21. De Groot S.R., Mazur P., *Non-equilibrium thermodynamics*, North-Holland publish. comp., 1969.

22. Drazin P.G. and W.H. Reid, *Hydrodynamic stability*, Cambridge Univ. Press, 1981.

23. Eliezer, S., Gathak, A., and Hora, H., *An introductionto equations of states, theory and applications,* Cambridge Univ. press, Cambridge, 1986.

24. J.L. Ericksen [E], Introduction to the thermodynamics of solids, *Applied Math. and Math. Comput.*, 1991, **1**.

25. Farwig, R., *Stationary solutions of compressible Navier-Stokes equations with slip boundary condition*, Commun. In Partial Differential Equations, **14**(11), 1989, 1579–1606.

26. Fereisl E., *Dynamics of viscous compressible fluids*, Oxford Univ. Press, Oxford 2003.

27. Feireisl, E. and Novotny, A., *Weak sequential stability of the set of admissible variational solutions to the Navier-Stokes-Fourier system*, SIAM J. Math. Anal. 37 2005, no. 2, 619–650.

28. Feireisl, E. and Novotny, A., *Large time behaviour of flows of compressible, viscous, and heat conducting fluids*, Math. Methods Appl. Sci. 29 2006, no. 11, 1237–1260.

29. Fichera, G., *Boundary value problems of elasticity with unilateral constraints*, Hand. der Phys., vol. VIa/2, S. Flugge 1972, 347–424.

30. P.C. Fife, *The Benard problm for general fluid dynamical equations and remarks on the Boussinesq approximation*, Indiana Univ. Math. J., **20** 1970, 303–326.

31. R. Finn, *On the equation of capillarity*, J. Math. Fluid Mech., **3** 2001, 139–151.

32. R. Finn, Luli, G.K., *On the capillary problem for compressible fluids*, J. Math. Fluid Mech., **9** 2006, 87–103.

33. J. Frehse, M. Steinhauer, W. Weigant, *On the stationary solutions for 2-D viscous compressible isothermal Navier-Stokes equations*, J. Math. Fluid Mech. **13**, 55–64, 2011.

34. Frolova, E., and Padula, M., *On existence of steady solutions of a horizontal layer of heavy, inhomogeneous incompressible fluid*, European Journal of Fluid Mechanics B/Fluids **23**, 665–679, 2004. Fro-Pad

35. Friedrichs, K.,O.: *On the boundary value problems of theory of elasticity and Korm's inequality*, Annals of Mathematics, ser.II, **48** (1947), 441–471.

36. Galdi, G.P., *An Introduction to the Mathematical Theory of the Navier-Stokes Equations*, Vol.1, Springer-Verlag, vol.38, 1994, corrected second printing 1998.

37. G. P. Galdi and M. Padula, *A new approach to energy theory in the stability of fluid motion*, Arch. Rational. Mech. Anal. **110**, 1990, 187–197.

38. Galdi, G.P., Novotny' A., and Padula M., *On the Twodimensional Steady-State Problem of a Viscous Gas in an Exterior Domain*, Pacific J. Math. **179**, 1997, 65–100.

39. Gallavotti, G., *Foundations of fluid dynamics*, Springer-Verlag, New-York, 2002.

40. Gilbarg, D., and Finn, R., *Three dimensional subsonic flows and asymptotic estimates for elliptic partial differential equations*, Acta Math., **98**, 1957, 265–280.

41. Gobert, J., *Une inegalite' fondamentale de la theorie de l'elasticite'*, Bull. Soc. Royale Sci. Liegi, **31**, 1962, 182–191.

42. P. Helluy and Golay, F., et al. *Numerical simulation of a wave breaking*, Math. Modelling Num. Anal. **39** 2005, 591–607.

43. Graffi D., *Il teorema di unicita' nella dinamica dei fluidi compressibili*, J. Rational Mech. Anal. **2**, 1953, 99–106.

44. Graffi D., *Il teorema di unicita' per le equazioni del moto dei fluidi compressibili in un dominio illimitato*, Atti Acc. delle Sci., Bologna **7**, 1960, 1–8.

45. G. Guidoboni & B.J. Jin, On the Nonlinear Stability of Marangoni-Benard Problem with Free Surface in the Boussinesq Approximation, Math. Models Methods Appl. Sci., Vol. 15, No.1, 2005, 1–22

46. Guidoboni G., Guidorzi, M., and Padula, M., *Uniqueness of Hopf solutions to a fluid-structure interaction model*, J. Math. Fluid Mech., **13**, 2011.

47. Guidoboni G., and Padula, M., *On the Benard problem*, Trends in Partial Differential Equations on Mathematical Physics, Obidos 7–10 june, 2003, J.F. Rodrigues, G. Seregin, J.M. Urbano eds., Progress in nonlinear differential equations and their applications, vol. 61, Birkäuser, Basel Boston Berlin, 2005, p. 137–148.

48. Guidorzi, M., Padula, M. *Approximate solutions to the 2-D unsteady Navier-Stokes system with free surface*, Hyperbolic problems and regularity questions, M. Padula & L. Zanghirati eds., Trends in Mathematics, Birkauser 2007, 109–120.

49. Guidorzi, M., Padula, M., and Plotnikov, P., *Hopf solutions to a fluid-elastic interaction model*, M^3 vol.18, 2008, 215–269.

50. B.J. Jin, *Existence of viscous compressible barotropic flows in a moving domain with upper free surface via Galerkin method*, Ann. Ferrara Univ. sez.VII, **49**, 2003, 48–71.

51. B.J. Jin & M. Padula, *On existence of nonsteady compressible viscous flows in a horizontal layer with free upper surface*, Comm. Pure Appl. Anal. **1**(3), 2002, 370–415.

52. B.J. Jin and M. Padula, *Steady flows of compressible fluids in a rigid container with upper free boundary*, Math. Ann. **329**, 2004, 723–770.

53. Joseph D.D., Stability of fluid motions, Springer Verlag, voll. I, II, Berlin, 1976.

54. T. Hagstrom and J. Lorenz, *All-time existence of classical solutions for slightly compressible flow*, SIAM J. Math. Anal., Vol. 29, 1998, 652–672.

55. B. Helffer and O. Lafitte, *Asymptotic methods for the eigenvalues of the Rayleigh equation for linearized Rayleigh-Taylor instability*, Univ. Paris-Sud. Math. 2424, 2002, 1–49.

56. J. G. Heywood and M. Padula, *On the Existence and Uniqueness Theory for steady compressible Viscous flow, Fundamental directions in Mathematical Fluid Mechanics*, Advances in Mahematical Fluid Mechanics, Birkhäuser Verlag, Basel-Boston-Berlin, 2000, 171–188.

57. Kellogg, B. & Kweon, J.R., Compressible Navier-Stokes equations in a bounded domain with inflow boundary condition, SIAM J. Math. Anal. **28**, 1997, 94–108.

58. Korn, A., *Eigenschwingungen eins elastischen Körpers mit ruhender Oberflache*, Akad. der Wissensch., Munich, Math-phys. Kl., Berichte, **36**, 1906, 351–401.

59. N. Kopachevsky and S.G. Krein, *Operator approach to linear problems of hydrodynamics*, vol.I, II, Operator theory: advances and applications, **128**, Birkhauser Verlag, 2001.

60. Kobayashi, T., and Shibata, Y., *Decay estimates of solutions for the heat-conductive gases in an exterior domain in R^n*, Comm. Math. Phys., **200**, 1999, 621–659.

61. Iooss, G., and Padula M., *Structure of the linearized problem for compressible parallel fluid flows*, Annali dell'Universita' di Ferrara, sez.VII, **43**, 1998, 157–171.

62. G. Iooss, and M. Rossi, *Nonlinear evolution of the bidimensional Rayleigh-Taylor flow*, Europ. J. Mech. B/Fluids 8, 1989, 1, 1–22.

63. Ladyzhenskaja, O.A., *The mathematical theory of incompressible fluid flow*, Gordon and Breach Sci. Publbl., 1969.

64. M.V. Lagunova, Solonnikov V.A., *Nonstationary problem of thermocapillary convection*, S.A.A.C.M., **1**, 1991, 47–72.

65. L.D. Landau & E.M. Lifshitz, *Fluid Mechanics, Pergamon press*, Quarterly J. Appl. Math., **69** 2011, 569–601.

66. P.L. Lions, *Mathematical Topics in Fluid Mechanics*, vol. 2, Compressible Models, Clarendon Press, Oxford Science Publications, Oxford, 1998.

67. Liu, B. and Kellogg, B., *Discontinuous solutions of the linearized steady state, viscous, compressible flows*, J. Math: Anal: Appl. 1997.

68. U. Massari, M. Padula, and S. Shimizu, *On nonlinear instability of capillary equilibrium figures*, on Quarterly J. Appl. Math., **69**, 2001, 569–601.

69. A. Matsumura and T. Nishida, *Initial boundary value problems for the equations of motions of compressible viscous and heat-conductive fluids*, J. Math. Kyoto Univ., **20**, 1980.

70. A. Matsumura and T. Nishida, *Initial boundary value problems for the equations of motions of compressible viscous and heat-conductive fluids*, Comm. Math. Phys., **89**, 445–464, 1983.

71. A. Matsumura and T. Nishida, *The initial value problem for the equations of viscous and heat-conductive fluids*, Proc. Jpn. Acad. Ser.A, **55**, 337–342, 1979.

72. A. Matsumura and M. Padula, *Stability of the stationary solutions of compressible viscous fluids with large external potential forces*, Stab. Appl. Anal. Cont. Media, **2**, 1992, 183–202.

73. J. Nash, *Le probleme de Cauchy pour les equations differentielles dun fluide generale*, Bull. Soc. Math. France **90**, (1962).

74. Nishida, T., *Equations of fluid dynamics- Free surface problems*, Comm. Pure Appl. Math. **39**, 1986, 221–238.

75. T. Nishida and Y. Teramoto, *On the Navier-Stokes flow down an inclined plane*, J. Math. Kyoto Univ. **33**, 1993 3. 78–801.

76. T. Nishida and Y. Teramoto, *On Navier-Stokes flow on a threedimensional horizontal plane*, Lect. Notes Pure Appl. Math., M. Decker Publish, 2001, in the press.

77. Novotny, A., and Padula, M., and Penel, P., *A remark on the well posedness of the problem of a steady flow of a viscous barotropic gas in a pipe*, (with A. Novotny', P. Penél), 1994 Comm. Partial Diff. Eq., **21**, 1996, 23–34.

78. S.A. Nazarov, A. Novotny, and K. Pileckas, *On steady compressible Navier-Stokes equations in plane domain with corners*, Math. Ann. **302**, 121–150 (1996).

79. Nishida T., Teramoto Y., Yoshihara H., *Hopf bifurcation in viscous incompressible flow down an inclined plane*, Journal of Mathematical Fluid Mechanics, Vol. 7, 2005, 29–71

80. A. Novotny, and K. Pileckas, *Steady compressible Navier-Stokes equations with large potential forces via a method of decomposition*, Math. Meth. Appl. Sci. **21**, 1998, 665–684.

81. Nishida T., Teramoto Y., Yoshihara H., *Navier-Stokes flow down an inclined plane: Downward periodic motion*, J. of Math. Kyoto University, Vol.33, No.3, 1993, 787–801.

82. Novotný, A., *Steady flows of viscous compressible flows L^2 approach*, SAACM, **3** 1993, 181–199.

83. Novotný, A., *Some remarks to the compactness of steady compressible isentropic Navier-Stokes equations via the decomposition method*, Comment. Math. Univ. Carolina, **37** 1996, 1–38.

84. Novotný, A. & M. Padula, *L^p-approach to steady flows of viscous compressible fluids in exterior domains*, Arch. Rational Mech. Anal. **126** 1994, 234–297.

85. Novotny, A., and Padula, M., *Physically reasonable solutions to steady compressible Navier-Stokes equations in 3-D-exterior domains I ($v_\infty = 0$)*, J. Math. Kyoto, **36**, 1996, 389–422.

86. Novotny, A., and Padula, M., *Physically reasonable solutions to steady compressible Navier-Stokes equations in 3-D-exterior domains II ($v_\infty \neq 0$)*, Math. Ann., **370**, 1997, 1–51.

87. Orr, W.McF., Proc. Roy. Irish Acad. (A), **27** 1907, 69

88. Oseen, C.W., *Neuere Methoden und Ergebnisse in der Hydrodynamik*, Leipzig, Akad. Verlagsgesellschaft M.B.H., 1927.

89. Padula, M., *On the uniqueness of viscous compressible steady flows*, Trends in Appl. of Pure Math. to Mech. Vol. IV, ed. Brilla, Pitman, London, 1981, 186.

90. Padula, M., *An existence theorem for steady state compressible Navier-Stokes equations*, Proc. Meeting "Waves and Stability in Continuous Media", Univ. Catania, 276–280, 1981.

91. Padula, M., *Existence and uniqueness for steady compressible flow*, Proc. Meeting "Dinamica dei Continui Fluidi e dei gas ionizzati, Univ. Trieste, 1982, 237–258.

92. Padula, M., *Uniqueness theorems for steady compressible heat-conducting fluids: bounded domains*, Rend. Accad. Naz. Lincei, **74**, 1983, 380.

93. Padula, M., *Uniqueness theorems for steady compressible heat-conducting fluids: exterior domains*, Rend. Accad. Naz. Lincei, **75**, 1983, 56.

94. Padula, M., *Non-linear energy stability for the compressible Bénard problem*, Boll. U.M.I. (6) **5-B**, 1986, 720.

95. Padula, M., *Existence and uniqueness for viscous steady compressible motions*, Arch. Rat. Mech. Anal., 97, 1987, 89–102.

96. Padula, M., *On the decay of the kinetic energy for compressible flows*, Proc. Symp. "Waves and Stability in Continuous Media" Laterza Ed. Bari 1986.

97. Padula, M., *Energy instability methods in fluid dynamics*, Proc. Meeting "Energy stability and convection" Pitman Research Notes in Mathematics, Longman, London, 1987.

98. Padula M., *On existence and uniqueness of non homogeneous motions in exterior domains*, Math. Z., **203**, 1990, 581–604.

99. Padula M., *Stability properties of regular flows of heat-conducting compressible fluids*, J. Math. Kyoto Univ., **32**, 1992, 401–442.

100. Padula, M., *A representation formula for steady solutions of a compressible fluid moving at low speed*, Transport Theory Statistical Physics, 21(1992), 593–614.

101. Padula, M., *On the exterior steady problem for the equations of a viscous isothermal gas*, Comm. Math. Carolinae, 34, 1993, 275–293.

102. Padula, M., *Steady flows of barotropic viscous fluids*, Quaderni di Matematica, **1**, 1997.

103. M. Padula, *On the exponential stability of the rest of a viscous compressible fluid*, Proc. of *Society of Trends in Application of Mathematics to Mechanics*, Nice 25–29 may 1998.

104. Padula, M., *On the exponential stability of the rest state of a viscous compressible fluid*, J. Math. Fluid Mech. **1**, 1999, 1–16.

105. Padula M., *On direct Lyapunov method in continuum theories*, Nonlinear problems in mathematical physics and related topics I, in honor of prof. Ladyzhenskaya, eds. Sh. Birman, S. Hildebrandt, V.A. Solonnikov, O.A. Uralsteva, Kluwer Academic/Plenum Publisher, New York, Boston, Dordrecht, London, Moscow, 2002.

106. M. Padula *Free work and control of equilibrium figures* Ann. Univ. Ferrara, sez.VII, **49**, 2003, 375–396.

107. M. Padula *Free Work Identity and Nonlinear Instability in Fluids with Free Boundaries* Proceedings of *Recent Advances in Elliptic and Parabolic Problems*, Proceedings of the International Conference, Hsinchu, Taiwan University, C-C Chen & M. Chipot & C.S. Lin, 16–20 february 2004, World Sci. 2005, 203–214.

108. M. Padula *Free work and control of equilibrium configurations* Progress in Nonlinear Differential Eqautions and their Applications, **61**, Birkhäuser, 2005, 213–223.

109. M. Padula, *The role of free work identity in viscous compressible flows*, Int. J. Computers & Math. with Appl. **51**, 2006, 849–856.

110. M. Padula, *On nonlinear stability of MHD equilibrium figures*, New directions in mathematical fluid mechanics, Advances in Math. Fluid Mech., 2009, 301–331.

111. M. Padula & K. Pileckas, *Existence and asymptotic behaviour of steady flow of a viscous barotropic gas in a pipe*, Ann. Mat. Pura ed Appl., ser. IV, **172**, 1997, 191–218.

112. M. Padula & M. Pokorny *Stability and decay to zero of the L^2 norms of perturbations to a viscous compressible heat conductive fluid motion exterior to a ball*, J. Math. Fluid Mech., vol. 3, 2001, 317–408.

113. Padula M., & V.A. Solonnikov *On Raileigh-Taylor stability*, Annali dell'Universita' di. Ferrara, vol. 45, 2000.

114. M. Padula & V. A. Solonnikov, *On the global existence of nonsteady motions of a fluid drop and their exponential decay to a uniform rigid rotation.*, Proceeding of *On Fourth European Conference on Elliptic and Parabolic Problems*, Rolduc and Gaeta 2001, World Scientific, New Jersey-London-Singapore-Hong Kong, 2002, 180–203.

115. M. Padula & V. A. Solonnikov, *On stability of equilibrium figures of a uniformly rotating liquid drop in n-dimensional space*, Proc. On Navier-Stokes Equations and Applications, RISM Kyoto, january 2006.

116. M. Padula & V. A. Solonnikov, *A simple proof of linear instability of rotating liquid drops*, Annali dell'Universita' di Ferrara, **54**, 2008.

117. M. Padula, *Nonlinear stability of liquid flow down an inclined plane*, J. Fluid Mech., preprint University of Ferrara.

118. M. Padula & V. A. Solonnikov, *On the global existence of nonsteady motions of a fluid drop and their exponential decay to a uniform rigid rotation*, Quaderni di Matematica, vol. 10, 2002, 185–218.

119. K. Pileckas, Zajaczkowski W., *On the free boundary problem for stationary compressible Navier-Stokes equations*, Comm. Math. Phys., **129**, 1990, 169–204.

120. P.I. Plotnikov & J. Sokolovski, *On compactness domain dependence and existence of steady solutions to compressible isothermal Navier-Stokes equations*, preprint de l'Inst. E. Cartan 2002/35

121. P.I. Plotnikov & J. Sokolovski, *Stationary solutions of Navier-Stokes equations for diatomic gases*, Russiam Math. Surveys, **62** 2007, 561–593; number 3, 2007, 567–608.

122. P.I. Plotnikov & E.V. Ruban & J. Sokolovski, *Inhomogeneous boundary value problems for compressible Navier-Stokes and transport equations*, J. Math. Pure et Appl., 2007.

123. Pyi Aye & T. Nishida, *Heat convection of compressible fluid*, Recent developmets in domain decomposition methods and flow problems, Math. Sci. and Apll., **11**, ed. by Fujita, Gakkotosho, Tokyo, Japan, 1998, 107–115.

124. Reynolds, O., Phil. Trans. Roy. Soc. London (A), **186** 1895, 123

125. Russo R., *On the uniqueness of viscous compressible fluid motions in unbounded domains*, Meccanica, **13**, 1978, 78–82.

126. P. Secchi, A. Valli, *A free boundary problem for compressible viscous fluid*, J. Reine Angew. Math., **341** 1983, 1–31.

127. Sedenko V.I., & Iudovich V.I., *Stability of steady flows of a perfect incompressible fluid with a free boundary*, P.P.M., **42**, 1978, 1049–1055.

128. Serrin J., *On the uniqueness of compressible fluid motions*, J. Rational Mech. Anal. **3**, 1959, 271–288.

129. Sohr H., The Navier-Stokes Equations, Springer 1990.

130. Solonnikov, V.A., *Solvability of initial value problem for the equations of motion of a viscous compressible fluid*, J. of Soviet Math., **14**, 1980, 1120–1133.

131. Solonnikov, V.A., *Solvability of an evolution problem of thermocapillary convection in an infinite time interval*, Lecture Notes in Num. Appl. Anal., **12**, 1991, 211–228.

132. V.A. Solonnikov, *On the stability of axisymmetric equilibrium figures of rotating viscous incompressible liquid*, Algebra Anal. **16** 2004, 317–336.

133. V.A. Solonnikov, *On the stability of nonsymmetric equilibrium figures of rotating viscous incompressible liquid*, Interfaces and free boundaries **6**, 2004, 461–492.

134. V.A. Solonnikov, *On instability of axially symmetric equilibrium figures of rotating viscous incompressible liquid*, Zap. Nauchn. Sem. POMI **318**, 2004, 277–297.

135. V.A. Solonnikov, *On instability of equilibrium figures of rotating viscous incompressible liquid*, J. Math. Sci. **128**, 2005, 3241–3262.

136. V.A. Solonnikov, *Letter to the editor,* J. Math. Sci. **135** No 6, 2006, 3522–3528.
137. Solonnikov, V.A., A. Tani, *Free boundary problem for the compressible Navier-Stokes equations with surface tension,* Constantin Caratheodory: an international tribute, T.M. Rassias ed., World Sci. Publ., 1991, 1270–1303.
138. Solonnikov, V.A., A. Tani, *Evolution free boundary problem for equations of motion of viscous compressible barotropic liquid,* Lect. Notes in Math., **1530**, Heywood, Masuda, Rautmann, Solonnikov eds., 1992, 30–55.
139. Spiegel E., *Convective instability in a compressible atmosphere I,* Ap. J., **141**, 1965, 1068–1087.
140. Tanaka, K., and Shibata Y., *On the steady flow of compressible viscous fluid and its stability with respect to initial disturbance,* J. Math. Soc. of Japan, **55**, 2003, 797–826.
141. A. Tani, *On the first initial boundary value problem for compressible viscous fluids,* J. Math. Soc. of Japan, **55** 2003, 797–826.
142. A. Tani, *On the free boundary problem for compressible viscous fluid motion,* J. Math. Kyoto Univ., **21**, 1981, 839–859.
143. C.A. Truesdell, *A first course in rational continuum mechanics,* **71**, Pure and Applide Mathematics, Academic Press inc., 1991.
144. Unno, W., Kato, S., and Makita M., *Convective instability in polytropic atmosphere,* Publ. Astr. Soc. Japan, **12**, 1960, 192–220.
145. A. Valli, *On the existence theorem for compressible viscous fluids,* Ann. Math. Pura Appl. **130**, 1982.
146. A. Valli, *On the existence of stationary solutions to compressible Navier-Stokes equations,* Ann. Inst. H. Poincaré, Anal. Non Linéaire 4, 1987, 99–113.
147. A. Valli, *Navier-Stokes equations for compressible fluids: global existence, qualitative properties of solutions in general case,* Comm. Math. Phys. **103**, 1986.
148. Vol'pert A.I., Hudjaev S.I., The Cauchy problem for composite systems of nonlinear differential equations, Math. Sb., **87**, 1972, 504–528.
149. J. A. Whitehead and M. M. Chen, *Thermal instability and convection of a thin fluid layer bounded by a stably stratified region,* J. Fluid Mech. **40**, 1970, 549–576.
150. Yih Chia-Shun, *Dynamics of nonhomogeneous fluids,* The Macmillan series in advanced mathematics and theoretical physics, New York, 1965.

Index

Absolute Temperature, 26
Absolutely Continuous Function, 10
Action-Reaction Theorem, 19
Arnold, 62
Autonomous System, 55
Auxiliary Function, 113, 115, 165, 205
    Decay Isothermal Fluid, 72
    Time Asymptotic Decay, 128–130, 190,
        214, 219
    Uniqueness Barotropic, 126
    Uniqueness Free Boundary, 187
    Uniqueness of Steady Flow, 127
    Uniqueness Polytropic Fluid, 218

Balance
    Momentum, 18
    Rotational Momentum, 18, 19
    Stress at Boundary, 39
Banach Space, 9
Basic Motion, 55
Basic Steady Flows, 77
    Rest State, 140
Benard Problem, 197
Bernoulli Equation, 67
Bochner Integrals, 9
Boundary Conditions, 36, 37
    Adherence, 38
    Capillary Effects, 40
    Deformable, 80
    Deformable Known, 37, 38, 44
    Deformable Unknown, 37, 39, 44
    Dirichlet, 43
    Elastic Membrane, 42
    Free Boundary, 37
    Free Regular, 42
    Impermeability, 38

Impermeable, 38, 39
Linear Elastic Stresses, 42
Neumann, 43
Newton's Cooling Law, 43
No Slip, 38
Non Compact Boundary, 52
Non Homogeneous, 78
Rigid Impermeable, 38, 44
Rigid Moving Domain, 37
Rigid Porous, 37, 38, 44
Stress Free, 37
Temperature, 43
On Velocity, 38, 199
Viscoelastic Membrane, 42
Boundary Value Problem, 89, 198
    Polytropic Fluid, 200
Boussinesq Approximation, 197
BVP
    Barotropic Exterior Domain, 78
    Barotropic Fluid, 75
    Barotropic Inviscid Steady Flow, 67
    Exterior Domain, 117
    Free Boundary, 81
    Incompressible Inviscid Steady Flow, 62
    Incompressible Viscous Steady Flow, 61
    Polytropic, 85

Capillary Equilibrium Configurations, 60,
    140
Cartesian Representation
    Domain, 3
    Surface, 3
Cauchy Inequality, 7
Cauchy Postulate, 17
Cauchy-Noll Theorem, 19
Chauchy Tensor, 19

M. Padula, *Asymptotic Stability of Steady Compressible Fluids*,
Lecture Notes in Mathematics 2024, DOI 10.1007/978-3-642-21137-9,
© Springer-Verlag Berlin Heidelberg 2011

Clausius-Duhem Inequality, 30
Conditions at ∞ on Velocity, 40
Conservation of Mass, 17
Constitutive Equations, 22, 32
Continuity Equation
    Global Form, 17
    Local Form, 18
Convection, 197
Critical Point, 56
Curvature, 137

Density
    Material, 12
Deviatoric Stress Tensor, 20
Direct Method, 56
Dirichlet Method, 53, 64, 82
Dissipation Principle, 30
Domain
    Bounded, 37
    Exterior, 37
    Fixed, 12
    Material, 12
    Moving, 12
    Rigid Bounded, 45
    With Non-Compact Boundary, 37
Doubled Mean Curvature, 5, 80
Dual Space, 6
    L^p, 7
Dynamical Balance of Stresses, 40, 41
Dynamical Condition, 39

Embedding Inequality, 8
Empirical Temperature, 25
Energy
    Energy of Perturbation, 135
    Method, 54
    Barotropic Fluid, 49
    Equation, 107
    First and Second Variations, 64, 142
    First Law of Thermodynamics, 28
    Free Boundary, 159
    Functional of Perturbation, 208
    Helmholtz Free, 33, 49, 68, 80, 93, 94,
        124, 141
    Helmhotz Free, 70
    Incompressible Fluid, 48
    Inequality, 103, 171
    Internal Energy, 26
    Internal Energy Functional, 208
    Kinetic Energy, 15
    Kinetic Energy Functional, 208
    Method, viii, 53, 60

Modified Energy Equation, 66, 96
Modified Energy Functional, viii, 63,
    98, 111
Modified Energy Inequality, 99
Modified Energy Method, 53
Norm of Perturbations, 71
Potential, 142
Potential Energy Functional, 208
Surface, 141
Total Energy, 27, 68, 141
Total Energy Equation, 208
Total Energy Functional, 208
Energy Equation, 64, 82, 90, 94, 158
    Barotropic Fluid, 77
    Heat-Conducting Fluid, 50
    Internal Energy, 28
    Isothermal Fluid, 71
    Navier-Stokes Perturbed, 61
    Perturbation, 93, 95, 109, 202, 208, 209
    Perturbations to Rest, 76
    Viscous Incompressible Fluids, 48
Enthalpy, 32, 89, 94, 97
Entropy, 26
Equilibrium Configurations, 139, 140
Euler
    Incompressible Equations, 22
Euler Compressible
    Steady Equations, 67
Euler Equations
    Compressible, 21, 33, 34
    Compressible Steady, 66
    Compressible Unsteady, 66
    Incompressible, 21, 62
    Incompressible Homogeneous, 22
    Incompressible Inhomogeous, 21
    Incompressible Steady, 62
    Incompressible Unsteady, 63
Euler-Lagrange Equations, 144
Eulerian Descriptions, 11
Existence
    Steady Flow, 51
    Unsteady Regular Flow, 51
    Unsteady Weak Flow, 52
Exponential Stability, 149
Exterior Domains, 78

First Law of Thermodynamics, 27
    Local, 28
    Macroscopic, 28
Fluid
    Barotropic, 48
    Heat-conducting, 35
    Polytropic, 1, 50

Barotropic, 1, 34, 74
Compressible, 30
Heat-Conducting, 35
Homogeneous, 22
Incompressible, 21
Incompressible Homogeneous, 31
Isentropic, 34, 89
Isothermal, 1, 33, 71, 89
Isothermal Free Boundary, 74
Linearly Viscous, 24, 34
Perfect Gas, 32
Polytropic, 32, 50, 74
Fluid Equations
Heat-Conducting, 35, 50
Flux through a Surface, 11
Forces
Body, 16
Density per Unit of Mass, 16
Density per Unit of Volume, 16
Long-Range, 16
Non Potential, 100
Oriented Surface, 16
Potential and Large, 96
Potential Large, 92
Short-Range, 16
Specific, 16
Fourier Law, 28
Frame Indifferent, 14
Frames, 13
Free Boundary Value Problem, 138, 154
Free Initial Boundary Value Problem, 139
Free Work
Inequality, 98, 99
Test Function, 88
Free Work Equation, viii, 54, 109, 213
Barotropic Fluid, 91, 103, 110
Decay to Rest State, 97, 165, 167, 205,
  214
Isothermal Fluid, 72
Ordinary Differential System, 65
Polytropic Fluid, 205
Uniqueness of Steady State, 103
Uniqueness to Rest State, 205
Free Work Functional, viii

Gronwall's Lemma, 100, 112, 122, 170, 178,
  213, 217

Hölder Inequality, 6
Helmholtz Decomposition, 70
Hilbert Space, 6
History of a Function, 23

Hyperbolic System
First Order, 62
Second Order, 64

IBVP
Barotropic Exterior Domain, 198
Barotropic Fluid, 74, 91, 106
Barotropic Inviscid Unsteady Flows, 66
Exterior Domain, 119
Free Boundary, 81
Incompressible Inviscid Unsteady
  Flows, 63
Incompressible Viscous Fluid, 60
Incompressible Viscous Perturbations,
  61
Isothermal Fluid, 71
Isothermal Free Boundary, 80
Polytropic Fluid, 85, 199
Polytropic Perturbations, 213
Ideal Fluids, 20
Initial Boundary Value Problem
Barotropic Exterior Domain, 78
Initial Data Control, 58, 133, 149, 150
Loss, 59
Instability, 55–57, 80, 122, 149
Linear, 57, 149
Nonlinear, 83
Phase Change, 79
Interpolation Inequality, 7
Irreversible Processes, 29
Isotropic Fluid, 19

Kinematical Dissipation, 28
Kinematics, 9
Kinetic Energy, 15
Korn's Constant, 168
Korn's Inequality, 156, 164, 168, 176, 192

Ladyzhenskaja Inequality, 8
Lagrange-Dirichlet
  Method, 62
Lagrangean Description, 11
Laplace-Beltrami Operator, 5, 137, 144
Large Potential Forces, 83
Leading Stress, 41
Lebesgue Spaces, 5
Linearization Principle, 58
Lipschitz Property, 4
Loss of Initial Data Control, 84, 133, 134,
  150
Loss of Initial Data Control, 84

Lyapunov
  Functional, 59, 63, 68, 71, 95, 170, 208, 212
  Method, 59, 213
  Theorem, 59

Marginal Stability, 149
Mass, 15
Material Time Derivative, 10
Matrix of transformation, 13
Matrix Transformation of the Basis, 13
Mechanical Potential, 89
Mechanical Process, 22
Minkowski Inequality, 7
Modified Energy, 167
  Decay to Rest State, 99
  Equation, 73, 209
  Equivalence to Norm, 73
  Estimate, 213
  Functional, 65, 73, 99, 134, 208, 215
  Inequality, 111
  Isothermal, Free Boundary, 164
Modified Energy Equation, 97
Modified Energy Inequality
  Isothermal Fluid, 73
Momentum, 15

Natural Norms, viii, 64, 96, 205
Navier-Stokes Equations, 24
  Compressible, 35, 49
  Incompressible, 24, 47
  Incompressible Steady, 61
  Incompressible Unsteady, 60
  Isothermal, 71
Newtonian Fluids, 24
Niremberg Inequality, 8
Nonlinear Exponential Stability, 57, 114, 119
  Exterior Domain, 79
  Free Boundary, 83
  Isothermal Free Boundary, 164, 167
  Non Potential Forces, 78, 109
  Polytropic Fluid, 86, 213, 214
  Porous Boundary, 78
  Potential Forces, 98
  Rest State, 76, 96
  Steady Flow., 109
Nonlinear Stability, 54, 56, 69, 78, 82, 149, 158, 162
Normal Vector, 4

Orlicz Space, 75, 90, 91, 97, 98
Oseen Equations
  Compressible, 40

Perturbation, 55
Perturbation IBVP
  Barotropic Fluid, 107
Perturbations Equations, 60, 159
Phase Change, 32, 79
Phenomenological Laws, 32
Poincaré
  Constant, 168, 176
  Inequality, 73, 106, 117, 127, 136, 168, 174, 176
Polytropy Index, 89
Pressure, 20
Principle of Local Action, 23
Principle of Material Indifference, 23

Regularity Assumption I, 14
Regularity Assumption II, 14
Regularity Class, 77, 107
Rest State, 139
  Barotropic, 75
Resultant, 10
Reversal Flow, 3
Reynolds Transport Theorem, 11, 12, 94, 160, 165, 214
Rotational Momentum, 15

Schwartz Inequality, 6
Second Law of Thermodynamics, 28–30
  Local, 30
  Macroscopic, 29
Side Condition on
  Density, 44
  Density [Global], 45
  Density [Local], 44
  Density at ∞, 46
Simply Connected Domain, 3
Sobolev
  Inequality, 176
  Space, 117, 179
Sobolev Inequality, 8, 104, 117
Sobolev Space, 7
Specific Heat at Constant Volume, 32
Stability, 54
  Autonomus Systems, 56
  Barotropic Fluid, 106
  Barotropic Inviscid Flows, 66
  Characteristic Number, 151

Equilibrium Figure, 146
Exterior Domain, 79
Linear, 57, 149
Marginal, 57
Nonlinear Asymptotic, 55
Nonlinear in the Mean, 55
With Respect Initial Data, 61
Strong Local Lipschitz Property, 4
Surface Tension, 41

Taylor Polynomial, 69
Temperature, 25
The Principle of Determinism, 23
Thermal Conduction, 25
Thermal State, 26
Thermo-Kinetic Process, 26
Thermodynamical Motion, 26
Thermodynamical Variables, 25
Torque of pole $\mathbf{x}_O$, 10
Torques
    Body, 16
    Long-Range, 16
    Surface, 17
Trace of a Function, 7
Traction, 16

Transformation Matrix, 13
Transformation Rule, 13
Transport Theorem, 11
Type of Boundary, 37
Type of Domain, 37, 39

Uniqueness
    Rest State, Free Boundary, 155
    Polytropic Fluid, 86, 201
    Rest State, Isothermal Free Boundary,
        82
    Rest State, Potential Forces, 75
    Steady Flow, Ω Exterior, 117
    Steady Flow, Barotropic Exterior
        Domain, 79
    Steady Flow, Non Potential Forces, 77,
        100, 101
    Steady Flow, Porous Boundary, 78, 113
    Steady Flow, Potential Forces, 90

Vectorial Notations, 9

Young Inequality, 7

LECTURE NOTES IN MATHEMATICS

Edited by J.-M. Morel, B. Teissier; P.K. Maini

Editorial Policy (for the publication of monographs)

1. Lecture Notes aim to report new developments in all areas of mathematics and their applications - quickly, informally and at a high level. Mathematical texts analysing new developments in modelling and numerical simulation are welcome.

   Monograph manuscripts should be reasonably self-contained and rounded off. Thus they may, and often will, present not only results of the author but also related work by other people. They may be based on specialised lecture courses. Furthermore, the manuscripts should provide sufficient motivation, examples and applications. This clearly distinguishes Lecture Notes from journal articles or technical reports which normally are very concise. Articles intended for a journal but too long to be accepted by most journals, usually do not have this "lecture notes" character. For similar reasons it is unusual for doctoral theses to be accepted for the Lecture Notes series, though habilitation theses may be appropriate.

2. Manuscripts should be submitted either online at www.editorialmanager.com/lnm to Springer's mathematics editorial in Heidelberg, or to one of the series editors. In general, manuscripts will be sent out to 2 external referees for evaluation. If a decision cannot yet be reached on the basis of the first 2 reports, further referees may be contacted: The author will be informed of this. A final decision to publish can be made only on the basis of the complete manuscript, however a refereeing process leading to a preliminary decision can be based on a pre-final or incomplete manuscript. The strict minimum amount of material that will be considered should include a detailed outline describing the planned contents of each chapter, a bibliography and several sample chapters.

   Authors should be aware that incomplete or insufficiently close to final manuscripts almost always result in longer refereeing times and nevertheless unclear referees' recommendations, making further refereeing of a final draft necessary.

   Authors should also be aware that parallel submission of their manuscript to another publisher while under consideration for LNM will in general lead to immediate rejection.

3. Manuscripts should in general be submitted in English. Final manuscripts should contain at least 100 pages of mathematical text and should always include

   – a table of contents;
   – an informative introduction, with adequate motivation and perhaps some historical remarks: it should be accessible to a reader not intimately familiar with the topic treated;
   – a subject index: as a rule this is genuinely helpful for the reader.

   For evaluation purposes, manuscripts may be submitted in print or electronic form (print form is still preferred by most referees), in the latter case preferably as pdf- or zipped psfiles. Lecture Notes volumes are, as a rule, printed digitally from the authors' files. To ensure best results, authors are asked to use the LaTeX2e style files available from Springer's web-server at:

   ftp://ftp.springer.de/pub/tex/latex/svmonot1/ (for monographs) and
   ftp://ftp.springer.de/pub/tex/latex/svmultt1/ (for summer schools/tutorials).

Additional technical instructions, if necessary, are available on request from lnm@springer.com.

4. Careful preparation of the manuscripts will help keep production time short besides ensuring satisfactory appearance of the finished book in print and online. After acceptance of the manuscript authors will be asked to prepare the final LaTeX source files and also the corresponding dvi-, pdf- or zipped ps-file. The LaTeX source files are essential for producing the full-text online version of the book (see http://www.springerlink. com/openurl.asp?genre=journal&issn=0075-8434 for the existing online volumes of LNM). The actual production of a Lecture Notes volume takes approximately 12 weeks.

5. Authors receive a total of 50 free copies of their volume, but no royalties. They are entitled to a discount of 33.3 % on the price of Springer books purchased for their personal use, if ordering directly from Springer.

6. Commitment to publish is made by letter of intent rather than by signing a formal contract. Springer-Verlag secures the copyright for each volume. Authors are free to reuse material contained in their LNM volumes in later publications: a brief written (or e-mail) request for formal permission is sufficient.

Addresses:
Professor J.-M. Morel, CMLA,
École Normale Supérieure de Cachan,
61 Avenue du Président Wilson, 94235 Cachan Cedex, France
E-mail: morel@cmla.ens-cachan.fr

Professor B. Teissier, Institut Mathématique de Jussieu,
UMR 7586 du CNRS, Équipe "Géométrie et Dynamique",
175 rue du Chevaleret
75013 Paris, France
E-mail: teissier@math.jussieu.fr

For the "Mathematical Biosciences Subseries" of LNM:

Professor P. K. Maini, Center for Mathematical Biology,
Mathematical Institute, 24-29 St Giles,
Oxford OX1 3LP, UK
E-mail : maini@maths.ox.ac.uk

Springer, Mathematics Editorial, Tiergartenstr. 17,
69121 Heidelberg, Germany,
Tel.: +49 (6221) 487-8259

Fax: +49 (6221) 4876-8259
E-mail: lnm@springer.com